Social Issues in Sustainable Fisheries Management

MARE Publication Series

Volume 9

Series Editors

Svein Jentoft, *University of Tromsø, Norway*
Svein.jentoft@uit.no
Maarten Bavinck, *University of Amsterdam, The Netherlands*
J.M.Bavinck@uva.nl

The MARE Publication Series is an initiative of the Centre for Maritime Research (MARE). MARE is an interdisciplinary social-science network devoted to studying the use and management of marine resources. It is based jointly at the University of Amsterdam and Wageningen University (www.marecentre.nl).

The MARE Publication Series addresses topics of contemporary relevance in the wide field of 'people and the sea'. It has a global scope and includes contributions from a wide range of social science disciplines as well as from applied sciences. Topics range from fisheries, to integrated management, coastal tourism, and environmental conservation. The series was previously hosted by Amsterdam University Press and joined Springer in 2011.

The MARE Publication Series is complemented by the Journal of Maritime Studies (MAST) and the biennial People and the Sea Conferences in Amsterdam.

For further volumes:
http://www.springer.com/series/10413

Julie Urquhart • Tim G. Acott • David Symes
Minghua Zhao
Editors

Social Issues in Sustainable Fisheries Management

Springer

Editors
Julie Urquhart
School of Science
University of Greenwich
Chatham
United Kingdom

Tim G. Acott
School of Science
University of Greenwich
Chatham
United Kingdom

David Symes
University of Hull
Hull
United Kingdom

Minghua Zhao
Greenwich Maritime Institute
University of Greenwich
London
United Kingdom

ISSN 2212-6260 ISSN 2212-6279 (electronic)
ISBN 978-94-007-7910-5 ISBN 978-94-007-7911-2 (eBook)
DOI 10.1007/978-94-007-7911-2
Springer Dordrecht Heidelberg New York London

Library of Congress Control Number: 2013956960

Cover image: Julie Urquhart & Tim Acott

Printed on acid-free paper

Springer is part of Springer Science+Business Media (www.springer.com)

Preface

Social Issues in Sustainable Fisheries Management draws on the themes from a major international conference held at the University of Greenwich, London on 4–5 April 2011, titled "It's Not Just About the Fish". Over 100 delegates, including researchers, policy makers, fishermen (and fishermen's wives) and industry representatives from 16 countries, including Europe, Australia, Canada, Nigeria, Pakistan and the USA, gathered together to discuss the social and cultural impacts of marine fishing on coastal communities. Despite vastly differing socio-economic, political and geographical contexts, similar issues face inshore fishing fleets and coastal communities worldwide. A number of themes emerged out of the conference which form the backdrop for this book, including: the need for integration of local communities into the decision-making and management process; developing trust and cooperation between scientists, policy makers and fishers; stewardship and sustainable fisher livelihoods; the contribution of fishing to local identity and sense of place; and gender roles and women in fishing communities.

Through this volume we hope to take a step towards a sustainable development paradigm for fisheries management, which explicitly incorporates a range of social and cultural issues. The contributions largely draw on European examples, although the issues presented are relevant in the global text. The editors would like to thank all the authors for their contributions and the wide range of community stakeholders that contributed to many of the projects outlined in this volume.

Contents

Contributors

Tim G. Acott School of Science, University of Greenwich, Central Avenue, Chatham Maritime, Chatham, Kent ME4 4TB, UK

Rodney Anderson IRIS Consulting, 5 Golden Square, Soho, London W1F 9BS, UK

Ana Moura Arroz Azorean Biodiversity Group, University of the Azores, 9700-042 Angra do Heroísmo, Azores, Portugal

Kathleen Blanchard Intervale Associates, PO Box 172, Doyles, NL A0N 1JO, Canada

Easkey Britton School of Environmental Sciences, University of Ulster, Cromore Road, BT52 1SA, Coleraine, Co. Londonderry, Ireland

Urszula Budzich-Tabor FARNET Support Unit, 1040, Brussels, 38 rue de la Loi, Belgium

Jennifer Dawe Intervale Associates, PO Box 172, Doyles, NL A0N 1JO, Canada

Katia Frangoudes UMR AMURE, IUEM, Université de Brest,12 rue Kergoat, CS 93837, Cedex 3, 29238, Brest, France

José J. Pascual-Fernández Instituto Universitario de Ciencias Políticas y Sociales, Facultad de Ciencias Políticas y Sociales, Campus de Guajara, Universidad de La Laguna, 38205, La Laguna, Tenerife, Spain

Mike Fitzpatrick Coastal and Marine Resource Centre, University College Cork, Cork, Ireland

Rosalina Gabriel Azorean Biodiversity Group, University of the Azores, 9700-042 Angra do Heroísmo, Azores, Portugal

Giorgio Gallizioli Directorate General for Maritime Affairs and Fisheries, European Commission, 1049, Brussels, Belgium

Laura-Mars Henichart Fisheries and Aquaculture Sciences Centre, Agrocampus Ouest, Rennes, France

Marie Lesueur Fisheries and Aquaculture Sciences Centre, Agrocampus Ouest, Rennes, France

Steven Mackinson Centre for Environment, Fisheries and Aquaculture Science (Cefas), Pakefield Road, Lowestoft, Suffolk, UK

Tim Martindale Goldsmiths, University of London, 222 Croxted Road, London, SE24 9DG, UK

Begoña Marugán-Pintos Department of Social Analysis, Universidad Carlos III de Madrid, Calle Madrid, 126, Getafe 28903, Madrid, Spain

Enésima Mendonça Azorean Biodiversity Group, University of the Azores, 9700-042 Angra do Heroísmo, Azores, Portugal

Richard Morgan Centre for the Economics and Management of Aquatic Resources, Portsmouth Business School, University of Portsmouth, Portsmouth, UK

Alison L. Neilson Azorean Biodiversity Group, University of the Azores, 9700-042 Angra do Heroísmo, Azores, Portugal

Estera Onoakpovike Greenwich Maritime Institute, University of Greenwich, Old Royal Naval College, Park Row, Greenwich, London SE10 9LS, UK

David Symes Reader Emeritus, Department of Geography, University of Hull, Cottingham Road, Hull, Yorkshire HU6 7RX, UK

Marilyn Tyzack IRIS Consulting, 5 Golden Square, Soho, London W1F 9BS, UK

Julie Urquhart School of Science, University of Greenwich, Central Avenue, Chatham Maritime, Chatham, Kent ME4 4TB, UK

Katrien Vervaele Groenestraat 22, 8377 Zuienkerke, Belgium

Russell Wall Intervale Associates, PO Box 172, Doyles, NL A0N 1JO, Canada

Kate Bigney Wilner Dalhousie University, Halifax, NS, Canada

Ruth Williams Centre for Rural Economy, School of Agriculture, Food and Rural Development, Newcastle University, Newcastle upon Tyne, NE1 7RU, UK

Douglas Clyde Kongshøj Wilson Innovative Fisheries Management, Aalborg University Research CentreFibigerstræde 13, 66c, 9220, Aalborg Ø, Denmark

Minghua Zhao Greenwich Maritime Institute, University of Greenwich, Old Royal Naval College, Park Row, Greenwich, London SE10 9LS, UK

Chapter 1
Introduction: Social Issues in Sustainable Fisheries Management

Julie Urquhart, Tim Acott, David Symes and Minghua Zhao

Keywords Sustainable fisheries • Fisheries management • Sustainable development • Ecosystem approach to fisheries • Social and cultural values of fisheries • EU Common Fisheries Policy • Social issues

1.1 Sustainable Fisheries

Throughout the twentieth and early twenty-first centuries fisheries policies and management systems worldwide changed as commercial fish stocks declined and fishing pressure intensified. Increased technological developments have enabled fishing fleets to become highly mobile and efficient and, together with a growing demand for fish for human consumption, have contributed to 87% of commercial fish stocks being either fully exploited or overexploited (FAO 2012). Unsurprisingly, the focus of fisheries policy and management has been directed at biological and economic dimensions in the effort to bring stocks back to sustainable levels and to protect marine environments. However, in the broad global context, social and cultural issues in fisheries policy and management have largely been ignored. Yet omitting socio-cultural objectives has consequences for many fishing communities that today are struggling to deal with the implications of such decision making (Symes and Phillipson 2009; Urquhart et al. 2011). Examples may include increased levels of unemployment, outmigration, weaker community structures and economic difficulties, which are especially felt in

J. Urquhart (✉) · T.G. Acott
School of Science, University of Greenwich, Central Avenue, Chatham Maritime, Chatham, Kent, ME4 4TB, UK
e-mail: j.urquhart@greenwich.ac.uk

T.G. Acott
e-mail: T.G.Acott@greenwich.ac.uk

D. Symes
Reader Emeritus, Department of Geography, University of Hull, Hull, UK
e-mail: dg@dgsymes.karoo.co.uk

M. Zhao
Greenwich Maritime Institute, University of Greenwich, Old Royal Naval College, Park Row, Greenwich, London SE10 9LS, UK
e-mail: M.Zhao@greenwich.ac.uk

J. Urquhart et al. (eds.), *Social Issues in Sustainable Fisheries Management*,
MARE Publication Series 9, DOI 10.1007/978-94-007-7911-2_1,

Fig. 1.1 The sustainability triangle. (Adapted from Charles 1994; Boyd and Charles 2006; Connelly 2007)

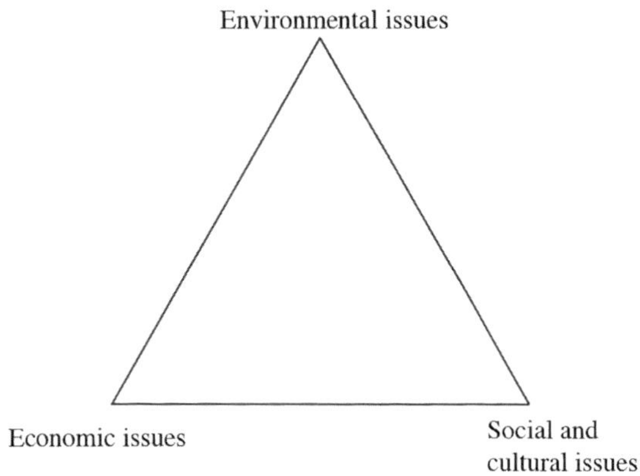

fisheries dependent communities (Scottish Government 2009). These examples have a strong socio-cultural dimension while biological and economic factors are important contributing factors. Consequently, it is increasingly being recognised that sustainable fisheries will only be achieved by integrating management and policy across biological, social and economic dimensions (FCR 2000; Forst 2009).

In this context, then, fisheries management needs to be understood in terms of recognising and respecting the environmental, economic and social impacts of policy and decision making. This mirrors broader societal trends in natural resource management that increasingly seek to understand sustainability from the perspective of the triple bottom line, i.e. social, economic and environmental dimensions. The issue of sustainability and fisheries is not new. In 1994, Anthony Charles reviewed sustainability concepts in the management of fisheries and analysed potential policy directions for the sustainable development of fisheries. Charles identified the need for management based on depleting fish stocks, the need to balance biological, socio-cultural and economic goals and the importance of controlling the rate of fish stock exploitation. From the 1950s conservation and rationalisation paradigms omitted consideration of social and cultural dimensions that are integral to understanding the idea of sustainable development. Moving towards a sustainable development paradigm for fisheries management entails thinking more explicitly about how marine fishing influences a range of social and cultural issues including community, wellbeing, identity, gender equality etc. The idea of linkages between these different dimensions has been depicted as a sustainability triangle (Charles 1994; Boyd and Charles 2006; Connelly 2007) (Fig. 1.1).

Figure 1.1 illustrates the idea of sustainability as negotiating a balance between environmental, economic and socio-cultural domains. Such conceptual mappings help reflection on the state of fisheries management and future directions for the industry. For instance, the conservation and rationalisation paradigms described by Charles (1994) pull fisheries management towards the lower left and upper corners of the triangle. To recognise humans as an integral element of this system greater concern needs to be given to the social and cultural element. However, despite the management of fisheries occurring against a backdrop of increasing national and international interest in sustainable development, this integration of social and cultural concerns has not perhaps received the attention that it should.

Two influential processes resulting in the emergence of the idea of sustainable fisheries were the UN Conference on Human Environment in 1972 and the UN Convention on the Law of the Sea in 1982 (see Garcia et al. 2003 for a detailed account). An internationally important milestone in the development of sustainability and fisheries management was the Committee on Fisheries (COFI) (March 1991) where there was a call for the development of new approaches that would lead to sustainable fisheries. The International Conference on Responsible Fishing (Cancun, Mexico, 1992) requested the FAO to prepare a code of conduct. This fed into the development of Agenda 21. Agenda 21 (published as a result of the UN conference on Environment and Development in 1992) included reference to fisheries and the marine environment in Chapter 17 (United Nations 1993). Agenda 21 was the attempt to identify an agenda to achieve sustainability in the twenty-first century. In Chapter 17 numerous issues relating to the sustainability of fisheries management are considered: "Fisheries in many areas under national jurisdiction face mounting problems, including local overfishing, unauthorized incursions by foreign fleets, ecosystem degradation, overcapitalization and excessive fleet sizes, underevaluation of catch, insufficiently selective gear, unreliable databases, and increasing competition between artisanal and large-scale fishing, and between fishing and other types of activities" (paragraph 17.71).

Specific recommendation is made in paragraph 17.74 (b) "to take into account traditional knowledge and interests of local communities, small-scale artisanal fisheries and indigenous people in development and management programmes". Agenda 21 has fed into national and international policy instruments in the context of developing more sustainable fisheries (Potts 2003). In response to these, and other initiatives, the Food and Agriculture Organisation (FAO) global Code of Conduct for Responsible Fisheries was adopted in 1995, stating: "The Code recognises the nutritional, economic, social, environmental and cultural importance of fisheries, and the interests of all those concerned with the fishery sector" (FAO 1995) and lays down principles for responsible fisheries. In 2002, the World Summit on Sustainable Development encouraged the application of an ecosystem approach to fisheries by 2010 (FAO 2003). FAO published their ecosystem approach to fisheries in 2003 (Garcia et al. 2003). It suggests an ecosystem approach is area-based, holistic and loosely cross-sectoral addressing a broader, society-wide stakeholder base in comparison to fisheries management.

Despite these treaties and initiatives calling for some recognition of social and cultural dimensions in fisheries management there is concern about the contribution that social science is making to fisheries management (Symes and Hoefnagel 2010). Surprisingly little is known about the processes of social representation of marine fishing, the issues faced and the social impacts of fisheries policy in the early twenty-first century (Symes and Frangoudes 2001). Several texts were published in the 1990s and early 2000s in response to the growing fisheries crisis in Europe (Crean and Symes 1996; Jentoft 1993; Symes 1998, 1999, 2000; Symes and Phillipson 2001) and further papers have addressed the management of fisheries, with critiques of policy and planning measures (Daw and Gray 2005; Steelman and Wallace 2001; Symes 2005; Symes and Hoefnagel 2010; Symes and Phillipson 2009). Further work consists largely of anthropological studies along the North Atlantic fringe (Acheson 2003; Symes and Frangoudes 2001) with ethnographic studies in Newfoundland, Maine and Northern Europe (Davis 1988, 2000; Jentoft 1993; Nadel-Klein 2000; van Ginkel 2001; Williams 2008), although there is also a growing interest in social-ecological approaches in fisheries (Wilson 2006; Ommer et al. 2011) and wellbeing (Coulthard et al. 2011; Britton and Coulthard 2013; Trimble and Johnson 2013). More recently, Chuenpagdee's (2011) edited book focuses on the particular issues faced by small-scale fisheries globally, including their contribution to livelihoods, community, employment and conservation.

To develop sustainable marine fishing communities, attention needs to be given to understanding the processes and relationships that are formed between environmental, economic and socio-cultural dimensions. Approaches such as ecosystems-based management and social-ecological systems assert that instead of seeing social issues as separate from ecological, they need to be incorporated in a more explicit and integrated way, recognising that humans are an integrated part of ecosystems (Berkes and Folke 1998; Kinzig 2001; Olsson and Folke 2001; Olsson 2003). Marine fishing is an activity that forms connections between a largely invisible undersea world and terrestrial coastal communities. The aim of this volume is to shed light on the relationships between the social, economic and ecological dimensions of marine fishing communities and fisheries management. Through a collection of essays, many of which are based on empirical case studies, the book examines issues in governance and co-management, social ecological knowledge, livelihoods, gender roles, social identity and cultural heritage. This collection is a timely contribution to the debate about the social and cultural values associated with marine fishing and provides some perspectives about how those values might be understood.

Firstly, however, in order to provide some context for the argument for the incorporation of social objectives, the following section focuses on the development of fisheries management and policy in Europe and provides an overview of the shifting management paradigms towards sustainable fisheries management. We focus on the European context, as the majority of contributions in this volume are from Europe. However, the issues faced by decision makers and fisheries communities worldwide are often similar and focus on how to sustainably manage our commercial fish stocks in order that they will continue to provide food and livelihoods for people both now and in the future.

1.2 Fisheries Management

According to Bromley (1991), a resource management regime, such as that governing fisheries, is defined as "a structure of rights and duties characterising the relationship of individuals to one another in respect of the particular environmental resource" (p. 22). At the core of its structure is the concept of property rights.

Over the past 40 years we have witnessed a remarkable transformation in the way we perceive the vast ocean commons in property terms from *res nullius*—a non-property or open access regime where management of the living resources could only be achieved through voluntary agreements—to a situation where the newly acquired 200 nautical mile exclusive fishing zones were treated as *res communis* to be managed by the coastal state for the benefit of the nation. More recently we have seen attempts to privatise use rights in fisheries through the introduction of individual transferable quota. Each of these transformations represents a significant change in the social construction of the resource and the institutional arrangements for its management. It is, therefore, surprising to find that in developing the theory of fisheries management no attention was given to the social context and, within the EU's Common Fisheries Policy (CFP), until very recently the fishing industry has been excluded from direct involvement in the policy process to decide the 'rights and duties' of the fishers.

1.2.1 Laying the Foundations for Modern Fisheries Management

Although the basic parameters of fisheries management were formulated as early as the mid 1950s, it was not until the late 1970s that the present institutional frameworks began to emerge as a consequence of the contagious spread of 200 nautical mile exclusive economic zones (EEZs). Seminal work by fisheries scientists Schaefer (1954, 1957) and Beverton and Holt (1957) had led to the generation of an integrated surplus production model "complex enough to accommodate the fluctuation of fish stocks … [and] simple enough to allow the scientist to predict the future size and yield of fish stocks for different catch regimes" (Holm 1996) that would permit estimation of maximum sustainable yield (MSY) for individual stocks and species.

Meanwhile Gordon (1953, 1954) was defining maximum economic yield (MEY) as a complementary reference point on the yield curve. When combined with Schaefer's work to create the classic bio-economic model, it laid the foundations for a system of fisheries management that survives, in modified form, to the present day. In simple terms it identified the biological and economic objectives of management. Later attempts (Charles 1988) to specify a social objective in the form of maximum social yield (MSocY) proved difficult to operationalise without compromising the biological and economic goals. Maximising the level of employment in fishing required intervention through subsidies to artificially lower production costs. This simply perpetuated the condition of resource over-exploitation and postponed the onset of sustainable fishing, thus jeopardising the livelihoods of the very people subsidies were intended to protect.

As the history of fisheries management in the North Atlantic subsequently demonstrated, the bio-economic model was insufficiently robust. Its weakness relates to the simplifying assumptions that permitted the model to function: fish stocks were held to be reasonably stable and to behave predictably under moderate levels of exploitation. It oversimplified the behavioural characteristics of different fish stocks (Caddy and Gulland 1985), ignored complex species interactions within marine ecosystems and took no account of the tendency towards instability within the ocean environment. It, therefore, committed fisheries managers to a reductionist view of marine ecosystems and was further compounded by an equally unrealistic representation of the fishing industry as rational 'economic man' aspiring to the collective achievement of MEY—in stark contrast to Hardin's 1968 'tragedy of the commons' thesis. Notwithstanding these flaws, successive generations of fisheries managers kept faith with the belief that the complexities and uncertainties surrounding the exploitation of fisheries could be reduced to a simple calculus for estimating the state of the stocks and identifying levels of surplus production to be harvested sustainably as 'total allowable catches' (TACs).

The task of management was to provide a framework of regulation to ensure the TAC was not exceeded. The problem, however, was that the fishers were left out of the management system: social objectives remained undefined; the realities of fishers' behaviour were omitted from the management model; and fishers' representatives were in most cases excluded from the policy community. This was to prove management's Achilles heel.

The genesis of fisheries management as a theoretical construct coincided with the post-war recovery of the fishing industry and a period of major expansion of Europe's fishing fleets. Despite the need for vigilance over the levels of exploitation of some commercial stocks, there was little sense of urgency about the unfolding situation in the 1950s and 60s. During that time, the industry—and especially the distant water fleets—underwent a remarkable technical transformation: the impact in terms of fishing capacity, sustained fishing effort and exploitation of stocks was dramatic. Apart from a 12 nautical mile zone of territorial waters, fishing was conducted in international waters and subject to fairly light regulation. In the north east Atlantic, international agreements were brokered by the North East Atlantic Fisheries Commission (NEAFC) established in 1959 after a number of false starts (Sen 1997). Its recommendations on closures, gear restrictions and TACs based on scientific advice from the International Council for the Exploitation of the Sea (ICES) made little impact on the growing problem of overfishing. Agreement was difficult; recommendations were binding only on those states that chose to accept them; and there was little enforcement.

Change was inevitable. It came about through unilateral declarations of 200 nautical mile EEZs—in effect the closure of the ocean commons—undertaken in the late 1970s by several coastal states fearful of the depletion of local stocks by distant water fleets. The idea spread rapidly: by the end of the decade virtually all North Atlantic coastal states had adopted the principle and its legitimacy was confirmed in the 1982 UN Convention on the Law of the Sea (Bailey 1996). As a result, the political geography of marine fisheries was completely transformed. It placed the burden of responsibility on the coastal state and gave expression to the idea of fisheries management as a science based management regime embedded in a state bureaucracy (Holm 1996). It also brought managers face to face with the realities of regulating a diverse, unorganised industry and the deficiencies of the management model.

1.2.2 *Facing the Challenges of Fisheries Management*

Closure of the commons created an opportunity to match the management system to the ecological, economic and social conditions of the coastal state's fisheries and the prevailing political philosophy. Even in the north east Atlantic there were quite pronounced differences ranging from the early adoption of rights based management in Iceland (Palsson and Helgason 1996; Eythorsson 1996), to the system of 'centrally directed consultation' in Norway (Hersoug and Rånes 1997) and the centralised command and control approach of the EU's Common Fisheries Policy. As most of the ensuing chapters deal with fisheries and fishing communities within the EU, this brief review of the challenges facing management is based largely on the experience of the CFP. A more detailed analysis of the approaches adopted in Norway, Faeroe and Denmark/EU is available in Gezelius and Raakjaer (2008).

The CFP is a bold but flawed experiment in multinational management of the fisheries within the combined EEZs of a union of independent states and is based on the principle of non-discrimination. Its decision making structure is unique comprising an independent Commission responsible for preparing policy proposals, together with a Council of Ministers drawn from all member states and an elected European Parliament sharing decision making responsibilities. In fisheries the normal procedures are set aside: the European institutions are granted exclusive competence in respect of stock conservation measures—the setting of annual TACs and technical conservation measures based on ICES advice—whereas in most areas policy formulation is shared with member states. Outcomes of EU decision making are presented as Regulations to be implemented directly in national law, rather than the softer Directives that allow member states some flexibility in the manner of implementation. Until recently decisions on stock conservation were the sole responsibility of the Council rather than the customary co-decision with the Parliament. Significant areas of management are reserved to the member state, including the all-important quota management system and regulation of inshore fisheries. The latter was the result of a temporary derogation that has, with time, become permanent and is perhaps the most important concession to the relevance of social issues in fishing. In theory, concerns over social justice in the allocation of fishing opportunities and fisheries dependency are primarily the responsibility of the member state (see Chap. 2). From the EU's perspective the social consequences of the CFP are treated as an externality to be dealt with by other policy areas or through local development projects that may be funded in part from the European Fisheries Fund (see Chap. 10).

To a degree the basic design of the CFP fell victim to the circumstances at the time of its negotiation in the late 1970s and early 80s—a period of respite from the fears of over-exploitation with the 'gadoid outburst' leading to recovery of cod and especially haddock stocks in the North Sea. The negotiating states were preoccupied with agreeing permanent allocation keys for distributing national shares of the TACs for the major commercial species; these keys formed the basis of 'relative stability' that eventually usurped the principle of equal access. Too little attention was paid to developing strategies for long-term management and the prevention of overfishing.

The test of the CFP's durability has been its ability to survive successive enlargements of the EU from nine, at the time of its construction, to 27 member states of which 22 are coastal states with active fishing industries and the extension of jurisdiction from the North Sea and adjacent waters to incorporate not only the Baltic Sea but also the Iberian seas and the archipelagos of the Azores, Madeira and Canary Islands. Judged a political success (Holden 1994; Nielsen and Holm 2007), despite issues of democratic deficit (Gray and Hatchard 2003), it has by its own account failed in its primary objectives of restoring EU fisheries to sustainable levels. According to the Commission (2007, p. 2) "30% of our stocks ... [are] outside safe biological limits ... [and] 80% of our stocks fished so intensively—above maximum sustainable yield—that the yield is reduced". At a time when scientists were demanding substantial reductions in TACs, fishing capacity was increasing, aided by the persistence of grant aid to enlarge and modernise the fishing fleets—a significant failure of political coordination. Equally damning evidence of process failure in the management of Europe's fisheries is the very high level of discards affecting some of the most valued and threatened commercial species (Scottish Government 2010), the result of an overly complex and increasingly restrictive regulatory system combining catch quota, days at sea and catch composition rules that make discarding in mixed demersal fisheries a legal obligation.

Employment in fishing has declined steadily partly as a result of the usual processes of substituting technology for labour and partly as a result of the essential contraction in the size of Europe's fishing fleets. More worrying is the growing difficulty in recruiting new crew members and the replacement of local labour by foreign contract workers—an indication of the way young people are reappraising fishing as a career choice and posing a threat to the social renewal of the industry.

These are symptoms of a failing policy. In the past the response of managers has been to treat the symptoms rather than the underlying causes, applying ever stricter limits to fishing activity. As Drummond and Symes (1996) argued such approaches are inappropriate and unsuccessful: inbuilt tendencies of the industry will breach the limits wherever they are drawn and so perpetuate unsustainability. For reasons of institutional inertia fundamental reform has proved difficult: the management system has remained broadly the same since its inception, revolving around a basic mechanism of TACs and national catch quota. Although the science of stock assessment has become more sophisticated and the use of target, threshold and limit reference points has added clarity to ICES advice, decision making is open to political abuse. In 16 out of the 19 years between 1987 and 2006, the final TACs for cod were set by the Council of Ministers above—usually well above—ICES recommendations (ICES 2006). The only significant change to the management process has been the introduction of long term management plans for the principal commercial species; thus far, however, even they have failed to circumvent the manipulative and destabilising effects of annual renegotiation of the TACs.

The CFP will also have to face two further, inter-related challenges in the immediate future: the elaboration of an ecosystem based approach to fisheries management that remains a largely unfulfilled promise in the current basic Regulation governing the conduct of the common policy; and the development of integrated marine management and marine spatial planning for which the CFP seems ill-prepared. Integrated management is likely to put the squeeze on fisheries, further reduce fishing opportunities and marginalise the social significance of fishing in much the same way, and for probably the same reasons, as Bigney Wilner outlines in Chap. 5 in respect of Nova Scotia.

The CFP has many critics (see for example Sissenwine and Symes 2007; Raakjaer 2009; EC 2009). Problems are addressed from a variety of perspectives—scientific, institutional, political—and may refer to quite specific issues including elaboration and prioritisation of policy objectives; failure to integrate local ecological knowledge and scientific stock assessments; complex systems of regulation; reluctance to embrace co-management or incorporate stakeholders in policy making structures; need to develop ecosystem based approaches as a coordinating policy framework; absence of rights based management, *inter alia*. Several reflect the absence of a clearly articulated social dimension in the way the CFP has been structured (see Chap. 4) and the invisibility of stakeholders in decision making, notwithstanding the establishment of stakeholder led Regional Advisory Councils (RACs) in 2003.

In their reflections on the CFP, Sissenwine and Symes (2007, p. 55) drew attention to the way in which it was regarded as "authoritarian and elitist in its unquestioning adherence to conventional fisheries science ... and remote, unresponsive and bureaucratic in its relations with industry". It had, in effect, lost the confidence of its client group and its very legitimacy was being challenged. To remedy the situation it was necessary to restructure the system so that European institutions are relieved of the burden of micromanagement and are able to focus on meta-governance functions (principles, objectives, strategies and targets), devolving responsibility for detailed management to regional organisations, member states and their fishing industries.

The results of the third decennial review of the CFP are still awaited. The orginal aspirations for a transformational reform of the system (Commission of the European Communities 2009) that would break away from the path dependent evolution of fisheries policy (Hegland and Raakjaer 2008) are likely to be frustrated by legal obstacles and a lack of appetite in the higher echelons of the system for radical change (Symes 2012). It is, therefore, hard to envisage any fundamental changes to fisheries management in the North Atlantic in the short to medium term. The quest for biological sustainability of commercial stocks will continue to dominate, though tempered by concerns for the overall health of the ecosystems that sustain them. The science underpinning stock assessments may be approaching its limits in terms of cost efficiency (Degnbol 1999). Coping with uncertainty, reinforced by the effects of environmental changes, will put pressure on management systems to be more adaptive. And the systems may undergo changes that give stakeholders more responsibility through the adoption of objectives led management and an implied reversal of the burden of proof (Lassen et al. 2008).

The example of the CFP demonstrates the obstacles to bringing about basic changes to well established systems. Although in some respects dysfunctional and impervious to pressures for fundamental change, the prospect of managing the complex and diverse fisheries of the EU's fishing zone outwith the framework of a common policy is difficult to contemplate. In areas like the Baltic and North Seas, seven or eight member states jointly exploit the fish stocks. Many of the key tasks currently undertaken by the Commission would still need to be undertaken in common, and such is the constriction of the EEZs of the individual states by median lines that few, if any, would benefit from a switch to coastal state management. Changes to the CFP are urgently needed but progress is likely to be slow and unspectacular.

1.3 Social Issues in Fisheries Management

The previous section outlined the complex political geography of fisheries management in Europe. It provides an example of the multitude of issues that are embedded within such policies and the difficulty of navigating a sustainable way forward through the range of competing factors. Policy and management of fisheries in Europe, and globally, have largely focused on biological and economic objectives (Clay and McGoodwin 1995; Symes and Hoefnagel 2010). However, with the FAO continuing to report an increase in the percentage of overexploited or fully exploited marine stocks (2012), the success of such policies must be called into question, especially considering the substantial social costs (Commission of the European Communities 2009; Symes and Phillipson 2009).

Understanding the 'social' in the context of fisheries management is a complex task. In popular media there is a tendency for fisheries and fishermen to be described in terms of global generalisations. Fishing is often portrayed as one thing, with fishermen blamed for pillaging the oceans. Such a view fails to distinguish between large-scale fishing activity involving ocean-going freezer trawlers capable of processing up to 150 t of fish per day and small under 10 m artisanal boats, often with just one crew, landing small quantities of fish per day caught within a few miles from harbour. Within this diversity of fishing activities lies a myriad of different social and cultural issues that are created as a result of the process of fishing. A move towards a sustainable development paradigm for fisheries management entails unpacking the social and cultural dimension of fishing and then integrating these issues into appropriate management and governance frameworks. This latter task is beyond the scope of this book, however the chapters presented in this volume do present a series of perspectives on social and cultural issues associated with fisheries management. Attention is needed on a range of social science approaches in fisheries management to help make visible impacts that might be missed from a narrower economic/biological paradigm.

Social issues in fisheries encompass a diversity of topics, including livelihoods, social cohesion, social innovation, social renewal, cultural values, sense of place and identity, education, wellbeing, equality, equity, dependency and spiritual values. These issues are often intangible and complex, requiring interdisciplinary perspectives, sometimes spanning quantitative and qualitative approaches, to reveal the range of social and cultural values that fishing provides to communities. For instance, fishing is important both for maintaining the identity and status of being a fisher for individuals, but is also important more broadly in terms of community identity and social cohesion, with fishing referred to as the 'glue' that binds the community together (Brookfield et al. 2005). Social values are often interrelated and interdependent. This is illustrated by considering the multiple factors that enable or constrain the process of social renewal in fishing communities. Social renewal refers to the process by which the next generation of fishers are created (e.g. succession and inheritance often based on kinship or neighbourhood relationships). Social renewal is influenced by a range of social and political issues. Firstly, the alleged privatisation of fishing rights results in fishing being a closed shop with very limited access to new blood, which deters new entrants. Secondly, social renewal is also influenced by factors relating to material standards of living and quality of life that can be enjoyed by fishers and their families. Further influences include demographic changes, such as an aging fisher population, the outmigration of young adults, reduced marriage opportunities, changes in social mobility, standards of education or training provision, housing, the range of alternative employment opportunities provided by local labour markets and the perception of fishing as an attractive way of life. In relation to most of these factors, the trends especially in remoter areas of the coastal periphery make it increasingly less likely that sons will simply follow fathers into the family fishing business. Indeed, today fishing is often less of an occupation of necessity or last resort but increasingly an occupation of choice where choice is circumscribed by negative images and uncertain future prospects surrounding fishing.

Understanding these types of social issues in the context of fisheries management is difficult and often requires drawing on a range of social science disciplines, such as anthropology, cultural studies, political science and human geography. Social science is characterised by methodological plurality yet fisheries management has tended to focus on a science-centred positivist paradigm. Thus, social science can offer a range of alternative ways of conceptualising social issues and problems, in order to contribute to our understanding of governance frameworks, the implications for the livelihood of people living in coastal towns, the hidden role of women in fisheries, influences on personal and community identity, sectoral linkages, social capital and demographic issues, for instance.

It is here that the contributions in this volume add to the debate. Through a range of disciplinary perspectives, we consider social issues that are pertinent to fisheries management. Many of the essays present empirical work that provides examples of methodological and conceptual approaches for unpacking social issues in fisheries management. The book is divided into five parts: issues in governance and co-management; issues in local ecological knowledge and scientific evidence; issues in fisher livelihoods; issues in gender roles and issues in social identity and cultural heritage.

1.4 Structuring the Argument

1.4.1 Governance and Co-management

Part 1 includes four chapters each of which probes the realities behind modern governance systems and policy approaches that are intended to underpin the sustainability of fishing activity through the incorporation of the principal stakeholders and an emphasis on participative management. Not all of these initiatives achieve their desired effects.

Highlighting recent advances in the theory of fisheries governance David Symes (Chap. 2) challenges the prevailing approaches to managing small-scale fisheries in parts of Europe. Using examples of recent policy proposals in the EU and UK, he argues that increasing emphasis on 'universal' systems of rights based management threatens to undermine attempts to build the resilience of small-scale fisheries through closer integration of the local social and ecological systems. By contrast, in Chap. 3, Mike Fitzpatrick celebrates the successful introduction of participative governance in the Celtic Sea herring fishery through the establishment of an informal management network—involving both scientists and stakeholders—to act as a problem solving forum, but warns of the need to address weaknesses in transparency, strategic objectives and the use of economic and social indicators if the early success is to be maintained.

Georgio Gallizioli (Chap. 4) switches the attention from specific governance issues to a broader overview of European fisheries policy and offers an insider's perspective on problems confronting the EU's obligations to provide employment and a fair standard of living for those engaged in fishing. Acknowledging the limited scope for action within the regulatory framework of the CFP itself, he suggests the more fruitful area for intervention lies in the European Fisheries Fund and its financial provision for local development initiatives to strengthen the economic base of fisheries dependent areas (see Chap. 10).

In Chap. 5 we cross the Atlantic and change the focus of attention from fisheries policy *per se* to the broader issue of policy discourse, power and integrated management. Kate Bigney Wilner provides an introduction to Canada's *Oceans Act* (1966) and its attempt to pioneer the concept and promote a participative approach to managing the multiple use of the marine resource field. But, as her case study of the Annapolis Basin reveals, the dominant role of federal agencies in the new institutional set up weakens the influence of local fishers and their representative organisations.

1.4.2 Local Ecological Knowledge and Scientific Evidence

Part 2 consists of two chapters that discuss the tensions between local knowledge and scientific evidence. Case studies from Canada and Europe suggest a strong argument for integrating local knowledge into decision-making and management of fisheries resources.

One of the ways in which fishers feel disrespected by modern governance systems is the sense that their vast store of local ecological knowledge (LEK) is ignored or set aside when it comes to assessing the state of stocks and manipulating the regulatory mechanisms. The two chapters in this section address this problem from different angles. In Chap. 6, Kathleen Blanchard et al. examine the idea of 'stewardship' as a means of engaging fishers and their experiential knowledge in the conservation of the threatened wolffish taken as bycatch off Newfoundland and Labrador. The benefits of building clear relationships with stakeholders, promoting a deeper awareness of the underlying science and an understanding of the fishers' concerns are borne out in increased levels of compliance. In the context of the EU, Mackinson and Wilson (Chap. 7) explore a more formal approach to solving the alienation of fishers from science based policy decisions as a result of exclusion and poor communication. Participatory action research incorporates fishers as collaborating partners in the design and conduct of scientific research. Respecting the differences between the partners and building a better understanding of the dilemmas that inevitably confront such partnerships are essential, but success will ultimately depend on developing an overall governance structure wherein stakeholders occupy a central role.

1.4.3 Sustaining Fisher Livelihoods

The social renewal of fishing and fishing communities is increasingly dependent on guarantees of sustainable livelihoods for fishers and their families in a very uncertain world. These guarantees can no longer rely on maintaining access to fishing opportunities for all those who wish to fish. At times of relatively scarce fishing opportunities the emphasis must switch to ensuring that value is added to local landings through effective marketing and distribution. The sustainability of fishing dependent areas especially in the remoter, less privileged parts of the coastal peripheries has come to rely more and more on diversifying local job opportunities and providing training in a broader range of employable skills.

Whether sustainable livelihoods are implicitly an issue for fisheries policy remains a moot point. What is beyond argument is the potentially severe social consequences of some policy decisions intended to improve the sustainability of depleted stocks or improve the profitability of fishing operations. In Chap. 8 Easkey Britton describes the loss of wellbeing experienced by individuals and households whose livelihoods were affected by the closure of the Lough Foyle salmon fishery. The impacts went deeper than the loss of a lucrative, seasonal source of income: for some it raised doubts about the prospects for future generations, though for others it prompted a more constructive view of how to manage inshore fisheries for sustainable livelihoods.

For many the solution to problems like those facing coastal communities around Lough Foyle lies in the diversification of local economies and the provision of alternative jobs that offer a different, and probably more secure, form of sustainable livelihood while choosing to remain living close to the sea. But for those who wish to secure their livelihoods through fishing the situation is more difficult. Opportunities for the diversification of fishing activities *per se* are strictly limited and becoming more so; moreover, the capital assets and skills associated with fishing are not readily transferable to other economic activities. This theme is explored by Richard Morgan et al. in their study of the Channel fisheries (Chap. 9) where diversification—in the sense of creating a multifunctional enterprise producing both commodity and non-commodity products—is quite rare. Limited attempts have been made to base diversification on deploying the vessel and seamanship skills in complementary activities (leisure and tourism, contract work) but even these can be constrained by legal and administrative obstacles.

The EU's approach to meeting the challenges of sustainable livelihoods in fisheries dependent areas is outlined by Urszula Budzich-Tabor in Chap. 10. The European Fisheries Fund focuses support on the fishing sector but also assists diversification of local economies disadvantaged by fisheries policy. The emphasis of the Axis 4 instrument is to strengthen links between the fisheries sector and the local community through creating synergy between different partners, diversifying fishing activities into pescatourism, recreational fishing etc. and adding value to local fish production (direct sales, new products, promotional activities etc.). Maintaining local ownership of the institutional framework (Fisheries Local Action Groups or FLAGS) and the individual projects they promote is an underlying aim of the Axis 4 initiative.

1.4.4 Gender Roles

One particular aspect of sustaining fisher livelihoods that has received increasing attention from social scientists over the last two decades or so is the undervalued and largely invisible roles of women within fishing and fishing communities. Fishing is popularly seen as a man's world. But this view conceals the huge importance of women's contributions to the viability of the fishing enterprise itself, the added responsibilities at home and in the community during the men's absence at sea and increasingly their roles as political activists. Arenas and Lentisco (2011) suggest that "Gender mainstreaming is not only a question of social justice but is necessary for ensuring equitable and sustainable human development. The long-term outcome of gender mainstreaming will be the achievement of greater and more sustainable human development for all" (p. 107).

A nuanced image of women's lives emerges from Katrien Vervaele's cameo portrait of Flemish fishermen's wives in Chap. 11. Written not by an academic scientist but by a local historian and freelance writer living in the coastal town of Oostende and based on extended interviews, her account reveals some of the emotional tensions and disturbed relationships that may sometimes occur in the families of distant water fishermen. To a degree the contributions of Flemish fisher wives seem less visible than those described in Chap. 13 and the disintegration of the fishing community more advanced with the widening of marriage catchments and the displacement of young fisher families from the quayside location to the suburbs or the rural hinterlands caused by differential house prices.

In Chap. 12 Katia Frangoudes and José Pascual-Fernandez trace the evolution of women's groups across Europe from their origins in the late 1940s to the surge of activity as a response to growing crises in Europe's fishing industries in the 1990s, and the increasing diversity of their aims from early concerns for living standards, health and safety at work to their involvement in fisheries governance as RAC members. Now well established as voluntary organisations, women's groups continue to face problems relating to legal status, funding and administrative capability. The perception of women's roles as undervalued and lacking formal recognition is borne out in Zhao et al.'s study (Chap. 13) that draws on interviews conducted in northern and south west England in 2010, covering a wide spectrum of women's work in fishing related activities as skippers, crew members, factory owners and process workers, as well as involvement in the fishing community as fisher wives and partners. Though relatively few in number, their active presence and their passionate commitment to the industry's sustainable future makes them increasingly visible.

1.4.5 Social Identity and Cultural Heritage

In the previous two sections, sustaining fisher livelihoods is largely—but not exclusively—concerned with material wellbeing. In treating the issue of women's roles, the examination moved perceptibly into the realms of social identity, status and prestige and so into a concern for personal and social wellbeing. Hitherto, these less tangible aspects of the social dimension of fishing have received far less attention from policy makers and, with a few notable exceptions, from fisheries social science. Today, their relevance, alongside sense of place and cultural heritage, is becoming increasingly influential in understanding the social renewal of fishing and fishing communities, as outlined by the four chapters in this section.

Firstly, in Chap. 14, Tim Acott and Julie Urquhart's account of the Channel fishing communities uses 'sense of place' to explore the material and perceptual relations that emerge from marine fishing activity. Sense of place which is derived from a way of life, social cohesion and a source of memory, inspiration, meaning and material expression translates into 'cultural ecosystem services' that can be developed to generate a more diversified, multifunctional fisheries based economy and ensure sustainable coastal communities.

Chapter 15 stays on England's south coast and focuses on sense of place in the use of heritage and maritime skills in the regeneration of the fishing experience in Newlyn, west Cornwall. Tim Martindale relates an example of 'heritage production' through the story of one man's attempt to rebuild a sailing lugger—a familiar feature of the local fishing scene in the nineteenth and early twentieth centuries—with the intention of establishing a sailing school that would offer not only a taste of an earlier fishing experience but also provide a means of personal development and the acquisition of life skills for local youth. A counterpoint to the story is the scepticism with which the project was received by some in the business community as to its practical relevance to the fishing industry's problems.

Moving northwards, Ruth Williams in Chap. 16 examines the impacts of industrial restructuring in north east Scotland on the lives of individuals and households that depend on fishing. Individual and collective identities formed around the distinctive nature of fishing and the symbolic values of belonging to a fishing community are being deconstructed. Visual images of change around the harbour, the closure of high street shops and decline in social events that once brought fishers and their families together are potent reminders that the fishing community, once the dominant influence, is becoming increasingly diluted.

For the final chapter in this section, it is a long haul, both geographically and culturally, from the urban fishing communities of north east Scotland to the Azores. In Chap. 17 Alison Neilson et al. provide an insight into how fishers in the Atlantic archipelago of the Azores perceive the sea, paying particular attention to how Azoreans identify themselves in relationship to the sea, their respect for the sea as a difficult environment in which to work and how conflicting interpretations of environmental justice reflect different forms of knowledge construction. Popular perceptions of environmental responsibility tend to neglect the narratives of those whose livelihoods are fashioned by a respect for the sea.

The final chapter of the book (Chap. 18) draws together the contributions in the edited volume by arguing for a more explicit inclusion of social issues if a sustainable development paradigm for fisheries is to realised. It suggests that one way of framing social issues may be through four social/cultural dimensions: institutional, discourse/knowledge, translation and methodological. This may provide a way of unpacking social issues and, through interdisciplinary approaches, including social, environmental and economic issues in fisheries management in a more integrated way.

1.5 Conclusion

This collection of essays, many of which are based on empirical studies into the social issues of fisheries management, is testament to the growing interest and demand for evidence on the social and cultural value of marine fisheries. Many of the chapters are geographically focused on Europe; the issues they raise and address are pertinent to fishing communities across the globe.

We set out at the beginning of this chapter to deconstruct the environmental, social and economic dimensions of fisheries management and to reflect on how humans are an integrated part of ecosystems. To introduce this edited collection, we have considered the issue of sustainability and its relationship to fisheries management, outlining the problem of a lack of consideration of social issues in fisheries policy. *Social Issues in Sustainable Fisheries Management* makes the argument for explicit recognition and incorporation of social and cultural objectives in fisheries policy. It provides evidence to illustrate the benefits of integrating local fishing communities into decision-making and adopting co-management approaches, as well as integrating local ecological knowledge with scientific evidence and capturing the broader cultural values that exist in fishing communities.

References

Acheson, J. M. (2003). *Capturing the commons: Devising institutions to manage the Maine Lobster Industry*. Lebanon, New Hampshire: University Press of New England.

Arenas, M. C., & Lentisco, A. (2011). Mainstreaming gender into project cycle management in the fisheries sector. Food and Agriculture Organization of the United Nations, Regional Office for Asia and the Pacific. Bangkok

Bailey, J. (1996). High seas fishing: Towards a sustainable regime. *Sociologia Ruralis, 36*(2), 189–200.

Berkes, F., & Folke, C. (Eds.). (1998). *Linking social and ecological systems: Management practices and social mechanisms for building resilience*. Cambridge: Cambridge University Press.

Beverton, R. J. H., & Holt, S. J. (1957). *On the dynamics of exploited fish populations*. London: Chapman and Hall.

Britton, E., & Coulthard, S. (2013). Assessing the social wellbeing of Northern Ireland's fishing society using a three-dimensional approach. *Marine Policy, 37,* 28–36.

Bromley, D. W. (1991). *Environment and economy: Property rights and public policy*. Oxford: Blackwell.

Brookfield, K., Gray, T. & Hatchard, J. (2005). The concept of fisheries-dependent communities. A comparative analysis of four UK case studies: Shetland, Peterhead, North Shields and Lowestoft. *Fisheries Research,* 72: 55-69.

Boyd, H., & Charles, A. (2006). Creating community-based indicators to monitor sustainability of local fisheries. *Ocean & Coastal Management, 49,* 237–258.

Caddy, J. F., & Gulland, J. A. (1983). Historic patterns of fish stocks. *Marine Policy, 7,* 267–278.

Charles, A. T. (1988). Fishery socioeconomics: A survey. *Land Economics, 64,* 276–295.

Charles, A. T. (1994). Towards sustainability: The fishery experience. *Ecological Economics, 11,* 201–211.

Chuenpagdee, R. (Ed.) (2011). World small-scale fisheries: Contemporary Visions. Delft: Eburon Academic Publishers.

Clark, C. W. (2006). *The worldwide crisis in fisheries: Economic models and human behaviour*. Cambridge: Cambridge University Press.

Clay, P. M. & McGoodwin, J.R. (1995). Utilizing social sciences in fisheries management. *Aquatic Living Resources,* 8, 203-207.

Commission of the European Communities (EC). (2007). *Reflections on Further Reform of the Common Fisheries Policy. Commission Working Document*. Brussels: CEC.

Commission of the European Communities (EC). (2009). *Green Paper: Reform of the Common Fisheries Policy. COM (2009) 163 final*. Brussels: CEC.

Connelly, S. (2007). Mapping Sustainable Development as a Contested Concept. *Local Environment, 12,* 259–278.

Coulthard, S., Johnson, D., & McGregor, J. A. (2011). Poverty, sustainability and human wellbeing: A social wellbeing approach to the global fisheries crisis. *Global Environmental Change, 21,* 453–463

Crean, K., & Symes, D. (1996). *Fisheries management in crisis*. Oxford: Fishing News Books, Blackwell Science Ltd.

Davis, D. (1988). 'Shore Skippers' and 'Grass Widows': Active and passive women's roles in a newfoundland fishery. In: J. Nadel-Klein & D. L. Davis (Eds.). *To work and to weep: Women in fishing economies.* St Johns, Newfoundland: Institute of Social and Economic Research, Memorial University of Newfoundland.

Davis, D. (2000). Gendered cultures of conflict and discontent: Living 'the crisis' in a newfoundland community. *Women's Studies International Forum, 23*(3), 343–353.

Daw, T., & Gray, T. (2005). Fisheries science and sustainability in international policy: a study of failure in the European Union's Common Fisheries Policy. *Marine Policy, 29,* 189–197.

Degnbol, P. (1999). The end of short-term prognoses or what constitutes valid biological knowledge as a basis for fisheries management. In D. Symes (Ed.), Multi-disciplinary research in fisheries management. Report of ESSFiN Workshop, Denmark 13–14 April (pp. 8–14). Hull: University of Hull.

Drummond, I., & Symes, D. (1996). Rethinking sustainable fisheries: The realist paradigm. *Sociologia Ruralis, 36*(2), 152–162.

Eythorsson, E. (1996). Coastal communities and ITQ management: The case of Icelandic fisheries. *Sociologia Ruralis, 36*(2), 212–223.

FAO. (1995). *Code of conduct for responsible fisheries*. Rome: Food and Agricultural Organisation of the United Nations

FAO. (2003). The ecosystem approach to fisheries. Issues, terminology, principles,
institutional foundations, implementation and outlook. FAO Fisheries Technical Paper.
No. 443. Rome 71 pp.

FAO. (2012). *The state of world fisheries and aquaculture*. Rome: Food and Agricultural Organization of the United Nations

FCR. (2000). An assessment of the socio-economic costs and benefits of integrated coastal zone management. European Commission: 61. Firn Crichton Roberts Ltd, G.S.o.E.S.U.o. Strathclyde.

Forst, M. F. (2009). The convergence of Integrated Coastal Zone Management and the ecosystems approach. *Ocean and Coastal Management, 52*(6), 294–306.

Garcia, S. M., Zerbi, A., Do Chi, T., & Lasserre, G. (2003). The ecosystem approach to fisheries. Food and Agricultural Organisation, Technical paper No. 443, p. 71.

Gezelius, S. S., & Raakjaer, J. (2008). *Making fisheries management work: Implementation of policies for sustainable fishing*. Dordrecht: Springer.

Gordon, H. S. (1953). An economic approach to the optimum utilization of fishery resources. *Journal of the Fisheries Board of Canada, 10,* 442–457.

Gordon, H. S. (1954). The economic theory of a common property resource: The fishery. *Journal of Political Economy, 62,* 124–142.

Gray, T., & Hatchard, J. (2003). The 2002 reform of the Common Fisheries Policy's system of governance-rhetoric or reality? *Marine Policy, 27*(4), 545–554.

Hardin, G. (1968). The tragedy of the commons. *Science, 162,* 1243–1248.

Hegland, T. J., & Raakjaer, J. (2008). Recovery plans and the balancing of fishing capacity and fishing possibilities: Path dependence in the Common Fisheries Policy. In S. S. Gezelius & J. Raakjaer (Eds.), *Making fisheries management work: Implementation of policies for sustainable fishing*. Dordrecht: Springer.

Hersoug, B. & Rånes, S.A. (1997). What is good for the fishermen, is good for the nation: co-management in the Norwegian fishing industry in the 1990s. *Ocean and Coastal Management.* 35, 2-3, 157-172.
Holden, M. (1994). *The common fisheries policy*. Oxford: Blackwell.
Holm, P. (1996). Fisheries management and the domestication of nature. *Sociologia Ruralis, 36*(2), 177–188.
International Council for the Exploration of the Sea. (2006). http://www.ices.dk/committee/acfm/comwork/report/2006/oct/cod-347d.pdf.
Jentoft, S. (1993). *Dangling lines-the fisheries crisis and the future of coastal communities: The Norwegian experience. St John's, Newfoundland.* Canada: Institute of Social and Economic Research, Memorial University of Newfoundland.
Kinzig, A. P. (2001). Bridging disciplinary divides to address environmental and intellectual challenges. *Ecosystems, 4,* 709–715.
Lassen, H., Sissenwine, M., Symes, D., & Thulin, T. (2008). *Reversing the burden of proof for fisheries management: Managing commercial fisheries within sustainable limits*. Copenhagen: ICES.
Nadel-Klein, J. (2000). Granny baited the lines: Perpetual crisis and the changing role of women in Scottish fishing communities. *Women's Studies International Forum, 23*(3): 363–372.
Nielsen, K. N., & Holm, P. (2007). A brief catalogue of failure: Framing evolution and learning in fisheries resource management. *Maritime Policy, 31*(6), 669–680.
Olsson, P. (2003). Building capacity for resilience in social-ecological systems. PhD, Stockholm University.
Olsson, P., & Folke, C. (2001). Local ecological knowledge and institutional dynamics for ecosystem management: A study of Lake Racken Watershed, Sweden. *Ecosystems, 4,* 85–104.
Ommer, R., Perry, I., Cochrane, K. L., & Cury, P. (Eds.) (2011). World fisheries: A social-ecological analysis. London: Wiley-Blackwell.
Palsson, G., & Helgason, A. (1996). Property Rights and Practical Knowledge: the Icelandic Quota System. In K. Crean & D. Symes (Eds.), *Fisheries management in crisis* (pp. 45–60). Oxford: Blackwell.
Potts, T. W. (2003). Sustainability indicators in marine capture fisheries. PhD, University of Tasmania.
Raakjaer, J. (2009). *A fisheries management system in crisis: The EU common fisheries policy.* Aalborg: University Press.
Schaefer, M. B. (1954). Fisheries dynamics and the concept of maximum equilibrium catch. 50th Annual Proceedings of the Gulf and Caribbean Fisheries Institute 1953, 1–11
Schaefer, M. B. (1957). Some considerations of population dynamics and economics in relation to the management of commercial fisheries. *Journal of the Fisheries Board of Canada, 14,* 669–681.
Scottish Government. (2009). *The inquiry into future fisheries management: Interim report: The European commission's green paper on the future of the common fisheries policy*. Edinburgh: The Scottish Government.
Scottish Government. (2010). *The future of fisheries management in Scotland. Report of an independent panel*. Edinburgh: The Scottish Government Office.
Sen, S. (1997). The evolution of high-seas fisheries management in the North-east Atlantic. *Ocean and Coastal Management, 35*(2–3), 85–100.
Sissenwine, M., & Symes, D. (2007). *Reflections on the common fisheries policy. Report to the General-Directorate for Fisheries and Maritime Affairs of the European Commission*. Brussels: CEC.
Steelman, T. A., & Wallace, R. L. (2001). Property rights and property wrongs: Why context matters in fisheries management. *Policy Sciences, 34,* 357–379.
Symes, D. (Ed.) (1998). *Property rights and regulatory systems in fisheries*. Oxford: Fishing News Books, Blackwell Science Ltd.
Symes, D. (Ed.) (1999). *Alternative management systems for fisheries*. Oxford: Fishing News Books, Blackwell Science Ltd.

Symes, D. (Ed.) (2000). *Fisheries dependent regions.* Oxford: Fishing News Books, Blackwell Science Ltd.
Symes, D. (2005). Altering course: Future directions for Europe's fisheries policy. *Fisheries Research, 71,* 259–265.
Symes, D. (2012). Regionalising the common fisheries policy: Context, content and controversy. *Maritime Studies, 11*(6).
Symes, D., & Frangoudes, K. (2001). The social organisation and reproduction of inshore fishing. In D. Symes & J. Phillipson (Eds.), *Inshore fisheries management.* Dordrecht: Kluwer. (2001).
Symes, D., & Hoefnagel, E. (2010). Fisheries policy, research and the social sciences in Europe: Challenges for the 21st century. *Marine Policy, 34,* 268–275.
Symes, D., & Phillipson, J. (Eds.) (2001). *Inshore fisheries management.* Dordrecht: Kluwer.
Symes, D., & Phillipson, J. (2009). Whatever became of social objectives in fisheries policy? *Fisheries Research, 95*(1), 1–5.
Trimble, M., & Johnson, D. (2013). Artisanal fishing as an undesirable way of life? The implications for governance of fishers' wellbeing aspirations in coastal Uruguay and southeastern Brazil. *Marine Policy, 37,* 37–44.
United Nations. (1993). *Agenda 21: Programme of action for sustainable development.* New York: United Nations
Urquhart, J., Acott, T., Reed, M., & Courtney, P. (2011). Setting an agenda for social science research in fisheries policy in Northern Europe. *Fisheries Research, 108,* 240–247.
van Ginkel, R. (2001). Inshore fishermen: Cultural dimensions of a maritime occupation. In D. Symes & J. Phillipson (Eds.), *Inshore fisheries management* (pp. 177-193). Dordrecht: Kluwer Academic Publishers.
Williams, R. (2008). *Changing constructions of identity: Fisher households and industry restructuring* (p. 248). Newcastle: School of Agriculture, Food and Rural Development, Newcastle University. (PhD).
Wilson, J. A. (2006). Matching social and ecological systems in complex ocean fisheries. *Ecology and Society,* 11(1): 9.
van Ginkel, R. (2001). Inshore fishermen: Cultural dimensions of a maritime occupation. In D. Symes & J. Phillipson (Eds.), *Inshore fisheries management.* Dordrecht: Kluwer.

Part I
Issues in Governance and Co-Management

Chapter 2
Finding Solutions: Resilience Theory and Europe's Small-scale Fisheries

David Symes

Keywords Small-scale fisheries • Resilience • Co-governance • European Union

2.1 Introduction

The failings of conventional fisheries management are by now all too familiar, yet the search for real solutions appears to be making relatively little headway. In the case of the European Union's Common Fisheries Policy (CFP), for all the promises of fundamental change that launched the 2012 reform process, the outcomes are but pale shadows of the ambitions highlighted in the Green Paper (EC 2009). Why? The underlying reasons are becoming increasingly clear: institutional inertia that owes much to the rigid legal framework of the EU constraining the scope for reform of the decision making system; a lack of political will for fundamental institutional change, with some member states seemingly unwilling to risk the wider European project over an issue as insignificant as sustainable fisheries; and a lack of credible alternatives to the current approach. All of this points to a real difficulty in breaking away from a path dependent course of action that had characterised earlier attempts at reforming the CFP (Gezelius and Raakjaer 2008). There may also be a more basic issue, namely that those who control the destiny of the CFP have failed to grasp the true nature of the problems that beset fisheries and marine environmental policy and the dangers implicit in continuing to follow the well-trodden path.

Against the background of this somewhat fatalistic and perhaps simplistic analysis, significant progress has been made in the conceptualisation of sustainable fisheries management. Over the past decade the social sciences have been active in the quest for alternatives to the often dysfunctional systems of management that have characterised modern industrial fisheries. Their research has taken a number of different directions offering, for example, a more developed understanding of the nature of fisheries/marine environmental issues that render them difficult to contain within a simple notion of 'management'; alternative models of decision making; or a bold new paradigm for the stewardship of natural resources. As a result several

D. Symes (✉)
Department of Geography, University of Hull, Cottingham Road,
Hull, Yorkshire HU6 7RX, UK
e-mail: dg@dgsymes.karoo.co.uk

J. Urquhart et al. (eds.), *Social Issues in Sustainable Fisheries Management*,
MARE Publication Series 9, DOI 10.1007/978-94-007-7911-2_2,

new concepts have entered the discourse on fisheries policy where 'governance' is now often preferred to 'management' and where 'participatory decision making' is regarded as the norm and 'regionalisation' is becoming part of the received wisdom of sustainable fisheries. To a degree, however, such developments may be cosmetic—part of a 'progressive' jargon that attempts to mask a largely unreconstructed view of how fisheries should be handled.

The aim of this chapter is to test the waters of the CFP to gauge how far alternative solutions to managing the complex issues of small-scale fisheries might be feasible. First three complementary developments in the conceptualisation of fisheries and their management are examined, focusing on the design of the governing system. Then the current governing systems at EU and member state levels are dissected to assess how far they are capable of accommodating the alternative approach. The analysis suggests that the existing management frameworks are becoming increasingly inhospitable to fundamental change, and the conclusion briefly considers the implications both for the future of small-scale fisheries and for the viability of alternative solutions in general.

2.2 Key Developments in Theory and Practice of Fisheries Policy

2.2.1 Reformulating the Nature of Problem Solving in Fisheries

Fisheries—especially when considered in the context of marine ecosystem sustainability—are commonly faced with problems which are difficult to define and separate from other wider issues and, therefore, difficult to effect permanent solutions. The term 'wicked problem' used to describe such situations was first coined nearly 40 years ago by Rittel and Webber (1973) in relation to planning theory and practice. As Jentoft and Chuenpagdee (2009) have argued, this is particularly relevant to fisheries where coastal ecosystems are typically diverse, complex and dynamic and where it is not always clear from a fisheries perspective what the problems are, what their underlying causes might be and how far the outcomes of management solutions may be unexpected and potentially perverse.

At a time when the major issues relating to fisheries management were considered to be essentially biological and concerned largely with population dynamics, it may have appeared possible to get away with linear projections and technical fixes (TACs and quota; mesh size regulations etc). These, however, tended to unravel as a consequence of uncertainties in the science of stock assessment relating to the measurement, processing and modelling of the data. Bio-economic modelling may have further compounded the problem by ascribing an unrealistic measure of certainty to human behaviour. As most social scientists will appreciate, many of the problems relating to fisheries and their management are unique in time and space and their solutions, involving judgement and choice, are likely to be contested. Defining the issues to be tackled is itself a wicked problem. Problem definition, goal formulation and solution finding are all likely to be challenged and the conflicts that arise both within and outwith the governing system may never be fully resolved. Stakeholders, often with widely differing perceptions of the problems and their solutions, form part not only of the system to be governed but also, through participative governance, the governing system itself. Though constant consultation, negotiation and reframing of the issues may help to reconcile some of the differences, most policy decisions will ultimately be based on political choice rather than scientific reasoning.

The lessons that follow from this analysis are that little progress will be made towards the goal of sustainable fisheries while we continue to adopt a simple, reductionist and instrumental approach, applying expert knowledge in the form of technical fixes delivered in a top-down, command and control system of decision making. Addressing wicked problems requires greater subtlety mediated through a broad governance approach rather than recourse to the much narrower managerial toolbox. Governing functions must be developed; local knowledge and experience (both ecological and socio-cultural) should be blended with formal evidence from both the natural and social sciences; and stakeholders need to be fully engaged in the governing process through communicative action. Some of these lessons may have already been learned, but so far the only evidence is to be found in the rhetoric of policy proposals rather than in practical solutions.

2.2.2 Improving the Governance of Fisheries

The seeds of interactive governance were being sown in the 1980s with the shift from state control to neoliberal and more inclusive systems of governance following the realisation that the state was unable to cope with increasingly complex societal problems (Rhodes 1996; Chuenpagdee 2011). Jentoft and Chuenpagdee (2009) acknowledge the important contribution of Kooiman et al. (2005) in developing the theme of governance and, in particular, elaborating the notion of interactive governance as a core component in tackling the wicked problems of fisheries and coastal governance. Kooiman et al.'s *Fish for Life: Interactive Governance for Fisheries* was the culmination of several years' involvement in articulating the concept of governance and, more specifically, a major collaborative social science project initiated in 2001.

Governance is viewed as a task shared by public and private actors alike, with the boundaries between the public and private domains becoming increasingly blurred. At the heart of interactive governance is the recognition of three distinct 'orders of governance': first order tasks involve the identification of everyday problems and are, therefore, closely akin to what we recognise as management; second order responsibilities are largely concerned with institutional arrangements; and third order or meta-governance functions involve elaborating the values, principles and criteria that guide policy making. According to Kooiman et al. (2005) too much attention has in the past been paid to the end stages of decision making and the means of delivering fisheries policy (first and second order functions) and too little was given to refining the values and principles on which rational decision making should be based.

Kooiman et al. (2005) started from the assumption that existing forms of policy making are too reliant on narrowly defined, static policy communities with limited knowledge applied to simple world models that are no longer fit for the purpose of governing diverse, complex and dynamic ecological and social systems. They argue for a broadening of the policy community and an open, interactive decision-making process capable of integrating a wide range of views. Built on ideas of inclusivity of representation, interactive learning and partnership building, interactive governance remains rooted in the principles of rationality, efficiency and performance (in terms of effectiveness, legitimacy and moral responsibility) which must be consistently applied often in the face of hard choices.

Scale is an important consideration both in terms of the efficacy of the governing system and also the complementarity of interactions between different levels of governance (supranational, national, regional and local). Interactive governance is expected to be most effective at the local level when dealing with smaller, less complex systems-to-be-governed, compared to the larger regional or national scales where not only are the issues more complex but the range and number of actors is also much greater. However, where decision making is too narrowly structured in terms of the definition of the problem and the range of stakeholders involved, the solutions are less likely to favour resilience of the overall system (Chuenpagdee 2011).

What is missing from Kooiman et al.'s thesis is a detailed route map for the implementation of interactive governance, although a much slimmer companion volume (Bavinck et al. 2005) does offer a practitioner's guide for use in small-scale, Third World fisheries. So far, the application of interactive governance has been mainly confined to providing an 'analytical lens' for judging the efficacy of governance arrangements for small-scale MPAs in developing countries (Chuenpagdee 2011). It would appear, however, that the EU's Green Paper (2009) has at least internalised one important lesson in its insistence that in a reformed CFP the Commission's own role should focus on meta-governance functions rather than the micro-management of EU fisheries.

2.2.3 A New Paradigm for Natural Resource Management

Whereas Jentoft and Chuenpagdee (2009) and Kooiman et al. (2005) were looking to effect changes within the institutional set up for modern fisheries governance—albeit radical changes to the ways in which we formulate fisheries problems and reach decisions concerning their solutions—Berkes (2010a) is in essence seeking to replace the existing management approach with a new paradigm for the stewardship of natural resources. There are close parallels between Berkes' most recent exposition of resilience theory and notions of wicked problems and interactive governance discussed above, notably in the underlying critique of fisheries management and the need to forge robust partnerships between stakeholders and those responsible for governing the use of natural resources.

Resilience theory is certainly not a new idea. Its origins can be traced back to Holling's (1973) definition of resilience as the capacity of an ecological system to absorb disturbance and reorganise while undergoing change. During the 1970s and 80s there was a major shift from a 'balance of nature' to a 'dynamic ecosystem' paradigm based on increased understanding of their diversity, complexity and uncertainty. In coastal fisheries uncertainty is endemic to the ecological and social systems involved. The integration of ecosystem and social system perspectives on resilience owes much to Berkes and Folke (1998), Adger (2000) and Berkes et al. (2003). The value of the most recent intervention (Berkes 2010a) lies in the advocacy of resilience as the basis for a fundamental reconceptualisation of resource management that can provide a good fit for recent thinking on "property rights, participation, interactions of institutions at multiple levels and experimentation in adaptive management and interactive governance" (p. 13).

Berkes' view of resilience is prefaced by a trenchant critique of conventional notions of natural resources and their management. Criticisms are directed towards the commodification of nature and its domination by a management élite using positivist and reductionist science in pursuit of false certainties and spuriously simple technical solutions, the consequent disempowerment of local communities and the erosion of local control over resource use. Conventional management approaches tend to reduce the inherent diversity that characterises the ecological and social systems associated with coastal fisheries, thus damaging the systems' resilience, making them more susceptible to crisis and less able to recover and self organise in response to natural perturbation (Berkes 2010a). More specifically, Adger (2000) alleges that specialisation and privatisation—two hallmarks of modern management—will reduce the social cohesion through which individuals and social groups adapt to environmental and economic change.

Despite theoretical advances emphasising the intrinsic uncertainties of marine ecosystems, recent developments in EU fishing policy (revival of MSY, long term management plans such as the cod recovery plan) suggest that outmoded equilibrium-based ecosystem models and mechanistic approaches to management still hold sway. Ultimately, these could prove dangerously counterproductive, reducing both the natural variability of the ecosystem and the ability of fishers to come to terms with uncertainty through flexible fishing patterns.

As the basis of a new paradigm for fisheries management, resilience theory envisages interlinked ecological and social systems responding to uncertainty, threshold effects and change through adaptive behaviour rather than abortive attempts by management to arrest or divert the course of change. The management goals are, therefore, framed not in terms of stabilising yields (MSY) and maximising economic revenues but focused more on building the resilience of the ecological and social systems and maintaining the flexibility of fishing operations in an increasingly uncertain world. While resilience is an inherent property of natural ecosystems (and pre-industrial society), it is essentially a learned skill in modern societies. It relies on the accumulation and sharing of local ecological knowledge, collective experience, participatory forms of institutional learning and a willingness to keep open the option of innovative response in pursuit of adaptive forms of management.

2.2.4 Evaluation

None of the three key concepts discussed above—wicked problems, interactive governance and resilience theory—are examples of radical new thinking. They each represent a stepwise progression of much earlier innovations developed in different fields. More importantly, they represent a convergence of ideas stemming from a philosophical aversion to the styles of fisheries governance that have unwittingly contributed to the largely unsustainable state of the world's developed fisheries and now threaten the future status of fisheries in the developing countries. Together they regard fisheries not as a self-contained policy area but one that is closely bound up with marine environmental conservation, on the one hand, and social/community sustainability, on the other. Reformulation of fisheries as a wicked problem and the application of resilience theory to its solution could help to reconfigure the notion of precautionarity and give substance to the ecosystem approach to fisheries management, much celebrated in policy rhetoric but still lacking in real achievement.

All three concepts come together to acknowledge the interdependence of ecological, economic and social systems, call for adaptive rather than prescriptive forms of problem solving that are tailor made to the conditions of the given fishery, place emphasis on the incorporation of stakeholders as full partners in the decision making process, and look to all these attributes to enrich institutional learning. There is, however, a sense in which the new approaches may run the risk of putting too much emphasis on the local (specific) nature of fisheries problems at the expense of their generality and universality. We must be careful, therefore, not to overindulge a localist perspective. On the contrary, we must remain fully alert to the importance of scale and the need for coherence both vertically across different scales of governance and horizontally between neighbouring administrative areas.

In sum, the three concepts begin to outline a refreshingly different framework for thinking about fisheries and offer a serious challenge to more conventional views about their management. The question that the second half of the chapter addresses is whether such a framework can deliver practical benefits particularly in a complex, highly structured system of governance like the EU. Berkes (2010b) acknowledges that the most likely setting for a resilience based approach is 'small-scale, community-based fisheries' in developing countries. How far this setting can be extended to include the small-scale coastal fisheries of Europe is a moot point. Thus far, the alternative framework remains only an outline and the three concepts are essentially philosophical contributions to the debates on fisheries governance. As 'works in progress' they have still to evolve into operational forms, though we can

Table 2.1 The alternative framework for small-scale fisheries management

Objectives	Structures and process	Regulatory approach
secure the resilience of local social and ecological systems	local, stakeholder led organisations	based on principles of 'parametric management' (Wilson and Kleban) and reliant on shared experience and understanding of fish stock behaviour in the local ecosystem
maintain broad range of fishing opportunities	integrated action re ecological and social systems – guided by precautionary, ecosystem based approach – informed by scientific advice incorporating local ecological knowledge	flexible use of 'technical conservation measures' (MLS; spatial/temporal closure of grounds; gear regulations etc.)
generate flexible, adaptive responses to changing economic and environmental conditions	open and transparent interactive dialogue among all stakeholders to define problems and scope solutions	avoidance of measures that limit flexibility of fishing operations (e.g. 'privatised' catch quota; effort limits)

begin to sketch out the basic requirements of the governing systems for small-scale fisheries based on the studies by Jentoft and Chuenpagdee (2009), Kooiman et al. (2005) and Berkes (2010a). The results presented in Table 2.1 may seem a far cry from the systems of management that characterise the CFP but perhaps not all that far removed from the surviving elements of local fisheries management still to be found in parts of the EU.

2.3 Small-scale Coastal Fisheries

2.3.1 Problems of Definition

The wicked nature of the problems associated with managing coastal fisheries derives from the diversity of coastal ecosystems, the complexity of the economic and social circumstances of their exploitation, and the often intricate relationships linking the ecological and social systems. It is compounded by the difficulty of defining what we mean by coastal fisheries and further exacerbated by the paucity of reliable information at all levels as to their size, structure and social significance. The problem of definition begins with the choice of either structural or spatial parameters—that is whether we attempt to define coastal fisheries on the basis of scale of enterprise, using the surrogate of vessel size (length, tonnage or engine power) or their location using distance from the shore (3, 6 or 12 nautical miles). The choice of parameter is likely to influence the style of management—whether coastal fisheries are managed as a subset of the fishing industry as a whole, or as a separate socio-ecological entity. Difficulties can arise where governments attempt to combine the two approaches as in the UK (see below).

However and wherever the boundary is drawn, a significant number of enterprises will be found to exhibit characteristics more in keeping with those on the other side of the line, as in the case of the 'super under 10 m' vessels in the UK fleet with fishing capacities well in excess of that normally associated with small-scale fisheries. Still further complications may arise over the inclusion or exclusion of recreational fisheries and different forms of mariculture, including the traditional cultivation and harvesting of inter- and sub-tidal shellfish beds and the more recent finfish farming. The latter is scarcely consonant with the popular image of 'small-scale enterprises' but important in the context of ecosystem effects.

Even without these complications, the bewildering economic, social and cultural diversity of small-scale fisheries may prove difficult to accommodate in an equitable system of management at the local scale. In the absence of reliable information as to the true nature of small-scale fisheries, there is a danger of being lulled into accepting a stereotype of polyvalent artisanal fisheries, combining several seasonal activities, deploying a variety of *métiers* (usually but not invariably static gears), engaging in small-scale commodity production and conditioned by distinctive life mode—when, in fact, the truth is often very different. There is also a tendency rather unwisely to present an image of small-scale fisheries as inherently eco-friendly, predicated as a function of vessel size and static gears in limiting habitat damage and, together with short trip distance, generating a modest carbon footprint. This view usually ignores the cumulative impact of large numbers of small vessels operating in a limited space.

Even more dangerous from a management perspective is to ascribe a set of economic drivers very different from those associated with large-scale fishing enterprises—to assume, in fact, that small-scale fisheries are the antithesis of so-called industrial fisheries. Most artisanal fisheries are integrated into the market economy either exploiting local opportunities or seeking wider niche markets for their high quality fresh fish and shellfish. True, there is a correlation between coastal fisheries and a small-scale, family based enterprise structure. But within that simple assertion there lurks a wide range of business objectives, personal aspirations, levels of technological sophistication, market awareness, participation rates (full-time, part-time, seasonal and occasional) and survival strategies. One self-evident truth is that small-scale fisheries tend to be dispersed among very large numbers of independent enterprises often lacking effective economic and political organisation and, therefore, exerting little political influence within the industry as a whole and playing a diminishing role in wider community politics.

2.3.2 Small-scale Fisheries and the Common Fisheries Policy

Practically everywhere one looks small-scale fisheries dominate the size structure of the world's fisheries. This is certainly true in the EU where around 83 % of the 85,000 strong fishing fleet is made up of vessels under 12 m overall length, but they account for only 10 % of gross tonnage and 35 % of aggregate engine power

(EC 2010a). Their contribution to overall landing value and to employment in the fisheries sector is more difficult to calculate, due to the paucity of comparable data and varying rates of participation. They are certainly significant at regional and local levels where in many of the remoter fisheries dependent areas the small-scale sector contributes an important source of local employment.

The European Commission's approach to small-scale fisheries is ambivalent. The CFP scarcely provides a sympathetic framework for the deliberation of wicked problems or for the successful incubation of resilience thinking. Not only is the EU's fishing zone far too large for effective governability, but the governing system enshrined in the CFP comes close to the archetypal centralised, command and control model reliant on a positivist and reductionist science and mediated by a management élite that Berkes (2010a) found inimical to the stewardship of natural resources—and 'a far cry' from the kind of thinking that resilience implies (Berkes 2010b, p. 55). In practice the CFP neatly sidesteps the issue of small-scale fisheries through a derogation from the 1982 regulation governing the conduct of the CFP that granted exclusive use of inshore waters (0–6 nm) to the coastal state's fishing industry. This allowed member states to assume much of the responsibility for management.

From an EU perspective, small-scale fisheries are viewed as a social rather than economic issue, associated more with the sustainability of coastal communities in the remoter parts of the EU's peripheral regions than with the overall performance in the fisheries sector. As Gallizioli (Chap. 3) makes clear the CFP is not seen as an instrument of social policy. Thus the CFP has made few direct concessions to the small-scale sector in providing protection from the effects of structural and geographical concentration in the commercial fishing industry. To an extent social issues are addressed through financial support from the European Fisheries Fund's Axis IV programme for improving the sustainability of coastal (and inland) areas with significant levels of fisheries employment through community led action (see Budzich-Tabor, Chap. 10). This involves a territorial rather than sectoral approach to sustainable development. Support is channeled through 'fisheries local action groups' (FLAGs) charged with devising and overseeing a strategy for strengthening the economic resilience and competitiveness of the local fishing industry through cooperation, partnership building and exploiting new niches in the marine ecosystem and local economy[1].

Somewhat surprisingly, the Commission chose to raise the issue of small-scale fisheries in the context of the 2012 reform of the CFP. In its Green Paper (EC 2009) the Commission raised the possibility of establishing differentiated management regimes for large-scale and small-scale sectors of Europe's fishing fleets, predicated on what might appear to be a false dichotomy. The suggestion was again linked to

[1] By the end of 2011 there were well over 200 FLAGs established across virtually all of the EU's coastal member states, with a particularly strong concentration in the Baltic states. Projects qualifying for financial support from the European Fisheries Fund were designed to strengthen the competitiveness of the fishing industry through adding value to local production, improving local marketing capability, diversifying fishing activities and increasing the integration of fisheries with other sectors of the local economy – notably tourism – *inter alia*.

the small-scale sector's "role in the social fabric and the cultural identity of the EU's coastal regions" (p. 14). Whereas the approach to managing the large-scale sector would have capacity reduction and economic efficiency as its central concerns using market based incentives (tradable fishing rights) to achieve these aims, the approach to small-scale fisheries management would be predicated on undefined social objectives, allocating non-tradable fishing rights to be used individually or through collective management schemes. Moreover, public funding would be available "to help the small-scale segment adapt to changing conditions in the wake of CFP reform" (p. 14)[2]. The Commission's view was that decisions concerning the small-scale sector should be taken as close as possible to the communities themselves—thus opening up the possibility of community-led management in which the seeds of resilience thinking might take root.

The notion of differentiated management regimes met with a mixed response in the consulting process that followed (EC 2010b). Although some member states lent support to the idea, there was a more widespread feeling that the choice of management system was best left to individual member states. No reference was made to the idea of differentiated management in the proposals laid before the Council of Ministers and the European Parliament in 2011 (EC 2011a). Repeated warnings were made of the need for "specific measures to help manage small-scale coastal fleets" (EC 2011a, p. 3) both as a general principle and as a caveat to the mandatory introduction of tradable fishing concessions as the principal instrument for managing overall fishing capacity and promoting a profitable industry. Privatisation of fishing rights has to be seen as an unfriendly act in the context of small-scale fisheries (Højrup 2011). It places limits on flexibility and adaptive behaviour and only those with capital or borrowing power can invest in the market for additional fishing rights. Moreover it limits the scope for future actions to transform the management system. Despite the adoption of notional time limits to such schemes, systems of privatised use rights acquire a permanence of their own, creating powerful vested interests and making it difficult for governments to afford the costs of compensating those who have invested heavily in the market for fishing rights.

Despite an initial commitment to fundamental reform, including a major redistribution of responsibilities between the European institutions on the one hand and the regions and member states on the other, the outcome of the reform process suggests a further strengthening of the present management framework (see Table 2.2). Few concessions have been made to the original ideas of regionalising the CFP

[2] Although proposals for the new Maritime and Fisheries Fund, 2014–2020 (EC 2011b) place renewed emphasis on integrated territorial development to reverse the decline of fisheries dependent communities, attention is also paid to the role of small-scale coastal fleets. In a sector where the majority of businesses are micro-enterprises with limited access to funding, special measures attracting higher levels of grant aid are proposed to support professional advice on business and marketing strategies and innovative development. Priority will be given to collective approaches building on existing social capital and permitting the attainment of critical mass for investment. The new fund will, for the first time, recognise the role of spouses in family fishing businesses through support for training and skill acquisition in fields of entrepreneurship and business management.

Table 2.2 The basic framework for managing the Common Fisheries Policy

Objectives	Structure and process	Regulatory approach
sustainable fisheries (MSY)	centralised decision making (Commission; Council; European Parliament)	long-term management plans for commercial species
efficient, profitable and competitive fishing sector	species based stock conservation informed by formal scientific advice (ICES)	output (TACs/quota) and input (days at sea) restrictions, supplemented by technical conservation measures (MLS; closures; gear regulations)
	limited engagement of stakeholders in management (POs) and advisory bodies (AC; RACs)	transferrable fishing concessions

(Symes 2012), delegating responsibility to member states and extending the participation of the industry in the management process. Indeed, with recommendations for the mandatory adoption of transferable use rights for all vessels over 12 m and for those under 12 m deploying towed gear, the CFP has deepened its attachment to a conventional, productivist view of management. This may be appropriate for industrial, offshore fisheries (though results may suggest otherwise) but it is questionable in the context of small-scale fisheries and inimical to the cultivation of a resilience approach.

2.3.3 Small-scale Fisheries in the UK

In order to gain deeper insight into the nature of management systems that address Europe's coastal fisheries, it is necessary to turn to the member states. Yet even at this level it is difficult to identify examples of local management systems dedicated to dealing with the conditions of small-scale fisheries (Symes and Phillipson 2001). Exceptions can be found in the *prud'hommies* of southern France (Frangoudes 2001) and the Spanish *cofradia* (Alegret 1998), both survivors of very much older systems, supplemented in France by the more recent hierarchical network of *comités de pêches* linking national, regional and local tiers of the administration. Throughout much of Europe, however, responsibility for coastal fisheries lies firmly in the hands of national administrations.

The UK offers something of a microcosm of the wider European situation, for the devolved administrations in Scotland, Wales and Northern Ireland have followed different pathways in the management of inshore waters (Phillipson and Symes 2010), providing a striking contrast between the devolved approach in England and Wales and the more centralised system in Scotland. In part this reflects differences in the overall importance of the inshore waters and the significance of the shellfish sector. As Table 2.3 suggests, shellfish production (excluding *Nephrops*) and the small-scale sector are more prominent in England and Wales, compared with either Scotland or Northern Ireland. The ensuing analysis focuses primarily on the evolving situation in England with its long established, locally devolved system of

Table 2.3 Relative importance of shellfish landings and under 10 m vessels in UK, by country of administration, 2008 . (Source: UK Sea Fisheries Statistics, 2008: Tables 3.2 and 2.2 (Defra 2009)

	England	Wales	Scotland	N. Ireland	UK
Landings by value £m					
Total landings	130.8	19.9	343.7	28.2	517.8
of which shellfish (excl. Nephrops)	68.0	15.5	56.1	3.2	142.8
%	52	78	16	14	28
Fishing fleet					
All vessels	3200	470	2213	204	6087
of which 10 m and under	2635	436	1505	147	4723
%	82	93	68	72	78

inshore management with the potential to incorporate a resilience-based approach but where ongoing changes could have the effect of closing the door to more adaptive forms of management in future.

2.3.4 Inshore Fisheries Management

In Scotland, where the political economy of fisheries has long been dominated by the more mobile offshore sector, responsibility for managing inshore waters (0–12 nm) has remained in the hands of the devolved administration in Edinburgh. Although wide powers were granted under the *Inshore Fishing (Scotland) Act*, 1984, actions were confined mainly to resolving local gear use conflicts with little evidence of measures to protect inshore stocks. Pressures for fundamental change to this reactive approach, involving more direct stakeholder involvement at local level (Symes & Ridgway 2003), resulted in proposals for 12 Inshore Fishing Groups (IFGs) responsible for developing local management plans, with an initial tranche of six pilot groups in 2009. IFGs fall short of the ideal for local integrated management in several respects. First, their executive committees are confined to commercial fishing interests; access to scientific and nature conservation expertise is provided through supporting advisory committees. Secondly, the composition of the executive committees sometimes favours non-local, nomadic fleet interests at the expense of local static gear fishermen. Finally, responsibility for implementing the management plans remains with Marine Scotland. The decision in 2012 to reduce the total number of IFGs to six, largely on financial and administrative grounds, further weakens their claim to reflect local management aspirations. Only in rare instances where local fishermen have succeeded in establishing a Regulating Order—as in the Shetland Islands—can it truly be said that a system of local inshore management has been implemented in Scotland.

By contrast, in England and Wales a governing system of devolved local management for inshore waters has been in place for well over a century. A network of Sea Fisheries Committees (SFCs) responsible for managing inshore fisheries, initially out to 3 nm but later extended to 6 nm, began to emerge after 1888. The resulting

12 Sea Fisheries Districts varied greatly in size, complexity and budgetary strength from the genuinely local to regional scale districts embracing as many as a dozen local authorities. Membership of the committees was divided equally between local authority councillors, representing the funding bodies, and stakeholders drawn largely from local commercial fishing interests. SFCs were tolerably well equipped with statutory powers for managing the fisheries, including both local byelaws and Regulating and Several Orders[3] (Symes 2002). One of the more remarkable features was their independent enforcement capability with a complement of land based and seagoing fishery officers and at least one vessel at their disposal.

There were, however, some significant limitations to their management functions, most notably the lack of emergency powers and an inability to cap fishing effort in the shellfisheries. As local regulation was subject to approval by the fisheries departments in London or Cardiff, changes in management were slow and the pace of adapting to changing conditions was reduced. The biggest challenge to SFCs came in the late 1990s with the added requirement of having to demonstrate that inshore fishing activities were conducted in an environmentally responsible manner. Although membership of the committees was extended to include local nature conservation interests, no additional resources or powers were made available. Only a handful of larger SFCs were able to employ their own scientific staff to carry out regular surveys of essential habitats in the inshore waters.

Pressure for reform of a system of inshore management that had remained substantially unaltered for almost 40 years came from within the SFCs, from independent observers (Symes 2002) and eventually from a government enquiry (Defra 2004). A new system of Inshore Fisheries and Conservation Authorities (IFCAs) entered into force in April 2011, which in some respects bore an uncanny resemblance to the one that preceded it (Phillipson and Symes 2010). Their geographical structure was little altered, except for Wales and North West England[4]. But there were significant changes including a redefinition of their remit in respect of both marine nature conservation and fisheries management; an extension of their powers; improved funding arrangements; and a shift in the balance of membership for the new authorities. This last change is seen by some in the industry as the most alarming: representation of the local authorities has been reduced and the seats allocated to local and national conservation interests increased at the expense of commercial

[3] Byelaws were used to limit the size of vessels operating within the district, regulate the design, size and number of towed gears deployed, effect local closures and vary national regulations governing minimum landing sizes to suit local circumstances, *inter alia*. Under the Sea Fisheries (Shellfish) Act, 1967, Regulating and Several Orders provided scope for more detailed management of shellfish beds, with Several Orders enabling division and allocation of beds to individual fishermen for the purpose of cultivation (restocking) and harvesting.

[4] The only significant change to the geography of inshore management came about as a result of the devolved Welsh administration's decision to unify arrangements in Wales and bring responsibility for inshore fisheries management 'in house', necessitating the amalgamation of districts in north west England. Elsewhere, the UK government decided against major rationalisation, preferring to maintain the identity of the previous SFCs and so retain established links within the local fishing industry.

fishing interests, leaving it open to speculation as to whether the underlying aims of IFCAs lie in the domain of fisheries or conservation management.

Through their very functions, structures and processes, IFCAs (like their SFC antecedents) appear well able to accommodate a resilience-based approach to inshore management. Indeed, many within those organisations would already claim to be implementing something very close to that approach. They are now geared to tackle local ecological and fisheries issues in a more integrated way. Their broad stakeholder base allows them to incorporate local ecological knowledge alongside more conventional forms of scientific advice in their decision making. And a genuine attempt at ecosystem-based fisheries management designed to suit local circumstances is now possible. What is required is for their actions to confirm the guiding principles of resilience thinking in coping with uncertainty through flexible and adaptive forms of management that will underpin the sustainability of local ecosystems and local inshore fisheries.

2.3.5 *Sectoral Management*

The future of small-scale fisheries, however, does not depend solely on the management of inshore waters. 70% of under 10 m vessel earnings are derived from non-quota species—mainly from shellfisheries within the 0–6 nm zone—but a significant minority of under 10 m vessels rely on landings of high value demersal fish subject to regulation under the CFP and managed through the UK's quota management system (QMS). The UK government is anxious to merge existing arrangements for quota management[5] into a unified system for all segments of the fleet built on a wealth-based approach that seeks to maximise the economic value of the UK's resource allocation through tradable fishing rights. The intersection of local (IFCA) and national (sectoral) management regimes casts doubt over the future direction of small-scale fisheries management in England.

After decades of inactivity, the opening decades of the 21st century have seen two initiatives for reforming the small-scale sector in England. Under the Sustainable Access for Inshore Fisheries (SAIF) initiative launched in 2009, a sector-led advisory group presented two contrasting options (SAIF 2010a, b). One involved the segregation of the small-scale sector, granting under 10 m vessels exclusive fishing rights within the 0–6 nm zone and extending the functions of IFCAs to include managing their quota entitlements. Despite doubts about the

[5] Under current arrangements, the UK's QMS is divided into three separate schemes: (a) The so-called sectoral quota scheme for vessels over 10 metre that are members of producer organisations (POs), in which the PO is responsible for managing the quota allocations of its member vessels, including quota swaps with other POs. An informal system of individual transferable quota is permitted within this scheme. Over 90% of UK quota is handled through sectoral quota management; (b) The quota allocations adhering to non-sector over 10 metre vessels (i.e. those outwith membership of a PO), managed by the relevant fisheries department; (c) Under 10 metre vessels that fish from a separate quota pool, managed by the fisheries department, usually on a monthly or bi-monthly catch limit basis.

ability of IFCAs, as currently constituted, to undertake this additional role, there was qualified support for this approach from within the small-scale sector. By contrast, the alternative strategy that found favour with the fisheries department and the National Federation of Fishermen's Organisations, lay in integrating the small-scale sector within a unified QMS with the owners of under 10 metre vessels opting either to join existing Producer Organisations (POs) or community quota groups (CQGs).

The government's own consultation on reforming the QMS in England (Defra 2011a) closely followed SAIF's second option. Seeking to remove the arbitrary division between under and over 10 metre vessels and dispense with government's direct involvement in quota management, the proposals confirmed the intention to establish a single QMS allocating individual transferable quota (ITQ) to those under 10 m vessels electing to join a PO and transferring the remaining under 10 m quota pool to establish CQGs[6]. Further proposals with implications for the small-scale sector included plans to exclude vessels that were either inactive or failed to take up their complement of fishing opportunities in order to extinguish the threats from dormant or latent fishing capacity. Rather more disturbing were suggestions to extend the QMS to include lobster and brown crab, key species for the small-scale sector.

One problem remains: the results of the consultation (Defra 2011b), based on a small uncontrolled sample, were inconclusive. There is considerable divergence of opinion across the industry as a whole, and within the small-scale sector, as to the most appropriate form(s) of managing small-scale fisheries. Implementation of the government's proposals will, therefore, have to await the results from piloting alternative solutions, including CQGs, incorporation within existing POs and a separate PO for under 10 m vessels.

Overall, the government's proposals and, to a degree, the small-scale sector's broad complicity, seem likely to usher in a radically different approach to that projected by the arrangements for managing inshore waters (see Table 2.4). The government is intent on rolling out a system of ITQ across the English fleet that will see the small-scale sector more deeply embedded in a system of management that has maximising the economic value of the UK's fishery resources as its core principle and individual tradable fishing rights as the indispensable mechanism. Its proposals will tie up most of the fishing opportunities available to the under 10s and so remove what remains of the flexibility of choice of fishing patterns, survival and development strategies that hitherto have underpinned the sustainability of the sector. The changes are also intended to commercialise and professionalise the small-scale

[6] One of the complications in reforming the QMS is the serious imbalance between existing capacity and quota allocations within the under 10 metre fleet, brought to light as a result of the Registration of Buyers and Sellers legislation introduced in 2007. Previously no formal records of the landings of under 10 metre vessels were available. The government has indicated its willingness to redistribute 3 % of the overall English demersal quota to help restructure the under 10 metre fleet as part of the proposals to create a unified QMS. Part of this would be allocated to vessels seeking membership of POs; the balance would be used to enhance the allocation of quota to establish the CQGs.

Table 2.4 Proposed framework for managing small-scale fisheries in England

Objectives	Structures and process	Regulatory approach
a) Inshore management • ensure exploitation of inshore fisheries conducted in a sustainable manner • balance achievement of economic and social benefits with protection of environment	local stakeholder led organisations (IFCAs) incorporating LAs, commercial and recreational fisheries and local and national conservation interests	use of byelaws and Regulating and Several Orders to provide effort restrictions and technical conservation measures
b) Sectoral Management • wealth based approach to ensure maximum value is extracted from available fishing opportunities	market led initiatives unified QMS quota managed through POs or community quota groups (CQGs)	Individual transferable quota (except for CQGs) with 'one way' value to prevent acquisition of under 10 m quota allocations by over 10 m vessels

sector by limiting the scope for part-time, seasonal and occasional participation in the fishery. They run counter to the interests of the small-scale sector and to the basic precepts of resilience thinking.

2.4 Conclusions

Despite the intellectual advances in our understanding of sustainable fisheries management outlined at the outset of the chapter, a gap clearly persists between theory and practice as evident from the ensuing analysis of European and UK small-scale fisheries management. The gap cannot be explained simply by reference to the long lead time in translating innovative theoretical designs into robust operational practice. In European fisheries management the problem appears to go much deeper: it relates, in part, to the intrinsic characteristics of the European political system and the difficulties in overcoming institutional inertia and the resistance of vested interests that have profited under the old regime.

The CFP is in the grip of incremental, path-dependent policy development where past decisions greatly influence the costs associated with future options. It becomes progressively more expensive—politically, organisationally and financially—and therefore more difficult to alter course. In the case of the CFP, path dependency is reinforced by a faith in positivist science to model social and natural causal choices, a belief in the ability of the state to manage these causal choices for the common good (Gezelius and Raakjaer 2008) and a shared commitment to output controls (TACs and quota) as the preferred means of delivering policy. Early attempts at reforming the CFP chose to accept these constraints. By contrast, the more ambitious 2012 reform agenda that sought to modify the centralised decision making system faced implacable legal challenges aimed at preserving the *status quo* (Symes 2012).

It was never very likely that the EU's Common Fisheries Policy would provide a fertile environment in which to cultivate the seeds of a resilience-based approach to stewardship of fishery resources. Its highly centralised and bureaucratic structures, allied to a strong neoliberal ethic, are inimical to a style of management grounded in the idea of fisheries and coastal governance as wicked problems, reliant on finding solutions through the adaptive behaviours of ecological and social systems and based on interactive learning and collaborative decision making. Instead the CFP seems more likely to continue its quest for sustainable fisheries through the rationales of science, managerialism and the market, ignoring the internal contradictions that inhibit the fishermen's ability to harvest the fishing opportunities fully and efficiently. The principles of non-discrimination and relative stability, allied to a seemingly inflexible decision-making system embedded in the European Treaties, leave little room for experimentation at the local level.

It might, therefore, seem more reasonable to assume that alternative forms of fisheries management which address issues at the local scale and in ways that suit local conditions would flourish in circumstances where the coastal state can exert more control over policy formulation, as in the case of inshore fisheries. Evidence from the UK, however, fails to lend unequivocal support for this argument. Small-scale fisheries that in the past have benefited in some degree from neglect by central government now find themselves being drawn ever more deeply into a mainstream, path dependent policy framework and the mélange of bureaucratic controls. Systems of local management that approximate to a resilience-based approach are being challenged by proposals to incorporate under 10 m vessels within universal systems of sectoral management that will increasingly expose them to the risks of privatisation. These risks are made all the more real by the absence of a unified response from the small-scale sector as a consequence of its highly diverse structures and aspirations.

Resilience theory is unlikely to form the template for managing the highly developed, industrial fisheries found throughout much of Europe. Nevertheless, its further elaboration and dissemination can serve two very useful functions in the European context. First, it can act as a mirror to reflect some of the cruder imperfections of the current approach where respect for the underlying ecological and social systems has been degraded. Secondly, it can buttress the argument that in sensitive coastal environments fisheries management must learn to cope with uncertainty and respond to turbulence by ensuring the continued flexibility and adaptability of small-scale, inshore fisheries.

References

Adger, W. N. (2000). Social and ecological resilience: are they related? *Progress in Human Geography. 24,* 347–364.

Alegret, J.-L. (1998). Property rights, regulating measures and strategic responses among the fishermen of Catalonia. In D. Symes (Ed.). *Property Rights and Regulatory Systems in Fisheries.* (pp 175–187). Oxford: Blackwell Science.

Bavinck, M., Chuenpagdee, R., Diallo, M., van der Heijden, P., Kooiman, J., Mahon, R. & William, S. (2005). *Interactive Governance: A Guide to Better Practice*. Eburon: Delft

Berkes, F. (2010a). Shifting perspectives on resource management: Resilience and the reconceptualisation of 'natural resources' and 'management'. *Maritime Studies, 9,* 13–40.

Berkes, F. (2010b). Rejoinder, *Maritime Studies. 9,* 55–58.

Berkes, F., & Folke, C. (Eds.). (1998). *Linking Social and Biological Systems*. Cambridge: Cambridge University Press.

Berkes, F., Colding, J., & Folke, C. (Eds.). (2003). *Navigating Social-Ecological Systems: Building Resilience for Complexity and Change*. Cambridge: Cambridge University Press.

Budzich-Tabor, U. (2014). Area-based local development—a new opportunity for European fisheries areas. In J. Urquhart, T. Acott, D. Symes & M. Zhao (Eds.) *Social Issues in Sustainable Fisheries Management*. Dordrecht: Springer.

Chuenpagdee, R. (2011). Interactive governance for marine conservation: an illustration. *Bulletin of Marine Science. 87,* 197–211.

Department for Environment, Food and Rural Affairs (2004). *Review of marine fisheries and environmental enforcement*. London: Defra.

Department for Environment, Food and Rural Affairs (2009). *United Kingdom Sea Fisheries Statistics 2008*. London: Defra.

Department for Environment, Food and Rural Affairs (2011a). *Consultation on reform of the English fisheries management arrangements*. London: Defra.

Department for Environment, Food and Rural Affairs (2011b). *Summary of responses to the consultation on domestic fisheries management reform in England*. London: Defra.

EC (Commission of the European Communities). (2009). *Green Paper: Reform of the Common Fisheries Policy.* COM (2009) 163 final. Brussels: The European Commission.

EC (Commission of the European Communities) (2010a). *Facts and Figures on the CFP*. Brussels: CEC.

EC (Commission of the European Communities) (2010b). *Synthesis of the consultation on the reform of the Common Fisheries Policy.* SEC (2010) 428 final. Brussels: CEC.

EC (Commission of the European Communities). (2011a). *Communication from the Commission to the European Parliament, the Council, the European Economic and Social Committee and the Committee of the Regions: Reform of the Common Fisheries Policy.* COM (2011) 417 final. Brussels: CEC.

EC (Commission of the European Communities) (2011b). *Proposal for a Regulation of the European Parliament and of the Council on the European Maritime and Fisheries Fund.* COM (2011) 804 final. Brussels: CEC.

Frangoudes, K. (2001). France. In D. Symes & J. Phillipson (Eds.), *Inshore Fisheries Management.* (pp. 139–153). Oxford: Blackwell Science.

Gallizioli, G. (2014). The social dimensions of the Common Fisheries Policy: a review of current measures. In J. Urquhart, T. Acott, D. Symes & M. Zhao (Eds.), *Social Issues in Sustainable Fisheries Management*, Dordrecht: Springer.

Gezelius, S., & Raakjaer, J. (Eds.). (2008). *Making Fisheries Management Work: Implementation of Policies for Sustainable Fishing*. Dordrecht: Springer.

Holling, C. S. (1973). Resilience and stability of ecological systems. *Annual Review of Ecology and Systematics. 4,* 1–23.

Højrup, T. (2011). *The Common Fisheries Policy has to recognise the need for common goods for coastal communities: common community quotas for sustainable life modes in coastal fisheries. The alternative to privatization of fishing rights in the home waters of Europe*. Ferritslev: Havbåde

Jentoft, S., & Chuenpagdee, R. (2009). Fisheries and coastal governance as a wicked problem. *Marine Policy. 33,* 553–560.

Kooiman, J., Bavinck, M., Jentoft, S., & Pullin, R. (Eds.). (2005). *Fish for Life: Interactive Governance for Fisheries*. Amsterdam: Amsterdam University Press.

Phillipson, J., & Symes, D. (2010). Recontextualising inshore fisheries: the changing face of British inshore fisheries management. *Marine Policy. 34,* 1207–1214.

Rhodes, R. A. W. (1996). The new governance: governing without government. *Political Studies. 44*(4), 652–657.

Rittel, H. W. J., & Webber, M.M. (1973). Dilemmas in a general theory of planning. *Policy Sciences. 4,* 155–169.

Sustainable Access to Inshore Fisheries Advisory Group (2010a). *Steps towards sustainable inshore fisheries: position paper.*

Sustainable Access to Inshore Fisheries Advisory Group (2010b). *Reform of the English inshore fishing fleet: Final report—key findings and recommendations.*

Symes, D. (2002). *Inshore fisheries management in England Wales: facing up to challenges of the 21st century.* English Nature research report 448. Peterborough: English Nature.

Symes, D. (2012). Regionalising the Common Fisheries Policy: context, content and controversy. *Maritime Studies. 11*(6).

Symes, D., & Phillipson, J. (Eds.). (2001). *Inshore Fisheries Management.* Dordrecht: Kluwer.

Symes, D., & Ridgway, S. (2003). *Inshore fisheries regulation and management in Scotland: Meeting the challenge of environmental integration.* Edinburgh: RSPB Scotland and Scottish Natural Heritage.

Wilson, J.A. & Kleban, P. (1992). Practical implications of chaos in fisheries. *Maritime Anthropological Studies,* 5(1), 67–75.

Chapter 3
From Boom and Bust to Local Stewardship: A Governance Benchmark for Celtic Sea Fisheries Management

Mike Fitzpatrick

Keywords Fisheries governance • Governance benchmarking • Fisheries co-management • Ecosystem approach to fisheries management • Celtic Sea Herring

3.1 Introduction

In 2004 six fishing vessels participated in the Celtic Sea herring fishery and the spawning stock biomass (SSB) was estimated to be 29.000 t. This was very close to the lowest SSB on record, when the stock had previously collapsed and was closed for five years from 1977 to 1982. The stock in 2004 was predominantly composed of 1 and 2-year-old fish and the overall feeling within the management advisory forum was that another complete closure of the fishery was a strong possibility. The total first sale value of the fishery was approximately € 250,000. This was a serious decline for a fishery which in earlier years was targeted by over 100 vessels and seasonally employed over 1500 people in processing factories alone (Molloy 2006).

Fast-forwarding to 2012, we find that the stock has made an excellent recovery back to historically high levels. The total allowable catch (TAC) has increased by over 300 % in four years but is constrained by highly precautionary fishing mortality rules prescribed under the recovery plan jointly developed and agreed by scientists and industry through the management advisory forum. Approximately 72 Irish vessels ranging in size from 10 to 45 m currently participate in the fishery. Over this period the local management forum has persisted and matured and attitudinal changes with regard to long-term decisions and trade-offs between markets and sustainable management are apparent.

This successful stock recovery presents an obvious contrast with the majority of European fisheries over the same period. This chapter, while focusing on the Celtic Sea herring fishery, utilises a governance benchmarking exercise to compare three Irish fisheries with differing success levels in stock recovery and varying governance profiles. First, the following section presents a general introduction to the Celtic

M. Fitzpatrick (✉)
Coastal and Marine Resource Centre, University College Cork, Cork, Ireland
e-mail: mike.fitzpatrick@ucc.ie

J. Urquhart et al. (eds.), *Social Issues in Sustainable Fisheries Management*, MARE Publication Series 9, DOI 10.1007/978-94-007-7911-2_3,

Sea herring fishery and its management history. This is followed by a governance benchmarking assessment of the fishery and two others for comparative purposes. Next is an analysis of how the local management forum, the Celtic Sea Herring Management Advisory Committee, has impacted on governance and the outlook for implementation of an ecosystem approach; and finally there are some conclusions on what can be learned from the successes and problems encountered in the governance of this fishery and the usefulness of the governance benchmarking approach.

3.2 The Fishery and its Management

The Celtic Sea herring fishery is a single species pelagic fishery predominantly targeted by Irish fishing vessels off the South coast of Ireland in International Council for the Exploration of the Sea (ICES) areas VIIj, VIIg and the southern part of VIIa (Fig. 3.1). The fishery is predominantly an inshore one and is conducted by a diverse fleet of vessels ranging from under 10 m multi-purpose inshore vessels up to modern 50 m pelagic vessels equipped with refrigerated seawater tanks. It has traditionally been a very important fishery for both the fleet and processing sectors in the south of Ireland although landings in the last 10 years have been well below their previous peaks and the length of the fishing season has also significantly decreased.

The fishery has in recent years been exploited almost entirely by Ireland with small reported catches by other nations. The only other significant players involved in the fishery are Dutch vessels and Dutch owned vessels registered in France and Germany. It is essentially a single species fishery.

The history of the fishery over the past 50 years has been one of an alternating boom and bust cycle (see Fig. 3.2). The TAC in 2010 was increased by 70 % over the 2009 figure and in 2011 increased by a further 30 %. In 2011 the rebuilding plan achieved its aim of maintaining SSB above the precautionary biomass level, B_{pa} for the third consecutive year, and the parameters of a long term management plan have been agreed and await ratification by the European Commission. Discussions have focused on the optimal balance between fishing mortality (F), stock biomass (SSB), total catch (TAC) and constraints on annual TAC fluctuations all of which are aimed at minimising the risk of stock collapse. Under the current management regime the fishing mortality rate is at its lowest estimated level in the past 50 years.

3.2.1 Current Management Institutions and Approaches

In 2001 the ICES advice for the Celtic Sea herring stock recommended a cut from the previous year's TAC of 20,000 t to a precautionary level of 6,000 t for 2002. This was mainly based on a poor age profile for the stock which showed an over dependence on juvenile fish. Although eventually the scientific advice for the stock was amended and the TAC was set at 13,000 t, stakeholders in the fishery were concerned enough to establish a Celtic Sea Herring Management Advisory Committee (CSHMAC) in 2001.

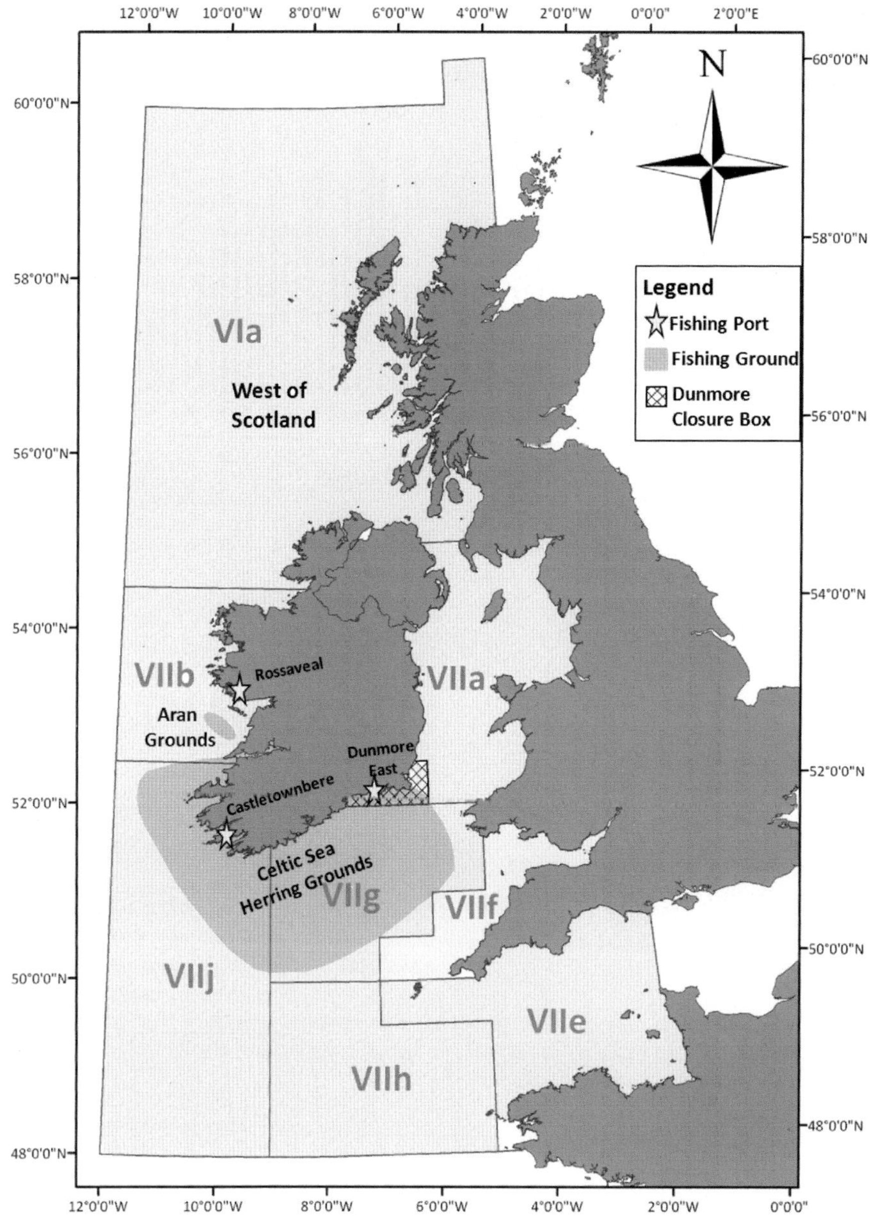

Fig. 3.1 ICES areas in Irish waters and Celtic Sea and Aran fishing grounds

The committee consists of representatives of fishermen, processors, scientists and control authorities. The Committee was established with the overarching goal of sustaining annual catches of 20,000 t and to rebuild the stock if necessary to achieve this. Another strong objective was to improve the partnership between industry and scientists.

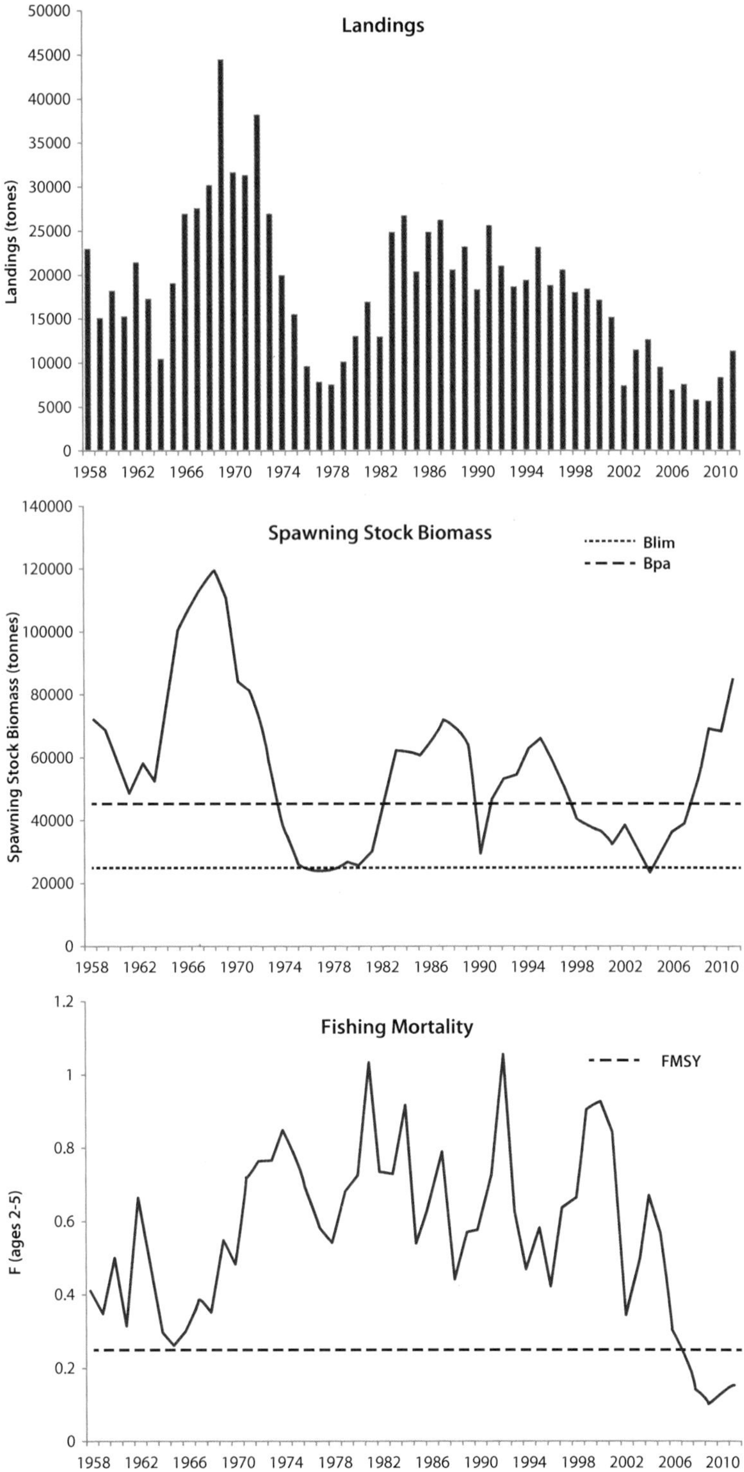

Fig. 3.2 Landings, spawning stock biomass and fishing mortality in the Celtic Sea herring fishery since 1958. (Source: Marine Institute Stock Book 2011)

In 2005 the Committee was officially recognised as an advisory committee by the Irish fisheries minister and tasked with providing advice to the minister and managers from the fisheries department. Although officially only advisory, following ministerial recognition the committee has found that more of its advice has been accepted and the partnership between industry and science has strengthened. In this sense the management of the fishery could be considered to represent an informal version of co-management.

One of the most significant measures taken was the closure for several years between 2002 and 2006 of a large area off Dunmore East known as the Dunmore Box (Fig. 3.1) where herring spawning took place and where fishing effort had previously been concentrated. This was aimed at reducing catches of small first time spawning herring. However, despite this initiative the TAC continued to decline so in 2007 a rebuilding plan was developed by the CSHMAC in conjunction with scientists from the Marine Institute. The rebuilding plan set a very low fishing mortality level, allowed for a small-scale fishery with a guaranteed quota allocation and strengthened the annual closure of the spawning area. In 2011 the stock was deemed to have recovered and from 2012 a long term management plan (LTMP) will replace it. The LTMP also sets a very low fishing mortality level (well below the fishing mortality estimated to achieve maximum sustainable yield, F_{MSY}) and retains the closure of the spawning area.

Another significant development in the fishery within the past decade has been the strengthening of control and enforcement in both legislative and operational terms. These changes have been driven mainly by the introduction of the pelagic weighing regulations and the establishment of an independent fisheries control agency. These factors have increased confidence in the precision of the scientific assessment and the Marine Institute in their most recently published advice state that "under the current management regime the quality of the catch data has improved" (Marine Institute 2011).

3.3 Governance Benchmarking

Grafton et al. (2007) describe "ineffective and inappropriate governance" as the number one cause of negative marine ecosystem outcomes. This is certainly echoed in the top five failings of the CFP identified in the European Commissions Green Paper (2009) which were all governance related. Grafton et al. suggest that a governance benchmarking exercise can identify underlying causes of unsustainable fishing and steps towards implementing an ecosystem approach. The idea and process is similar to governance profiling described in Juda and Hennessey (2001) and the governance baseline approach outlined by Olsen et al. (2009). The governance benchmarking assessment evaluates how current governance arrangements may impact on the implementation of an ecosystem approach in the Celtic Sea herring and uses two other fisheries, Aran Ground *Nephrops* and Celtic Sea mixed demersal fisheries, for comparative purposes.

The *Nephrops* fishery on the Aran Grounds in Area VIIb is a well-established fishery that has been exploited since the mid-1970s but has been exclusively an Irish fishery since around 1988. Currently there are 12 large whitefish vessels (> 15 m) and another 8 smaller, weather dependent vessels in the fleet. The majority of these vessels fish from the port of Rossaveal on the west coast of Ireland. Landings of *Nephrops* from the Aran Grounds in recent years have been around 700–900 t. Currently a single TAC is applied to the overall Area VII *Nephrops* fishery, which includes stocks in the Irish Sea, Porcupine Bank, SW Ireland and the Celtic Sea in addition to the Aran Grounds. Despite the use of a variety of technical measures the gear used is still largely unselective for the target species, *Nephrops*, as well as the most common by-catch species such as haddock and hake. There is a single target species for the fishery; it operates in a well-defined inshore area and the participating vessels predominantly land into one port and through one co-operative. The major problem in the fishery is discarding of fish and small *Nephrops*, which have been observed as being high.

The mixed demersal fishery in the Celtic Sea area (centred on ICES Areas VIIg and VIIj) is a highly diverse fishery targeting mainly cod, haddock and whiting, involving a large number of vessels from Ireland, France, the UK and Belgium, ranging in size from 10 to 40 m and fishing with a variety of gears including otter trawls, beam trawls, gillnets and Scottish seines. Currently, the fishery is managed by TACs and quotas. In addition there is a seasonal closure during cod spawning of three ICES statistical rectangles in Areas VIIg and VIIf that has been in place since 2005 as well as a range of gear-based technical measures. Discarding is believed to be considerable for all species driven *inter alia* by restrictive TACs and poor gear selectivity. The current scientific advice for the major whitefish stocks in this area is uncertain. In comparison to the Celtic Sea herring and Aran *Nephrops* fisheries, this is much more problematic with governance arrangements complicated by the mix of target species, fleets, gears and national management structures. There are emerging positive examples of co-operation across fleets in the fishery. The seasonal closure currently in place is the result of a transnational industry initiative and there are active discussions between industry and scientists, facilitated through the North Western Waters Regional Advisory Council (NWWRAC) in developing a long term management plan for whitefish in the area.

The criteria used for governance benchmarking are derived from a number of sources. The primary source is the five key governance principles identified from the literature by Grafton et al. (2007): accountability, authority and responsibility; transparency; incentives; risk assessment and management; and adaptability. These are supplemented with the principles of the Ecosystem Approach to Fisheries Management (EAFM) and marine management summarised from a number of the most commonly cited and relevant policy documents and journal papers (see Table 3.1). The criteria are also inclusive of the principles used in a European Commission White Paper on Good Governance including: participation, openness, accountability, coherence and effectiveness (EC 2001). These five principles are also contained in Art. 2 of the current CFP regulation (EC 2371/2002) and the European Commission Green Paper on CFP Reform (2009). Juda's (1999) interpretation of integration included integration between natural and social sciences as a desirable step towards including multi-disciplinary perspectives and this facet is incorporated in the benchmarking criteria. Table 3.2 lists the criteria used and summarises the benchmarking scores for each of the three fisheries.

Table 3.1 Principles of the ecosystem approach summarized from literature

Principle	FAO[1]	UNGA[2]	CBD[3]	MSFD[4]	CCAMLR[5]	Australian EBM[6]	ICES[7]	WWF[8]	Costanza *et al*[9]	Juda[10]
Maintain ecosystem integrity & function	√	√	√	√	√	√	√	√	√	√
Broad stakeholder participation	√	√	√	√		√	√	√	√	√
Apply the precautionary principle	√	√	√	√	√		√		√	√
Use adaptive management			√	√		√	√	√	√	√
Ensure spatial & management compatibility	√	√	√				√		√	
Use a sectorally integrated approach	√	√	√				√			√
Use a social-ecological systems approach	√	√				√		√		
Utilise broad knowledge base		√	√			√				
Use incentives	√								√	√
Devolve management to lowest appropriate level			√	√						

[1] FAO 2003, [2] UNGA 2006, [3] Convention on Biological Diversity 1998, [4] EC 2008, [5] CCAMLR 1982, [6] Scandol et al. 2005, [7] ICES 2005, [8] Ward et al. 2002, [9] Costanza et al. 1998, [10] Juda 1999

Table 3.2 Results of the governance benchmarking assessment

Category	Criteria	Benchmark Score*		
		Celtic Sea Herring	Aran-Nephrops	Celtic Sea De-mersal
Objectives	Are there clear strategic objectives?	1	1	1
	Are there clear operational objectives?	2	1	1
	Has a long-term management plan been agreed and implemented?	3	0	1
Responsibility	Is there accountability for decisions and outcomes?	2	2	2
	Are there clearly defined roles and responsibilities?	2	1	1
	Are independent management assessments used?	2	2	2
	Are social performance indicators used?	0	0	0
	Are economic performance indicators used?	1	1	1
	Are ecosystem performance indicators used?	1	1	1
Transparency	Is the decision-making process transparent to non-participants?	0	0	0
	Is the research process collaborative?	3	2	3
Participation	Is there a formal or informal co-management process?	3	1	1
	Are a broad spectrum of stakeholders involved in management?	2	1	1
Incentives	Are there incentives to avoid by catch & habitat damage?	2	0	1
	Is a rights-based-management system used?	0	0	0
	Is there strong enforcement of the rules?	4	3	3
Adaptive Management	Is in-season adjustment to management possible?	1	0	0
	Is there a real-time closure option	2	1	1
	Is fishers tacit knowledge utilized?	2	2	2
Integration	Is there an integrated institutional framework?	1	1	1
	Is there integration between natural and social sciences?	2	1	1

*Benchmark scores: *4*-Governance element fully in place; *3*-Governance element mostly satisfied, but not yet fully operationalized; *2*-Governance element partially satisfied, but further development is required; *1*-Governance element is not satisfied, but steps towards its development are in place; *0*-Governance element missing in the fishery

For the purpose of summarising and communicating the results of the governance benchmarking exercise a five-point grading system used by Grafton et al (2007) is employed as it provides a simple visual indication of the degree to which the criteria have been operationalised. In order to facilitate score comparison across the case study fisheries a numeric notation is used to indicate performance rather than the alphabetic one used by Grafton et al.

The governance benchmarking results are based on 20 interviews with a range of participants in the management of the assessed fisheries. The interviewees included two inshore fishermen, four fishermen operating traditional "dry-hold"[1] vessels, three skippers of larger refrigerated sea water vessels, two representatives of fishermen's organisations, two scientists responsible for the assessment of the stocks, two managers of fish processing plants, two salesmen for fisheries cooperatives, a fisheries protection officer, a director of a responsible fishing certification scheme and a director of a marine environmental NGO. The benchmarking is also based on observations made while attending approximately 60 meetings of CSHMAC and on my own experiences of working within Irish fisheries governance as both an employee of a fishermen's representative organisation and afterwards as an observer.

3.3.1 Discussion of Benchmarking Results by Category

3.3.1.1 Objectives

Clear and prioritised management objectives are essential to developing and measuring the success of management plans (Pascoe et al. 2009). The current CFP objectives are cursory and extremely high level (Symes 2009) and accordingly offer very little to guide strategic planning at the fishery level. Although the CFP specifies that it should satisfy environmental, social and economic objectives the European Commission itself criticises the lack of priority setting between objectives and the fact that "There are no clear indicators and yardsticks that could provide more concrete guidance or to help measure policy achievements" (EC 2009).

It is obviously difficult to see how operational objectives can be set in the absence of higher level strategic ones and accordingly this criteria scores poorly. In the Celtic Sea herring fishery there has been an indigenous attempt to set long-term objectives firstly through the recovery plan and now through the agreed long term management plan (LTMP). However these objectives are narrowly focused on biological or stock targets and the long term plan does not have any formal status as it has not at the time of writing been assessed by STECF. The fact that the CSHMAC is advisory rather than a statutory management forum with limited ability to make some significant decisions also makes it difficult to give a higher grade to this criterion.

The drafting of long-term management plans has been incentivised by a ruling from the European Commission (EC 2011), which prescribes a highly precautionary TAC setting in the absence of an LTMP. The use of LTMPs should have an additional benefit of reducing the level of political horse trading at December council meetings. Of the three fisheries assessed here only Celtic Sea herring has a locally agreed LTMP. The NWWRAC is currently developing a LTMP for Celtic Sea demersal

[1] "Dry-hold" vessels store their herring catches in the traditional way, mixed with ice in lockers or compartments in the fish hold. Their numbers have decreased rapidly over the past 10 years due to increasing completion from vessels which can store their fish for longer in refrigerated sea water (RSW) tanks.

fisheries but it is still a draft and has not yet been subjected to any scientific assessment. To date there have been no attempts to develop a management plan for the Aran *Nephrops* fishery. The Irish Marine Institute Stock Book for 2011 makes the following recommendation: "There are no explicit management objectives or a management plan for Nephrops stocks in VII. FSS recommends that management objectives be established and that management plans be developed with stakeholders and implemented for fisheries catching Nephrops" (Marine Institute 2011).

The difficulty with not having clear policy objectives is illustrated in a recent review of management arrangements for Irish herring fisheries, instigated in 2011 by the Irish fisheries minister at the request of some industry representatives unhappy with existing arrangements. Written submissions from all interested parties were sought and a Ministerial proposal was produced in response. Due, in particular to objections to the proposed restriction on future access to the fishery, a public meeting to discuss the issues was organised in January 2012. In terms of consultation the process was fair, all interested parties were given two opportunities to make written submissions and one to air their views at a public meeting. However, the problem remains that the Minister's criteria on which to base his decision are essentially arbitrary. There is no national policy on allocation of fisheries quotas in general nor on pelagic fisheries specifically. However, there is an *ad hoc* process of limiting access to pelagic fisheries underway. To date restricted access regimes for mackerel, horse mackerel, blue whiting, boarfish and herring have been established but in each case the allocation criteria has varied. Concerns about the basis for the Minister's decision are evident in the fact that at least one fishermen's organisation submitted a request to the Minister asking that he include, with his final decision, an explanation for the criteria used.

3.3.1.2 Responsibility

Interviews with participants in the governance system indicate that they generally feel that their roles are reasonably well defined but that accountability is very poorly structured which results in the 'blame game' being regularly played out between various governance parties. This is symptomatic of a poorly structured governance system, which is mirrored at national and European level. The creation of advisory bodies such as CSHMAC and the RACs at both levels does little to improve accountability as an advisory group can easily disown negative outcomes by provision of examples of their advice being ignored and their ignorance of the criteria used to make the final decision. Furthermore, an advisory committee usually has influence only on certain aspects of the management process. In terms of Schlager and Ostrom's (1992) hierarchy of decision-making, advisory committees operate mainly at the operational level, partly at the collective choice level but critically not at the constitutional level where the most fundamental decisions are taken.

On independent management assessment the fisheries score more highly. ICES evaluate scientific aspects of management, particularly with respect to precaution and periodically ICES Working Groups will nominate a problematic stock for a full audit. STECF also assess technical and economic aspects of management decisions and plans. However, there is no mandatory requirement, such as exists in the US under the Magnusson-Stevens Act, for a full management strategy evaluation.

The use of indicators with which to monitor the success of management plans is a definite problem area. At present biological or stock indicators are used comprehensively but from a governance perspective broader reference points and indicators are essential elements. However, this relates to the issue of clear objectives: without explicit social, economic and ecosystem objectives the use of indicators is rather pointless, except perhaps to produce a data set to act as a baseline to inform the assessment of success relative to some future objective. The use of ecosystem performance indicators should increase rapidly as a range of ecological data is now required under the Data Collection Framework, the Marine Strategy Framework Directive, the Habitats Directive and the Water Framework Directive.

The absence of social indicators reflects the fact that social objectives have been dealt with in an equivocal fashion in European fisheries governance (Symes and Phillipson 2009). Repeated references to issues such as "providing a fair standard of living for those who depend on fishing activities" in the basic CFP regulation (Council of European Ministers 2002) are not backed up with any explicit objectives or operational targets. This problem originated in early CFP negotiations where France and Italy had tried in 1960 and again in 1992 to have social objectives included in the CFP, specifically to have funding allocated to alleviate unemployment arising from shrinking fishing fleets but these attempts were unsuccessful due to concerns about increasing the Community budget (Holden and Garrod 1996).

3.3.1.3 Transparency

There is a highly opaque decision-making process in each of these three fisheries. A member of the general public would have extreme difficulty in getting information on how operational or strategic decisions were made. Whether management meetings are occurring at local, national or European level, very few of the negotiations are subject to public scrutiny. CSHMAC has recently developed an action plan to address this issue as it was raised during the assessment process for Marine Stewardship Council (MSC) certification.

The collaborative research process scores are better as there is a long-standing relationship between scientists and CSHMAC and improving levels of collaborative scientific initiatives in relation to Celtic Sea demersal fisheries. Some conflict with industry has set back attempts to build a science-industry partnership in the Aran *Nephrops* fishery but there has been a project attempting to utilise the knowledge of fishers participating in the fishery.

3.3.1.4 Participation

Participation of stakeholders in management is a key principle of the ecosystem approach, second only to the maintenance of ecosystem structure and function in terms of citation frequency in the EAFM literature. The 2001 EU White Paper on governance (EC 2001) lists participation as one of its five key components and states that "The quality, relevance and effectiveness of EU policies depend on ensuring wide participation throughout the policy chain—from conception to implementation."

Dubbink and van Vliet (1996) describe three governance levels, the macro-level of state and inter-state bureaucracy, the meso-level of civil and private organisations and the micro-level of individuals. The co-management and interactive governance perspectives emphasise that good governance requires a greater input from the meso- and micro- levels. Grafton et al., in their paper on governance benchmarking, also alludes to the same issue when describing the challenge of connecting higher-level ecological goals with day-to-day management decisions as the missing link in fisheries governance (Grafton et al. 2007).

Typologies of both participation (Arnstein 1969; Pretty 1995) and co-management (Sen and Raakjaer Nielsen 1996; Carlsson and Berkes 2005) have been developed, all of which loosely range from hierarchical, manipulative or non-participatory modes at one end through various degrees of consultation through to full delegation of power or self-mobilisation at the other end. If we accept the definition of co-management as "the sharing of power and responsibility between the government and local resource users" (Berkes et al. 1991) then none of the three fisheries exhibit what would strictly be classified as a formal co-management structure. However, in all but name, which denotes an advisory role, CSHMAC can be considered as an informal co-management process as the majority of its recommendations are implemented across many aspects of management.

Aran *Nephrops* and Celtic Sea demersal fisheries are more typical of the general Irish fisheries management framework in that they do not have a dedicated management forum and are centrally managed at a departmental level and as part of the NWWRAC sub-committee structures. As these fora are advisory or consultative the degree of sharing of rights and responsibilities is quite low and thus cannot be classed according to Berkes et al.'s definition as co-management.

In terms of breadth of stakeholder participation Irish fisheries management is quite restrictive being limited largely to representatives of state institutions (department officials, scientists and the state fisheries development board) and the fisheries organisations. The NWWRAC stakeholder profile is more inclusive with one third of membership being open to interested parties from outside the fishing industry. This division of power within the RAC structures has been criticised as creating an imbalance of power particularly for minority interests with less resources than well-funded fisheries organisations (Lutchman et al. 2009; Long 2010). The counter position to this has been described as the participation paradox: "the greater the number of actors, the smaller the role each plays, and the lesser the importance of traditional sectors" (Suarez de Vivero et al. 2008).

CSHMAC has recently increased the diversity of stakeholders with the invitation of an environmental NGO, representatives of the fisheries control agency and social scientists to attend meetings on a regular basis. The MSC certification process has incentivised, through recommendations on governance of the fishery, the formalisation of these expanded stakeholder arrangements.

3.3.1.5 Incentives

Economists have emphasised the importance of understanding the role incentives play in fisheries management for many years (Clark and Munro 1975; Hatcher 1997; Hatcher and Gordon 2005). In the past 20 years that emphasis has been expanded to accommodate complex systems theory and in particular the need to embed incentives within an ecosystem approach (Hanna 1998; Hilborn et al. 2005; Grafton et al. 2006; Charles 2006). Rights-based management has been identified as a key enabling factor for positive economic outcomes in common pool resource management contexts (Ostrom 1990; Grimur Valdimarsson and Metzner 2005; Bromley 2008; Costello et al. 2008).

However, the emphasis on incentives and rights-based mechanisms has not significantly penetrated the governance regime of the three Irish fisheries assessed here with the exception of the control regime. Unsurprisingly this produces a feeling among industry that the governance regime is all stick and no carrot. In all three fisheries there are significant disincentives for conservation actions due to the fact that all three fisheries are in either full or partial open access regimes.

In relation to avoiding bycatch and habitat damage the incentives again are all top-down which fishermen often perceive as a negative. The designation of Special Areas of Conservation under the Habitats Directive, in contrast to the closure of spawning boxes for herring and cod in the Celtic Sea, have not received much fishing industry support. This may be due to their permanent nature, dissatisfaction with the designation process or simply because they do not have a perceived benefit for their target species.

A recent announcement by the Fisheries Minister whereby additional quota will be given to fishermen using nets with an approved escape device for young fish in the Celtic Sea demersal fishery indicates a change in attitude towards the use of incentives to avoid bycatch.

It remains to be seen whether environmental certification can act as a strong driver of change in terms of incentives to avoid environmental damage. CSHMAC has asked, as part of the certification process, to develop an environmental impact plan by the MSC process auditors and this plan will address issues such as cetacean bycatch, the use of observers, protection of gravel spawning beds and other environmental impacts from the fishery.

3.3.1.6 Adaptive Management

Examples of 'active' adaptive management are few due, at least in part, to practical difficulties in designing management measures as experiments and also in attributing outcomes to measures adopted (Defeo et al. 2007; Walters 2007). Nevertheless, it is widely cited as being a crucial element of an ecosystem approach (Walters 1997; Olsson 2006; Armitage et al. 2009). 'Passive' adaptive management, which places a different emphasis on the learning aspect of the management process and does not require multiple simultaneous management strategy experiments, is probably a more pragmatic option. It incorporates the idea of addressing uncertainty through learning by doing, and is explicitly iterative. It is sometimes disparagingly described as *ad hoc* management but in fact adaptive management follows a planned and deliberate sequence of monitoring, assessment and design.

Aspects of adaptive management are being implemented in these fisheries: there is a trend towards increased use of real-time measures and fishermen's knowledge. However, there are some serious challenges to the application of adaptive management in the three fisheries. These include a persistent desire for stability, predictability and certainty by all stakeholders. Additionally, the explicit use of alternative management strategies, evaluation of their consequences and scenarios aimed at addressing uncertainty will require a change of mind-set and additional flexibility which does not necessarily fit with the current development of LTMPs. Such a planned and experimental approach more than likely requires an institutional maturity, which would have to be preceded by a period of co-management capacity building.

3.3.1.7 Integration

Poor scores on integration within the institutional framework are unsurprising given the disintegrated marine governance structures existing at Irish and European level currently. Despite the fact that there is now an Integrated Maritime Policy (EC 2007) and a Marine Strategy Framework Directive (European Council and Parliament 2008) both of which cover multiple industrial sectors the degree to which fisheries policy will be integrated particularly within the IMP framework is debatable (Juda 2007; van Hoof and van Tatenhove 2009; Rätz et al. 2010; Wakefield 2010).

At an Irish level the degree of disintegration is a concern. There is an inter-departmental co-ordination committee comprising the assistant secretaries of at least five different departments with marine responsibilities. There is some evidence of a move to improve this as a consultation that aims to develop an integrated Irish marine policy.

Juda (1999) states that "social scientists also have an essential role to play in the governance process since ecosystem-based governance addresses human behavior". In comparison, particularly with Nordic countries such as Norway and Denmark, Irish fisheries research and governance have not until recently included any significant role for social science or economics so integration between natural and social sciences has been almost non-existent. Moves to redress this are being made and current Irish research programmes include investigations of governance aspects of the ecosystem approach, the economics and socio-economics of Irish fisheries, and the collation and use of fishermen's tacit knowledge.

3.4 The Role of the Local Management Forum & EAFM Implementation

Three basic modes of fisheries governance have been described (Gray 2005; Symes 2006): top-down or hierarchical governance; self-governance which involves devolution of responsibility to the individual level; and co-governance involving a partnership between the state and user groups.

Few world fisheries systems correspond exactly with these ideal alternatives but instead contain elements of each to a greater or lesser degree. The governance benchmarking exercise has shown that overall Irish fisheries governance can be classified as a hierarchical or top-down system but one that shows a slight trend towards increasing incorporation of co-governance elements. Although the Minister and civil servants consult on the majority of issues with the fishing industry and the number of fora where such consultations take place has been increasing, stakeholder representatives are limited to an advisory role. Executive authority in all cases still rests exclusively with the Minister and his department officials.

A hierarchical system if it functions well is not necessarily negative. However, in the case of Irish fisheries the effectiveness of the hierarchical structure is compromised both by weak national policy making capacity and by serious legitimacy problems with the CFP. The first issue, that of weak national policy-making capacity, is well illustrated by the lack of a management framework for Irish inshore shellfish fisheries. Given that 73 % of vessels on the Irish fleet register are under 12 m, it is evident that good governance arrangements for the sector should be a priority. The most tangible fisheries recommendation in a recent Irish Department of Agriculture policy statement, Harvest 2020, is that "the implementation of a specific Inshore Fisheries Management framework should proceed as speedily as possible" (Department of Agriculture 2010). The fact that 40 years after Ireland entered a common European fisheries system it still lacks a management framework for the main fisheries sector within its exclusive competence is more of an indictment of its past governance regime than a laudable objective for the future.

The second issue with the hierarchical governance regime, that of the legitimacy of the CFP, is summarised in the report on a comprehensive Irish fisheries strategy review conducted in 2006, which concluded that the principal cause of conflict in Irish fisheries was the fact that "the EU Common Fisheries Policy, which the State is required to implement, is universally unpopular with the fishing industry" (Cawley et al. 2006). This legitimacy problem creates significant challenges for centralised policymaking and governability (Jentoft 2000; Chuenpagdee and Jentoft 2009), which strengthen the case for some further devolution or regionalisation.

Specifically in relation to the ecosystem approach the lack of policy direction at national level is compounded by shortcomings in the CFP. Under Art. 2.1 of the 2002 CFP (EC 2371/2002) there is a commitment made to the "progressive implementation of an ecosystem based approach to fisheries management". However, the lack of any definition of strategy, goals or indicators for implementation of an

EAFM within the 2002 CFP has been widely criticised (Sissenwine and Symes 2007; Lutchman et al. 2009; Symes 2009). The Commission itself has criticised its own progress on the ecosystem approach in the CFP Green Paper (EC 2009) where they find that "while direct references are made to adopting a precautionary and an ecosystem approach …. there are no clear indicators and yardsticks that could provide more concrete guidance or to help measure policy achievements". Simply put, there are no extant European or Irish fisheries policy drivers towards implementation of the ecosystem approach. This is a definite barrier to the implementation of an ecosystem approach as it constrains the capacity for change at lower levels.

The governance benchmarking exercise examines how these policy issues are manifested in three Irish fisheries. In general the fisheries do not score particularly highly but Celtic Sea herring does perform better overall. In terms of an average grade across all the criteria examined, the Celtic Sea herring fishery scores 2, indicating that governance elements are partially satisfied but further development is required. The other two fisheries, Aran *Nephrops* and Celtic Sea mixed demersal, do less well with an average score of 1, which indicates that governance elements are not satisfied, but steps towards their development are in place. The most significant differences between the fisheries were in relation to the existence of a long-term management plan and also the degree of management participation. In the case of Celtic Sea herring these two factors are intrinsically linked, as the presence of a dedicated management forum over a number of years created a platform for a strong industry-science partnership, which in turn facilitated the development of a long-term management plan.

In total, on 10 out of 21 criteria Celtic Sea herring scored better than either of the other two fisheries. Not all of these improved scores can be attributed to the presence of a co-management process; for instance, a higher score for control and enforcement reflects the fact that regulations governing pelagic fisheries are better defined and more prescriptive than for demersal fisheries. However, in the categories of operational objectives, accountability, broad stakeholder involvement, incentives to avoid bycatch and habitat damage, adaptive management and integration, much of the drive to improve these aspects has come through CSHMAC. It is a strongly held belief among those interviewed that both governance performance and biological stock status for Celtic Sea herring would be closer to those for the other two fisheries in the absence of a longstanding management forum.

3.5 Conclusions

Chuenpagdee and Jentoft, and others within the interactive governance school of thought, caution against approaching governance as a set of idealised performance indicators which are attainable within any system (Kooiman 2005; Chuenpagdee and Jentoft 2007). They advocate an examination of governability, which involves a detailed assessment of the interactions between the governing system and the system to be governed. This gives a more realistic measure of the capacity of a

Table 3.3 Building blocks and obstacles in moving towards an ecosystem approach in Irish fisheries

Building opportunities
Collaborative research initiatives
Increasingly effective control and enforcement
Example of some co-management success with Celtic Sea Herring
Top-down drivers towards development of Long Term Management Plans
Changing incentives and greater industry assumption of responsibility under MSC or other certification schemes
Increasing use of ecosystem indicators required under EU legislation
Obstacles
Opaque management process and decision-making criteria
Lack of clear strategic and operational objectives
Underuse particularly of social and also economic indicators
Participation is purely consultative for most fisheries and stakeholder field is narrow
Underuse of 'positive' incentives such as rights based management and incentives to minimise environmental impacts
Absence of an integrated framework
Adaptive management would require both a general mind-set and institutional change

given social-ecological system to attain good, but reachable, rather than ideal, but unattainable, governance goals. The governability approach recognises that many natural resource management processes are inherently political, are influenced by variable human and financial resource availability and that many governance performance indicators are contestable. This is evidenced in the on-going debate about the benefits of participation in resource management and whether a greatly expanded pool of participants enhances or inhibits the management process (Dubbink and van Vliet 1996; Mikalsen and Jentoft 2003; Suarez de Vivero et al. 2008). The reality is that the right level of participation, devolution, transparency etc., depends on the individual case and detailed contextual understanding is required to ensure good governance outcomes.

However, a governance benchmarking exercise is very useful as an intermediate or scene-setting stage for more detailed analysis (Adrianto et al. 2005). In this chapter an attempt has been made to put the crude mechanistic benchmarking scores into context with the history of the fishery and its management. To summarise the findings from this assessment Table 3.3 lists the significant building blocks and obstacles towards the implementation of fisheries ecosystem plans in the context of the fisheries assessed.

These opportunities and obstacles highlight the need for stronger policies which both facilitate and incentivise local management actions and which ensure that wider societal concerns are addressed within local management fora. CSHMAC has shown that local management initiatives can autonomously improve governance structures and, in the process, promote stock recovery and ameliorate conflict. However, when left to their own devices, and without strong policy direction, issues which may not rank highly on the priority list of the fishing and processing industries (for example, non-commercial food-web elements or the necessity for transparency and inclusiveness in decision-making) will inevitably not be reflected in management actions. Additionally, despite the informal co-management status that advisory committees may attain, their ability to address higher level decisions is limited. Accordingly, issues such as the setting of high-level objectives, the use of social and economic indicators, institutional integration and resolution of property rights issues lie outside their control and depend on policy makers at both the European and national levels to improve their performance.

This has implications for management of coastal fisheries in the wider European context. Coastal fisheries do not exist in a governance and biological vacuum; actions taken and stock levels are influenced by a complex web of interactions across varying ecological and institutional scales. Ideally, a form of multi-level governance is required. Multi-level governance has been defined as "the sharing of policy-making competencies in a system of negotiation between nested governments at several tiers (supranational, national, regional and local) on the one hand and private actors (e.g. NGOs, producers, consumers and citizens) on the other" (van Hoof et al. 2012). This multi-level governance would be informed by strategic policy directives aimed at ensuring that high-level sustainability objectives are achieved. At the local fishery level tailored and collaborative decision-making aimed at the long-term would be possible through a local management forum. Crucially, there should be one or more intermediate levels, such as the Regional Advisory Councils (RACs), where issues such as interactions between fleets from different member states and the possibility of scaling up responses to locally successful management initiatives would be discussed. While such a governance system would not be a panacea for all fisheries management problems, it would certainly address some of the prominent obstacles to implementation of an ecosystem approach.

References

Adrianto, L., Matsuda, Y., & Sakuma, Y. (2005). Assessing local sustainability of fisheries system: A multi-criteria participatory approach with the case of Yoron Island, Kagoshima prefecture, Japan. *Marine Policy, 29*(1), 9–23.

Armitage, D. R., Plummer, R., Berkes, F., Arthur, R. I., Charles, A. T., Davidson-Hunt, I. J., Diduck, A. P., Doubleday, N. C., Johnson, D. S., Marschke, M., McConney, P., Pinkerton, E. W., & Wollenberg, E. K. (2009). Adaptive co-management for social and ecological complexity. *Frontiers in Ecology and the Environment, 7*(2), 95–102.

Arnstein, S. R. (1969). A ladder of citizen participation. *Journal of the American Planning Association, 35*(4), 216–224.

Berkes, F., George, P., & Preston, R. (1991). Co-management. *Alternatives, 18*(2), 12–18.

Bromley, D. (2008). The crisis in ocean governance: Conceptual confusion, economic nonsense, political incoherence. *MAST, 6*(2), 29–33.

Carlsson, L., & Berkes, F. (2005). Co-management: Concepts and methodological implications. *Journal of Environmental Management, 75*(1), 65–76.

Cawley, N., Murrin, J., & Bric, O., R. (2006). *Steering A New Course: Strategy for a Restructured, Sustainable and Profitable Irish Seafood Industry 2007–2013*. Dublin, Department of Communications, Marine and Natural Resources.

CCAMLR . (1982). *Convention on the Conservation of Marine Living Antarctic Resources*. Canberra, Australia.

Charles, A. (2006). Benefits and costs of implementing the ecosystem approach to fisheries. *Bergen Conference on Implementing the Ecosystem Approach to Fisheries*, Bergen, Norway, 26–28 September 2006.

Chuenpagdee, R., & Jentoft, S. (2009). Governability assessment for fisheries and coastal systems: A reality check, *Human Ecology, 37*(1), 109–120.

Clark, C. W., & Munro, G. R. (1975). The economics of fishing and modern capital theory: A simplified approach. *Journal of Environmental Economics and Management, 2*(2), 92–106.

Convention on Biological Diversity (1998). *Report of the Workshop on the Ecosystem Approach*. Malawi. UNEP/CBD/COP/4/Inf.9.

Costanza, R., Andrade, F., Antunes, P., den Belt, M. v., Boersma, D., Boesch, D. F., Catarino, F., Hanna, S., Limburg, K., Low, B., Molitor, M., Pereira, J., Rayner, S., Santos, R., Wilson, J., & Young, M. (1998). Principles for Sustainable Governance of the Oceans. *Science, 281*(5374), 198–199.

Costello, C., Gaines, S. D., & Lynham, J. (2008). Can catch shares prevent fisheries collapse? *Science, 321*(5896), 1678–1681.

Defeo, O., McClanahan, T. R., & Castilla, J. C. (2007). A brief history of fisheries management with emphasis on societal participatory roles. In Defeo, McClanahan & Castilla (Eds.), *Fisheries Management: Progress Towards Sustainability*, (pp. 1–21), London: Wiley-Blackwell.

Department of Agriculture, Food & the Marine. (2010). *Harvest 2020: a vision for Irish agri-food and fisheries*. Dublin.

Dubbink, W., & van Vliet, M. (1996). Market regulation versus co-management—two perspectives on regulating fisheries compared. *Marine Policy, 20*(60), 499–516.

European Commission. (2001). *European Governance—A White Paper*. COM (2001) 428 final. Brussels, pp. 35.

European Council (2002). Regulation (EC) No 2371/2002. OJ L 358, 31.12.2002, 59.

European Commission. (2007). *An integrated maritime policy for the European Union*. COM (2007) 575 final. Brussels.

European Council and Parliament (2008). *Directive 2008/56/EC of the European Parliament and of the Council of 17 June 2008 establishing a framework for community action in the field of marine environmental policy* (Marine Strategy Framework Directive).

European Commission. (2009). *Green Paper on Reform of the Common Fisheries Policy*. Brussels. COM (2009) 163.

European Commission. (2011). *Communication from the Commission concerning a consultation on Fishing Opportunities*. Brussels. COM (2011) 298.

FAO (2003). The ecosystem approach to fisheries. *FAO Technical Guidelines for Responsible Fisheries. No. 4, Suppl 2* pp. 112. Rome.

Grafton, Q., Kompas, T., McLoughlin, R., & Rayns, N. (2007). Benchmarking for fisheries governance. *Marine Policy, 31*(4), 470–479.

Grafton, R. Q., Arnason, R., Bjorndal, T., Campbell, D., Campbell, H. F., Clark, C. W., Connor, R., Dupont, D. P., Hannesson, R., Hilborn, R., Kirkley, J. E., Kompas, T., Lane, D. E., Munro, G. R., Pascoe, S., Squires, D., Steinshamn, S. I., Turris, B. R., & Weninger, Q. (2006). Incentive-based approaches to sustainable fisheries. *Canadian Journal of Fisheries and Aquatic Sciences, 63*(3), 699–710.

Gray, T. S. (2005). Theorising about participatory fisheries governance. In T. S. Gray (Ed.), *Participation in Fisheries Governance*. Dordrecht: Springer.

Hanna, S. S. (1998). Institutions for Marine Ecosystems: Economic Incentives and Fishery Management. *Ecological Applications, 8*(1), 170–174.

Hatcher, A. C. (1997). Producers' organizations and devolved fisheries management in the United Kingdom: Collective and individual quota systems. *Marine Policy, 21*(6), 519–533.

Hatcher, A. C., & Gordon, D. (2005). Further Investigations into the Factors Affecting Compliance with U.K. Fishing Quotas. *Land Economics, 81*(1), 71–86.

Hilborn, R., Orensanz, J. M., & Parma, A. M. (2005). Institutions, incentives and the future of fisheries. *Philosophical Transactions of the Royal Society B-Biological Sciences, 360*(1453), 47–57.
Holden, M. J., & Garrod, D. (1996). *The Common Fisheries Policy*. Fishing News Books.
ICES. (2005). Guidance on the application of the ecosystem approach to management of human activities in the European marine environment. *ICES Cooperative Research Report. No. 273*, pp. 22.
Jentoft, S. (2000). Legitimacy and disappointment in fisheries management. *Marine Policy, 24*(2), 141–148.
Juda, L. (1999). Considerations in developing a functional approach to the governance of large marine ecosystems. *Ocean Development & International Law, 30*(2), 89–125.
Juda, L. (2007). The European Union and Ocean Use Management: The Marine Strategy and the Maritime Policy. *Ocean Development & International Law, 38*(3), 259–282.
Juda, L., & Hennessey, T. (2001). Governance profiles and the management and use of large marine ecosystems. *Ocean Development and International Law, 32,* 43–69.
Kooiman, J., Bavinck, M., Jentoft, S., & Pullin, R. (2005). *Fish for Life: Interactive Governance for Fisheries.* Amsterdam, Amsterdam University Press.
Long, R. (2010). The role of regional advisory councils in the european common fisheries policy: Legal constraints and future options. *The International Journal of Marine and Coastal Law, 25*(3), 289–346.
Lutchman, I., Grieve, C., Des Clers, S., & De Santo, E. (2009). *Towards a reform of the Common Fisheries Policy in 2012—A CFP Health Check.* IEEP, London, pp. 80.
Marine Institute (2011). *The Stock Book: Annual Review of Fish Stocks in 2011 with Management Advice for 2012.* Rinville, Oranmore, Co. Galway, Ireland, Marine Institute, pp. 502.
Mikalsen, K. H., & Jentoft, S. (2003). Limits to participation? On the history, structure and reform of Norwegian fisheries management. *Marine Policy, 27*(5), 397–407.
Molloy, J. (2006). *The Herring Fisheries of Ireland, 1900–2005*, Marine Institute.
Olsson, P., Gunderson, L. H., Carpenter, S. R., Ryan, P., Lebel, L., Folke, C., & Holling, C. S. (2006). Shooting the rapids: navigating transitions to adaptive governance of social-ecological systems. *Ecology and Society, 11*(1).
Ostrom, E. (1990). *Governing the commons: the evolution of institutions for collective action.* Cambridge University Press.
Pascoe, S., Proctor, W., Wilcox, C., Innes, J., Rochester, W., & Dowling, N. (2009). Stakeholder objective preferences in Australian Commonwealth managed fisheries. *Marine Policy, 33*(5), 750–758.
Pretty, J. N. (1995). Participatory learning for sustainable agriculture. *World Development, 23*(8), 1247–1263.
Rätz, H.-J., Dörner, H., Scott, R., & Barbas, T. (2010). Complementary roles of European and national institutions under the Common Fisheries Policy and the Marine Strategy Framework Directive. *Marine Policy, 34*(5), 1028–1035.
Scandol, J. P., Holloway, M. G., Gibbs, P. J., & Astles, K. L. (2005). Ecosystem-based fisheries management: An Australian perspective. *Aquatic Living Resources, 18*(3), 261–273.
Schlager, E., & Ostrom, E. (1992). Property-rights regimes and natural resources: A conceptual analysis. *Land Economics, 68*(3), 249.
Sen, S., & Raakjaer Nielsen, J. (1996). Fisheries co-management: A comparative analysis. *Marine Policy, 20*(5), 405–418.
Sissenwine, M., & Symes, D. (2007). *Reflections on the Common Fisheries Policy*. The European Commission, pp. 76.
Suarez de Vivero, J. L., Mateos, J. C. R., & del Corral, D. F. (2008). The paradox of public participation in fisheries governance. The rising number of actors and the devolution process. *Marine Policy, 32*(3), 319–325.
Symes, D. (2006). Fisheries governance: A coming of age for fisheries social science? *Fisheries Research, 81*(2–3), 113–117.

Symes, D. (2009). Reform of the European union's common fisheries policy: Making fisheries management work. *Fisheries Research, 100*(2), 99–102.

Symes, D., & Phillipson, J. (2009). Whatever became of social objectives in fisheries policy? *Fisheries Research, 95*(1), 1–5.

Ward, T., Tarte, D., Hegerl, E., & Short, K. (2002). *Policy proposals and operational guidance for ecosystem-based management of marine capture fisheries*. World Wildlife Federation, pp. 83.

United Nations General Assembly. (2006). *Report on the work of the United Nations Open-ended Informal Consultative Process on Oceans and the Law of the Sea at its seventh meeting.* UNGA, pp. 28.

Valdimarsson, G., & Metzner, R. (2005). Aligning incentives for a successful ecosystem approach to fisheries management. *Marine Ecology Progress Series, 300,* 286–291.

van Hoof, L., van Leeuwen, J., & van Tatenhove, J. (2012). All at sea, regionalisation and integration of marine policy in Europe. *Maritime Studies, 11*(9).

van Hoof, L., & van Tatenhove, J. (2009). EU marine policy on the move: The tension between fisheries and maritime policy. *Marine Policy, 33*(4), 726–732.

Wakefield, J. (2010). Undermining the Integrated Maritime Policy. *Marine Pollution Bulletin, 60*(3), 323–333.

Walters, C. J. (1997). Challenges in adaptive management of riparian and coastal ecosystems. *Conservation Ecology, 1*(2).

Walters, C. J. (2007). Is Adaptive Management Helping to Solve Fisheries Problems? *Ambio, 36*(4): 304–307.

Chapter 4
The Social Dimensions of the Common Fisheries Policy: A Review Of Current Measures

Giorgio Gallizioli

Keywords Common Fisheries Policy • European Fisheries Fund • Fisheries dependent communities • Integrated Maritime Policy • Temporary cessation of fishing activities • PESCA programme

4.1 Introduction

The third reform of the Common Fisheries Policy (CFP) offers an opportunity to reflect on the achievement of fundamental objectives of the policy that, as the European Regulation R.2371/2002 reminds us[1], are threefold: to ensure sustainable economic, environmental and social conditions for the sector. In the 2009 Green Paper (EC 2009), the Commission highlighted the shortcomings of the current CFP and gave an indication of possible solutions by tackling, in particular, the environmental and economic aspects. The proposals for the reform of the CFP confirmed those orientations[2].

This chapter will not comment further on the current reform, but will rather aim to give a retrospective view on what has been done through the CFP with regard to one of its fundamental objectives: the social dimension. I intend to focus on initiatives that aim at helping those in the fishing industry, and fishing communities more broadly. Given its specificities, the analysis will be restricted to the harvesting sector, thereby excluding the other sectors, which traditionally make up the fisheries sector (processing and marketing of fish products, aquaculture and other related

[1] All legal texts referred to in this chapter are the European Union's R(egulations), Dir(ectives) or Dec(isions).

[2] EUROPEAN COMMISSION (EC), COM(2011) 416, 417, 424, 425. Also the Communication EC, COM(2011) 804 (European Maritime and Fisheries Fund: EMFF), as amended by COM(2013)245, is a part of the reform package.

The views expressed in this chapter are those of the author alone and do not necessarily represent the views of the European Commission.

G. Gallizioli (✉)
Directorate General for Maritime Affairs and Fisheries,
European Commission, 1049, Brussels, Belgium
e-mail: giorgio.gallizioli@skynet.be

J. Urquhart et al. (eds.), *Social Issues in Sustainable Fisheries Management*,
MARE Publication Series 9, DOI 10.1007/978-94-007-7911-2_4,

activities). It is, however, important to bear in mind that for every person employed in the harvesting sector at least three additional jobs are created in related economic activities (FAO 2010).

The goal of this short analysis is to consider what has been accomplished by the CFP over more than 30 years of the policy's existence and to seek to provide an assessment of its achievements regarding the socio-economic conditions of fishers.

4.2 The Legal Framework

The aims of the European Community (now the European Union, or EU) are those defined in the Treaty of Rome, and in the Acts that have subsequently amended it. The Treaty of Rome particularly emphasised farmers and the Common Agricultural Policy (CAP) was designed to ensure the optimum utilisation of labour and a fair standard of living for the agricultural community by increasing the individual earnings of those engaged in agriculture. This objective of the CAP remains and is still stated with identical wording today, in Article 39 of the consolidated version of the Treaty on the Functioning of the EU (TFEU), also know as the Lisbon Treaty (EU 2010). The only difference from the original provisions of the 1957 Treaty, albeit very significant in the present context, is that the term "agricultural shall be understood as also referring to fisheries".

Farmers (and fishermen formerly under the CAP in the 1970s, and subsequently under the CFP) are a category of professionals to whom the Treaty pays particular attention. Indeed, the Lisbon Treaty underlined the mission entrusted to the EU of establishing a 'social market economy, aiming at full employment and social progress' (Article 3.3) (Schmitt 2010). It is important to, therefore, examine what measures have been taken to provide support for farmers and fishers in order to fulfill the objectives laid down in the Treaty. It is important to note that such special treatment should also be extended to the communities to which these persons belong, that is to the social fabric, comprising people sharing the same values and way of living who reside in a specific area.

4.3 The Problems of Definition

In order to assess the measures adopted in favour of fishers and the fishing community, it is first important to provide some definitions. In the CFP (as in other legislation, for example the recent International Labour Organization (ILO) Convention concerning work in the fishing sector), 'fishers' refers generically to all persons engaged in a professional occupation on board any fishing vessel used for commercial exploitation of fisheries resources. The definition is thus derived from having a professional activity on board a fishing vessel, rather than on a precise activity. In general, there are no specific requirements regarding a minimum percentage of income derived from fishing as a test of eligibility for EU funding though fishing should be seen to be the main source of income. Indeed, under the European Fisheries Fund (R.1198/2006) the pursuit of multiple jobs is actually encouraged. This definition, therefore, covers a rather broad range of jobs and so not only encompasses those workers on board vessels which are actually engaged in the capture of wild fish, but also workers carrying out related and auxiliary activities providing they perform such activities on board a vessel equipped for fishing. Duties and wages may differ, but the qualification of 'fisher' is extended to workers who do not actually harvest fish. The scope will be even broader under the new European Maritime and Fisheries Fund (EMFF), as fishermen may be charged with activities concerned the protection of the marine environment or the collection of waste. The reverse is also generally true in that harvesting fish along the shoreline does not qualify the worker as a 'fisher'. This may result in adverse consequences for some categories of worker who harvest living marine resources. In order to tackle this shortcoming, the new proposal extends the definition to persons professionally harvesting marine organism without a vessel[3].

Nor is the definition of a 'fishing community' straightforward. It remains unspecified, although it is generally understood to include, besides fishers and their families, all those living in an area where fishing activity, and those activities related to fisheries, have a particular economic, social, cultural and historical value. Due to economic development, such communities may have evolved over time, to the point where they lose their original features so that the activity of catching fish becomes essentially an historical reference (EC 2011f). A useful example of this wider definition may be found in the Magnuson-Stevens Fishery Conservation and Management Act, which states that "(T)he term 'fishing community' means a community which is substantially dependent on or substantially engaged in the harvest or processing of fishery resources to meet social and economic needs, and includes fishing vessel owners, operators, and crew and United States fish processors that are based in such community.".

The intention that these communities should be given special treatment, has been restated several times in EU legislation, starting with Council Resolution of 3 November 1976 (the 'Hague Resolution') whereby "The Council recognizes … there are regions in the Community … where the local communities are particularly dependent upon fishing and the industries allied thereto. … in applying the CFP, account should also be taken of the vital needs of these fishing communities"[4]. Some years later, in its Resolution of 30 May 1980, on the Common Fisheries Policy, the Council referred to a community 'dependent upon fishing' and stated that the CFP should ensure "fair distribution of catches having regard, most particularly, to traditional fishing activities, to the special needs of regions where the local populations are particularly dependent upon fishing and the industries allied thereto".[5]

---

[3] Subject to Member State recognition (Art. 3.2 (6), COM(2011)804).

[4] EC, Official Journal of the European Union (OJ) C 105/1981, p. 1.

[5] EC, OJ C 158/1980, p. 2

It is, however, not easy to define precise indicators for measuring a degree of dependency on fishing activities and related industries of a community or region. In fact, a real dependency on fishing activity, meaning the dependency of a community on the wealth generated from the quantity and value of catches landed at a port and processed in factories located a short distance from that port, is difficult to calculate. Nowadays, the question of dependency is even more debatable, as the contribution of the harvesting sector to local economies in Europe is very limited, save for a few exceptions of small coastal municipalities located in rather isolated, less developed areas, that have been unable to reconvert their economy.

In fact, specific indicators have never been set by the European law and the qualification as a fisheries area is left to each Member State. However, it is commonly accepted that the concept of dependency on fishing activity includes aspects other than the landing and processing of catches, so as to comprise also the history of the community, its tradition, culture, urban landscape and social fabric. If a stricter definition were to be adopted whereby a fishing community only exists where the fishery sector constituted the mainstay of the local economy, so that a drastic reduction in the harvesting activity would impact on the whole local economy causing its decline (Love 2010), this would only apply to a very few areas in EU 27. As statistical analyses show, out of 128 NUTS-2 coastal regions in the EU[6], less than twenty have a significant ratio of dependency on the fisheries sector (EP 2007).

Legislation also uses the term of 'fisheries area', that, in this context, means a location where fishers live, not where they operate. Finally, the term 'region' does not necessarily refer to an administrative entity recognised as such under national law and may encompass more than one administrative entity.

4.4 Evolution of Employment in EU Marine Fisheries

It is a notoriously difficult task to calculate the number of active fishermen across the EU. In fact, because of disparate methodologies employed by individual Member States, there are no accurate statistics on the number of persons working as professional fishers in the EU. By using figures published by the European Commission, it was estimated that some 300,000 full- and part-time fishermen could have benefitted from EU financial intervention in 1990 when the EU comprised 15 Member States only but some 150,000 in 2008 (with 27 Member States) (EC 2006). After having increased during the 1970s and 80s, the employment in marine capture fisheries has more than halved over the last 20 years. Clearly, statistics do not cover the situation of informal work relationships in fisheries, which are not unusual in this sector. Although on a world scale, the number of fishers has increased over the same period, the steady decline in the number of those employed in the European harvesting sector, should not come as a surprise, since this is the natural trend of economic development that promotes a shift towards secondary and tertiary sectors.

[6] NUTS—Nomenclature of territorial unit for statistics (R.1059/2003).

The decrease in the number of professional fishers is attributable to the reduction of the number of the vessels, a consequence of the technological development and the depletion of fishing stocks. In addition, fishing, and other traditional maritime occupations, is become less attractive. They are seen today as unrewarding, dangerous and not family friendly. The EU is seeking to counter this negative trend through specific measures aimed at restoring the reputation of these occupations. Some of these measures are considered in the Integrated Maritime Policy, launched in 2007 (EC 2007).

4.5 Measures Intended to Benefit Fishers and Their Communities

This aspect—the central one in this chapter—concerns the measures the EU has put in place in order to fulfill the mission assigned to it by the Treaty, in supporting fishermen and their communities. Initiatives in the fisheries sector have been taken since the 1970s, when changes to the Law of the Sea and vested interests of some EU Member States triggered legislation regarding fish products and fisheries production structures, under the Chapter of the Treaty of Rome devoted to agriculture.

Given the overall perspective of this volume in focusing primarily on social issues, the measures under consideration relate principally to the allocation of fishing rights and the financial support for improving living standards. Although legislation designed to improve the health and safety conditions on board fishing vessels or working conditions is pertinent, their goals are fundamentally different to those pursued by Article 39 of the TFEU, which refers to a fair standard of living and, therefore, mainly to economic conditions. Despite the fact that these aspects are now critical in the attempt to attract the younger generation into this traditional activity, the goals of improving health and safety conditions are not specific to fishermen, as they concern every category of worker and will not be treated here.

The analysis will consider only the transfer to Member States of Community financial resources earmarked for the achievement of objectives of a socio-economic nature. Furthermore, this analysis will be confined to aspects directly concerning people. Aid granted for production or processing infrastructure (vessels, ports, factories), which comprise the bulk of the EU contributions until now, lies outside the scope of this analysis. The forthcoming European Maritime and Fisheries Fund, scheduled to come into operation in 2014 relating to the provisions of the revised CFP, is expected to alter the order of priorities, as compared to the actions financially supported under current legislation.

The allocation of fishing quotas (or, in general, the so-called fishing opportunities) to the individual fisher or to producer organisations are, of course, of great importance. However, the system is built in such a way that the recipients are the

Member States and not distinct coastal regions (Wakefield 2009). Only Member States are entitled to distribute quotas to operators according to national law, political goals and practices. No coastal community can claim exclusive rights to stocks found in the waters adjacent to its coastline, although restrictions may apply to access by the fleets of other states[7]. Further, the criteria for establishing the quantities that can be allocated are mainly based on a rational exploitation of the available resources as recommended by scientific advice, rather than on considerations of a social nature. Finally, although quotas obviously have an economic value, there is no public money at issue.

4.6 Financial Support from the EU

Even when support for fleet adjustment, the underpinning of the marketing and processing sectors, promotion of aquaculture and costs associated with fishing in third country waters are included, the amounts earmarked in the EU budget for the fisheries sector have never been massive when compared to other sectors of the economy. However, they may appear in a very different light when the total amount is set against the number of professional fishers in the EU.

EU financial support within the CFP was never designed to create new jobs in the harvesting sector. Its purpose is primarily to facilitate the restructuring of the sector as a consequence of the excessive fishing capacity and the depletion of stocks. The goal of EU intervention over the last 25 years has been to achieve a balance between fishing capacity and available resources as the prerequisite for a profitable industry. The changes in the characteristics of the EU fleet as a result of technological progress and environmental concerns, have clearly had an impact on the number of jobs on board vessels and on the professional qualifications of crew members (EC 2002). On the other hand, the downtrend in the economic performance of fishing businesses has had an unfavourable impact on the revenues of fishermen and on the retention of workers in the sector. If stocks were exploited at maximum sustainable yield, the catching industry would be profitable, fishermen would obtain higher revenues and many more jobs would be created (World Bank & FAO 2009; NEF 2012). This is precisely the aim of the current CFP's reform.

Until now, EU measures have been designed, above all, to cushion the blow of the social consequences of the decline of fishing activity, whether the inactivity be a consequence of a conservation measure or the poor economic results of a fishing business. However, as the new reforms suggest, a different approach has been chosen and many of the earlier measures are to be discontinued because of perceived ineffectiveness. This change in policy comes at a time when unemployment rates in Europe are historically high.

[7] R2371/2002, Art. 17.

4.7 From Special Initiatives to Structural Funds for Fisheries

In 1971, Community funding from the agricultural budget was made available to the fisheries sector in order to develop a fishing sector which was in need of modernisation in the context of the European economy as a whole. When, during the 1980s, it became apparent that fishing capacity needed to be reduced, special measures were devised for scrapping vessels or limiting their activity. A premium was granted to vessel owners in exchange for a temporary or permanent reduction of production capacity. The amount was calculated on the basis of vessel tonnage and the number of 'laid-up' days. It was only in 1995, that measures of a truly socio-economic nature were introduced to aid those workers whose main occupation was as fishers on board operational sea-going fishing vessels[8].

For fishing vessel owners, the premium was an incentive to keep idle a production tool so as to reduce pressure on resources. For crew members, the measure was compensation for a situation where the employee is denied access to social welfare payments. We should also remember the widespread situation of share fishermen, who likewise may lack protection under national social security schemes in the event of cessation or insolvency of a fishing business[9]. The aid is particularly justified because these workers are exposed to the risk of not being able to find other jobs in the same region in which they live, as job opportunities for their skills are generally scarce. The measure was also an attempt to retain within the sector 'dynamic and skilled' workers[10]. Such compensation payments entail no additional costs for businesses that are often struggling for their own economic survival.

The situation of inactivity of vessels due to conservation measures has been extended to include the stoppage of activity following the termination of, or a delay in, the renewal of fisheries agreements with third countries[11], the compulsory abandonment of traditional fishing techniques imposed by a new law and, more recently, for a limited period, to the temporary cessation of fishing activities due to economic crises induced by the abrupt increase of fuel prices in 2008. Under this more generous scheme, part of the 'basic salary' of fishers employed on a vessel admitted into the scheme is charged to the relevant public authorities[12]. In addition to these indemnities, legislation has also provided an incentive to fishers to take early retirement, through a contribution to the pension fund. These measures have been extended under the framework of the Financial Instrument for Fisheries Guidance (FIFG)[13] (2000–2006) and the European Fisheries Fund (EFF)[14] (2007–2013). A further extension of these socio-economic measures under the provisions of the next EMFF is not envisaged by the Commission's proposal.

---

[8] R.2719/95.

[9] See Dir.2008/94 and EC, COM(2007)591, Reassessing the regulatory social framework for more and better seafaring jobs in the EU.

[10] R.2468/98, recital 11.

[11] Dec.95/451.

[12] R.744/2008.

[13] R.1263/1999.

[14] R.1198/2006.

In short, the measures currently available with the aim of ensuring an income for individual fishermen are: the compensation to fishers and owners of fishing vessels for the inactivity of their vessels, and aid granted to workers due to their departure from the fishing sector, including early retirement. Other types of aid are also gathered together under the umbrella of 'socio-economic measures'. They concern the enhancement of skills, within or outside the fishery sector, or facilitating the start-up of an entrepreneurial activity for a young fisher, though these are not necessarily specific to the fishing sector.

4.8 Assessment of the Socio-Economic Measures

Socio-economic measures, as they have been labelled in the legal acts, remained fundamentally unchanged during the two periods of financial programming for structural interventions from 1994 to 2006 (for further details see Suris-Regueiro et al. 2011[15]). Ex-post assessments of these schemes show that only compensation for temporary cessation of fishing activity as a fisher has had a significant effect, in terms of the members involved and the amounts disbursed. Conversely, actions aimed at the early retirement of those over 55 years, and the departure of workers from the sector, have been barely utilised. While the first scheme is well received, as it allows some flexibility to businesses, the latter actions are most probably unattractive when compared with existing labour market conditions and the public social schemes in force. In fact, the decrease of jobs in the harvesting sector is due essentially to the number of vessels removed from fleet registers and only partly replaced by new entries (usually more efficient and larger units). The data available does not allow calculation of ratios between jobs lost and tonnage or KWs withdrawn.

Conversely, it is has been shown that, during the period under consideration, public aid has created jobs in the aquaculture or processing, marketing or related sectors (ports, shipyards, financial institutions, administration etc.), which have offset the reduction in the number of jobs in the fisheries sector. However, it is not possible to trace those fishers that have left the harvesting industry to take up jobs elsewhere in the fishing sector. The scarcity of information available on the attendance of fishers on retraining courses, financed essentially by an unspecific Fund, as it is the European Social Fund, makes it difficult to assess the numbers of workers who, after retraining, remain within the fisheries sector but with a different occupation.

What is clear is that it is the small-scale sector that has been the most adversely affected by a reduction of fishing activity on grounds of resource conservation measures, or the need to improve the economic performance of the sector. Communities that are heavily dependent on fishing and where fishers are unable to

[15] 6 See also: EC, Evaluation ex post des programmes de l'IFOP pour la période 1994–1999, 2004; EC, Ex-post evaluation of the Financial Instrument for Fisheries Guidance (FIFG 2000–2006), 2010; EC, Interim evaluation on establishing EU financial measures for the implementation of the CFP and in the area of the Law of the sea, 2007–2013, 2010.

find other professional occupations in the local area or alternative sources of income (e.g. social welfare payments) face the risk of irreversible decline if they do not embark on diversification of their local labour markets.

4.9 From Harvesting Activity to the Coastal Community

In 1994, the Commission adopted a specific Community initiative for the fisheries sector, known as PESCA[16], in order to assist coastal regions facing loss of revenue due to the poor economic performance of the fishing industry. At that time, businesses were already struggling with limited catches and increasing costs induced, *inter alia,* by new rules on health and safety conditions on board vessels and stricter hygiene requirements for fish products in order to comply with standards in the single market.

250 millions Ecus were made available over the period 1994–1999 to regions deemed to be dependent on fisheries because of their relatively high employment rates in the sector and the added value of fish products within the local economy. The long list of regions eligible for such aid was created by the EU Member States. The initiative mirrored that concerning the rural world, known as LEADER (Liason Entre Actions pour le Development de l'Economie Rurale), operating from 1991 to 2006, and subsequently integrated into the current framework for rural development. The methodology involving a bottom-up approach, as well as the aim of facilitating the diversification of the activities, are identical. Although the two schemes—LEADER and PESCA—may have overlapped in some coastal regions, projects financed under LEADER did not include the fish harvesting sector.

The PESCA initiative was an attempt to redress a situation where, due to declining fish stocks, poor economic yields, new consumer habits or new patterns of trade, either the available employment opportunities were insufficient or the revenues generated were unattractive. In order to revitalise the local economies, the communities needed to become less dependent on the harvesting activity. In so doing, fisheries policy introduced elements of regional policy now referred to 'territorial cohesion'[17], by supporting projects for economic development not directly related to the fishing industry as in pesca-tourism, gastronomic tours, leisure activities, museums of the fishing history, craft goods. In particular, fish workers could benefit from training and vocational education with a view to finding jobs outside the fishing sector.

The same objective of promoting sustainable development and an improvement in the quality of life in coastal areas with a significant, though declining, fishing activity, prompted the revival in 2007, of PESCA-like initiatives, supported by the EFF, to be carried out under the initiative of Fisheries Local Action Groups (FLAGs). It should be noted that in this framework, fishers and harvesting industry are no longer seen as the drivers of economic development in a fisheries area.

[16] OJ C 180/94, p. 1.

[17] Art. 174 TFEU.

Nowadays, themes like technical innovation, networking, food quality, environmental protection, waste management, energy saving and alternative energy, are more popular. It is expected that some 300 such groups will be established under the current programme (see Chap. 11). Some € 567 million has been set aside for the sustainable development of fisheries regions across the EU, which corresponds to 13 % of the EFF budget. It is too early to evaluate this measure, which has raised a lot of interest and will surely be extended in the future. It should be kept in mind, however, that these projects, because of their content, have only a limited impact on the economy and cannot on their own overturn a situation of decline in the sector. Hundreds of such initiatives of this kind cannot fundamentally change the situation for those municipalities, amongst the 6,600 located on the European coast, where fishers are suffering severely from poor economic performance of the fishing industry. Such initiatives should, therefore, be considered no more than a complement to the local economy. However, what is of particular importance is the possible spillover of innovation, good practice and the transfer of know-how which may trigger new dynamics in areas facing similar difficulties. This is made possible particularly through the network that underpins the FLAGs.

It is in this wider context, that the contribution of women to the fishing industry and to the fishing community has been recognised and attracted EU financial support with pilot projects launched as early as the late 1990s[18]. It is well known that the presence of women on board fishing vessels is a rare occurrence, and, therefore, they generally do not qualify as fish workers. Yet, they contribute to fish production, not only as shellfish gatherers, but women play an important role in the processing and marketing of fish products (see Chaps. 12, 13 and 14). Furthermore, their involvement, often unnoticed, in activities within a family run business is crucial. The entitlement of spouses of self-employed workers to benefit from the general system for social protection is a long-awaited and deserved recognition of their contribution[19].

As this territorial-based approach is now the favoured avenue for the economic development of coastal areas, such initiatives seem certain to be extended under the new general framework for EU support from 2014 onwards and is currently under discussion in the Council of Ministers and in the European Parliament. In the new context, the need to insert coastal development into the wider Integrated Maritime Policy and the specific investments including maritime spatial planning, as well as the attainment of the targets fixed by the Europe 2020 Strategy[20], will be taken into account.

4.10 Governance

Over the period under analysis, financial allocations earmarked for implementing measures of a socio-economic nature have largely been underutilised. One of the reasons cited for the poor performance is the difficulty for those most likely to

[18] See OJ C 216/97, p. 31.

[19] Dir.2010/41.

[20] EC, COM(2010)2020.

benefit to voice their concerns when the measures are being elaborated. With regard to governance in the fisheries sector, important steps have been made in recent years and particularly since the 2002 reform of the CFP when Regional Advisory Councils (RACs) were introduced. It is, therefore, useful to recall the opportunities available to economic operators, workers and NGOs to make their voices heard when new measures are being prepared in Brussels. Consultation is an essential feature in the production of EU law and embedded in the Lisbon Treaty.

In the law-making process, apart from the possibility of engaging directly in public consultations, interested parties may be represented through several different organisations including the longstanding Advisory Committee on Fisheries and Aquaculture (ACFA)[21] established in 1971 and the Sectoral Social Dialogue Committee (SSDC) for marine fisheries in 1974[22]. Two other committees are of relevance: one is tasked with the implementation of the programme established in the context of the EU structural fund for fisheries, the Monitoring Committees appointed for each Member State in a local partnership approach, and the other the Regional Advisory Councils (RACs) created for each of five regional seas together with Councils representing the distant water and pelagic fisheries[23]. The RACs are entitled to submit recommendations and suggestions to the Commission and Member States, and to contribute to the achievement of sustainable economic, environmental and social conditions.

Due to its very composition, the RAC has a key role in representing the interests of the components of a fishing community in a multinational set-up. Those entitled to RAC membership includes vessel owners, small-scale fishers, employed fishers, producer organisations, processors, traders and other market organisations, women's networks, environmental organisations, aquaculture producers, consumers and recreational or sport fishers. Given their wide ranging stakeholder membership, RACs are today probably the most powerful instrument for expressing stakeholder views currently in the hands of the fishing community (Long 2010). It is up to RACs, possibly with enhanced competences and administrative capacity, to interact with other actors with an interest in coastal zone management. Their effectiveness could be enhanced by stronger producer organisations and other influential social partners. Should the RAC be deemed inadequate to represent the fishing sector in the coastal management process, as recommended by the FAO's Code of Conduct for 'responsible fishing', and endorsed by the EU, changes can be made[24].

4.11 Conclusions

At the outset of the CFP in 1968, a document relating to the anticipated social pillar in the set of proposals elaborated by the Commission was not finally tabled. It was deemed that strengthening production and marketing structures would be more

[21] Established by Dec.71/128, revised in 1999.

[22] Established by Dec.74/441, revised in 1998.

[23] Dec 2004/585.

[24] See EC, COM(2011)425, Art. 54.

effective in sustaining fishing-dependent communities. However, measures to alleviate the emerging difficulties introduced in the late 1980s—and largely still in force today—have failed to secure a stable and sufficient income for fishers or the improvement of living conditions in fishing communities in parts of the EU. It is true that, where requirements for health and safety and improved working conditions have been complied with, significant improvements have taken place as a result of new legislation and related financial support. But a mechanism specifically designed to guarantee a stable and adequate income for all professional fish workers has never been achieved, in marked contrast to what has been done for farmers through the provisions of the EU Treaty. Clearly the two EU policies are quite differently implemented (Wakefield 2009).

This analysis has referred neither to measures adopted for improving the skills of fish workers, nor to actions focusing on the small-scale/artisanal fisheries and may in this respect be seen as incomplete. In fact, notwithstanding Article 41 of the TFEU which refers to measures in the spheres of vocational training within the CAP, other sectors of EU intervention are perhaps better suited than the CFP for dealing with programmes for vocational training or retraining of fishers. Measures relating to small-scale fisheries do exist as exemptions from certain regulations that apply to larger vessels and in higher rates of financial assistance than those which apply to larger vessels. However, in the absence of rules that can link a specific segment of the fishing fleet to a given coastal community, it cannot be argued that a special regime designed to sustain the revenues of small-scale fishers has been developed (see also Symes and Phillipson 2009). After three decades of the CFP, the EU fleet is composed of fewer, but larger and more powerful vessels, which employ less crew on board. What is more, fish workers are less and less connected to coastal communities partly due to the increasing numbers of non-EU nationals working on board EU fishing vessels. It can be argued that this trend owes more to uneven levels of economic development across Europe and beyond than to the direct effects of the CFP.

In summary, save for a few exceptions, the EU harvesting sector remains economically fragile and requires further restructuring. Studies carried out on the ex-post evaluation of the socio-economic measures, for the period 1994–2011, show that the amounts earmarked in the EU budget have generally been underutilised. At this stage a more detailed analysis is not possible as data collection is ongoing and for several initiatives only aggregate data are available. Only the compensation for temporary cessation of fishing activity appears to have been a qualified success with relatively large numbers of workers from a relatively small number of Member States benefitting from the financial provisions. Support for scrapping vessels could also qualify as a socio-economic measure to the extent that the vast majority of vessel owners in the small-scale fleet work on board their vessels; significantly, however, a number of national decommissioning schemes have specifically excluded vessels under 10 or 12 m.

When compared to other fishing nations, the EU's harvesting sector is today generally worse off than at the outset of the CFP, despite extensive modernisation. Competition for the use of the same marine areas from other growing European economic sectors (DG MARE 2012), and the constraints, imposed by the goal of good environmental status in the marine environment[25], make the future of Europe's fishing industry uncertain. The fishing industry will have a prosperous future only if it becomes part of an otherwise flourishing coastal community. Genuine fisheries-dependent communities are the exception. Dependency is more and more 'virtual' rather than 'real'. Today the image of being a fishing community has a value exceeding that of the landed catches (Brookfield et al. 2005). The future of coastal fishing communities depends more on economic development of the marine and maritime economy, than on the size of the harvesting sectors.

References

Brookfield, K., Gray, T., & Hatchard, J. (2005). The concept of fisheries-dependent communities. A comparative analyses of four UK case studies: Shetland, Peterhead, North Shields and Lowestoft. *Fisheries Research, 72*(1), 55–69.

DG MARE. (2012). *Blue growth: Scenarios and drivers for sustainable growth from the oceans, seas and coasts.* Report by ECORYS, Deltares & Oceanic Development for DG MARE. Rotterdam/Brussels. MARE/2010/01.

European Commission. (2002). *Action plan to counter the social, economic, and regional consequences of the restructuring of the EU fishing industry.* COM(2002)600.

European Commission. (2006). *Employment in the fisheries sector: Current situation.* (FISH/2004/4).

European Commission. (2007). *An integrated maritime policy for the European union.* COM (2007) 575.

European Commission. (2009). *Green paper: Reform of the Common Fisheries Policy.* COM(2009)163.

European Commission. (2011a). *Proposal for a Regulation of the European Parliament and the Council on the common organization of the markets in fisheries and aquaculture products.* COM(2011) 416.

European Commission. (2011b). *Communication from the Commission to the European Parliament, the Council, the European Economic and Social Committee and the Committee of the Regions. Reform of the Common Fisheries Policy.* COM(2011)417.

European Commission. (2011c). *Communication from the Commission to the European Parliament, the Council, the European Economic and Social Committee and the Committee of the Regions on external dimension of the Common Fisheries Policy.* COM(2011)424.

European Commission. (2011d). *Proposal for a regulation of the European Parliament and the Council on the Common Fisheries Policy.* COM(2011)425).

European Commission. (2011e). *Proposal for a regulation of the European Parliament and the council on the European Maritime and Fisheries Fund [repealing Council Regulation (EC) No 1198/2006 and Council Regulation (EC) No 861/2006 and Council Regulation No XXX/2011 on integrated maritime policy.* COM(2011)804.

European Commission. (2011f). Regional social and economic impacts of change in fisheries-dependent communities. (Fish/2006/06).

European Parliament. (2007). *Regional dependency in fisheries.* p. VI.

E. U. (European Union) (2010). Article 39, Consolidated versions of the Treaty on European Union and the Treaty on the Functioning of the European Union. *Official Journal of the European Union. C 83/2010, 53*, 36.

[25] Dir.2008/56.

FAO. (2010). *The state of the world fisheries and aquaculture 2010.* Rome: The Food and Agricultural Organization of the United Nations.
Long, R. (2010). The role of Regional Advisory Councils in the European Common Fisheries Policy: Legal Constraints and Future Options. *The international Journal of Marine and Coastal Law, 25,* 289–346.
Love, P. (2010). *Fisheries. While stocks last?* Paris: OECD.
NEF. (2012). *Jobs lost at sea.* New Economics Foundation, London
Schmitt, M. (2010). La dimension sociale du traité de Lisbonne. *Droit social*, 682
Suris-Regueiro, J. C., Varela-Lafuente, M. M., & Garza-Gil, M. D. (2011). Evolution and perspectives of the Fisheries Structural policy in the European Union. *Ocean & Coastal Management, 54*(8), 593–600.
Symes, D., & Phillipson, J. (2009). Whatever became of social objectives in fisheries policy? *Fisheries Research, 95*(1), 1–5.
Wakefield, J. (2009). Fisheries: A failure of values. *Common Market Law Review, 46*(2), 431–470.
World Bank & FAO. (2009). *The sunken billions. The economic justification for fisheries reform.* World Bank Publications. Washington.

Chapter 5
The Unfulfilled Promise of Integrated Management: How Policy Discourses Operate in Annapolis Basin, Canada

Kate Bigney Wilner

Keywords Discourse analysis • Policy discourse • Fisheries policy • Coastal communities • Critical political ecology

5.1 Introduction

In the Canadian context, the principles of integrated management (IM) are "ecosystem-based management, sustainable development, the precautionary approach, conservation, shared responsibility, flexibility and inclusiveness" (DFO 2002). Barriers to implementing the *Oceans Act* include coordinating inter-departmental change in a siloed environment and making sense of terms such as 'sustainable development' and 'shared responsibility' in a heavily industrialised nation characterised by a turn towards market-based environmental governance. An additional barrier is the soured relationship between certain government and community actors as a result of developments like the collapse of groundfish stocks in the 1980s and 1990s and the implementation of aboriginal rights with regard to natural resources. The text of the *Oceans Act* champions sustainable coastal communities and coastal economies, yet, to date, post-*Oceans Act* changes in marine and coastal governance that are evident have had few meaningful improvements in resource sustainability or on the lives of those living in coastal communities.

Discourses, as a combination of words and action, structure political struggles and, when powerful, capture debates. For instance, the 'tragedy of the commons' is one of the most well-known and powerful discourses in fisheries and oceans governance, informing widely applied solutions to the fisheries 'tragedy' like privatization of access in the form of individual transferable quotas. This discourse still structures thought and practice of fisheries management, while the language of integrated management also reveals underlying assumptions in particular about coastal

K. B. Wilner (✉)
Dalhousie University, Halifax, NS, Canada
e-mail: bigneyk@Dal.Ca

J. Urquhart et al. (eds.), *Social Issues in Sustainable Fisheries Management*, MARE Publication Series 9, DOI 10.1007/978-94-007-7911-2_5,

communities. This chapter examines discourses around, and the implementation of, integrated management in Canada's Bay of Fundy on the Atlantic coast. The context for this question is the resistance of a group of community research partners[1] to the term 'integrated management'. The project, called the Coastal CURA, was designed to support coastal communities as they engage with new models for integrated management. While integrated management appeared to answer certain critiques of current government practice, such as fragmentation, opacity, exclusive science, short-term time horizons and anti-democratic consultation processes, and even to resemble models of resource management that the community partners practice and promote, community partners resisted the terminology of integrated management.

This analysis of integrated management thus examines the underlying positions, epistemologies and assumptions behind the Canadian government's vision of integrated management, as compared with community critiques and visions of integrated management. These visions differ in imagining what sustainable communities look like and who ought to be involved in ensuring community sustainability. Which visions are captured in policy and in practice is in no small part due to power relations within communities and between communities and government resource managers. How different actors talk about problems and solutions both reflects and reinforces those power relations. A critical look at what discourses exist, how they are used and with what effects is one step in illuminating the power relations that help or hinder coastal community involvement in discussions which shape their futures.

This chapter has three components. First, the relationship between integrated management and community is explored. Next, a conceptual framework connecting political ecology, geography and policy studies is developed to focus attention on questions of space, power and discourse in integrated governance of coastal and marine space and resources. Third, this framework is applied to policy discourses around integrated management in the Annapolis Basin in Nova Scotia, Canada, to describe existing discourses and to demonstrate how their use privileges certain actors. The framework helps illustrate how the lack of attention to power relations in the shift to governance structures like integrated management further disempowered a coastal community in their attempts to wrest a measure of control over their livelihoods. The actors in this case study include clam harvesters, coastal NGOs, municipal, provincial and federal government departments and agencies and many others. A sustainable coastal community can and should include all of these actors, and integrated management can be a way to achieve this inclusion. Attention to the discursive power struggles in the Annapolis Basin provides one explanation for the failure of integrated management.

[1] Part of a Community-University Research Alliance funded by the Canadian Social Sciences and Humanities Research Council. For more information see www.coastalcura.ca.

5.2 Integrated Management[2]

Integrated management (IM) as a governance term typically refers to managing all human activity with an impact on marine or coastal ecosystems, such as tourism, shipping, oil and gas, recreation, industrial, residential, agricultural, energy production as well as fishing, by bringing representatives from those industries, and coastal citizens, called stakeholders, together with the state to coordinate management within a given area. Some of these are new industries, while some, like fishing, are usually long established. This, in principle, could alleviate stakeholder conflict and address the cumulative impact of traditional and new activities (Cicin-Sain and Knecht 1998). Other definitions of integrated management focus on citizen engagement in negotiating public policy (Bastien-Daigle et al. 2008), and/or scaling between local, regional and national levels of governance (GESAMP 1996). Most critiques of integrated management take its founding premises for granted and relate its challenges to 'implementation problems' (ICES 2007) inherent in 'scaling up' local initiatives to global problems (Agardy 2005), or to multiple governmental jurisdictional issues, including the rights of indigenous governments (Ricketts and Harrison 2007). However, the creation of new areas of management authority involves developing new governance bodies for that space, which creates new relationships within the territory, and defines what activities are permissible and not permitted. This involves creating, sustaining or altering power relations (Zimmerer 2006). The next section first outlines thinking on governance, then turns to how geographic thinking about space and scale are useful in attending to the oft-neglected political issues that exist in integrated management.

5.2.1 *Integrated Management as Governance*

The state—or the formal, elected, sovereign, centralised government—is no longer the single, or even the central, entity responsible for governance. Rather informal, decentralised and collective decision making structures are being recognised as sources of political power (Rhodes 1996). Integrated management thus acknowledges that the coasts and oceans are an arena valued by and requiring the input of all coastal stakeholders. The study of politics and policy now must consider the roles of state and of non-state actors and their interactions in, for example, new venues for stakeholder deliberation involving state and non-state actors, from private interests, to non-government civil society and environmental organisations, researchers and aboriginal groups (Griffin 2010). This is especially so because in practice, the promise of citizen engagement in the switch to governance has produced "a fuzzy terrain …, somewhere in-between, but articulating with, state and market,

[2] I use the term IM to represent integrated coastal and oceans management (ICOM) and integrated coastal zone management (ICZM). This chapter focuses on the human (rather than natural scientific) aspects of integrated management.

but irreducible to either; a terrain that was neither state nor private, yet expressing a diverse set of social activities and infused with all manner of social power relations, tensions, conflict and social struggles" (Swyngedouw 2005, p. 1996). Hajer and Wagenaar (2003) warn that these participatory spaces can exist in an 'institutional void' where there are few pre-given rules, which can bring loosened notions of responsibility, authority and accountability. This preponderance of uncodified space opens up terrain for conflict over access to decision-making and access to resources. As opposed to encoded democratic rules (when they are followed), "inclusion or exclusion, legitimacy … representation … accountability of [participatory] groups or individuals often take place in non-transparent ad hoc and context-dependent ways" (Swyngedouw 2005, p. 1999). For example, Griffin (2010) shows how regional fisheries bodies are places where powerful actors can maintain, manipulate and increase their power over less powerful actors within the governance regime. Alternately, market forces can dominate: competing with representation and democratic principles are "equally strong processes at work pointing in the direction of a greater autocratic governmentality … i.e. the democratic character of the political sphere is increasingly eroded by the encroaching imposition of market forces that set the 'rules of the game'" (Swyngedouw 2005, p. 1993). Government can also carry on as usual: Griffin (2010) finds from her study of the European Union that because regional committees are only intended to provide advice to the commission who makes policy, there is little evidence that stakeholders actually influence fishery decisions as a result. Decentralisation can allow the state to maintain its control over decision making, leaving intact the status quo (Griffin 2010).

This is, in part, because interactions within stakeholder bodies, between stakeholders and governments or stakeholders and experts are characterised by unequal power relationships: "to deny the existence of power struggles in a participatory approach like IM is unrealistic" (Bastien-Daigle et al. 2008, p. 120). Jentoft called the lack of attention to power relationships in fisheries management surprising (2007). A different facet of power is evidenced in governance arrangements characterised by decentralisation and "pluralisation of power and decision-making centres" (Abrahamsen 2004, p. 1459). This facet of power is not based on coercion, but on regulating the self: it "works through systems of knowledge and discursive practices to provide the meanings, norms, values and identities that not only constrain actors, but also constitute them" (Abrahamsen 2004, p. 1459). In this way integrated management is related to other neoliberal forms of governance, characterised by decentralisation, which establish a way in which the "conduct of conduct" is worked on at sites at a distance, literally and figuratively, from the state (Digeser 1992; Abrahamsen 2004). Technologies specific to integrated management include the use and alternation of space and scalar relations, such as through maps, as well as new governance institutions, and the supporting policies and programming that accompany legislative changes.

These technologies form the basis for new kinds of knowledge "that make some kinds of actions seem naturally more appropriate than others as an invaluable aid to the process of government" (Agrawal 2005, p. 224). These produce altered environmental subjectivities, i.e. how people think of the environment and their position within it. These changes resonate with Jentoft's assertion that management tools and systems "express a political position on relations of power, conflict and social justice" by distributing power and altering power relations (Jentoft 2007, p. 428). For instance, "management systems change the very perception of what it means to be a fisher, such that management systems are now considered a fact of life" (Jentoft 2007, p. 428). Fishing communities, clam harvesters, even government employees are formed by the technologies and practices of government. So for example, small-scale fishing may be considered deviant under integrated management, and those pursuing that livelihood would be encouraged through practices of integrated management to regulate themselves back into the modern economy.

5.2.2 Joined up or Multi-level Governance

Like other forms of "joined up" or decentralised governance, integrated management proposes a change in scalar relationships and responsibilities, moving local actors 'up' to larger arenas and national or transnational ones, e.g. shipping 'down' to engage with smaller scales. "Spatial regulation regimes are also social regulation regimes" that "reflect economic and political interests of proponents rather than some natural state in nature or society" (Nichols 1999, p. 390). For instance, Mansfield (2005) calls the declaration of the 200 mile exclusive economic zone an exercise of sovereignty—"asserting the ocean as national space"—therefore, a dimension of scalar politics (p. 469). Another example is the dominance of the global scale as an explanatory agent. The global scale has been naturalised as the place where certain environmental problems exist, such as climate change, or the fisheries crisis. This "serves to disembody the causes and consequences of such problems, and their construction as such, from practices and politics taking place at a multitude of sites and scales of governance" (Bulkeley 2005, p. 883). As a result, "there is little consideration of the possibilities that the governance of global environmental issues might emanate from the 'bottom up'" (Bulkeley 2005, p. 883). Preoccupation with the process and study of globalization also runs the risk of removing the state from critical attention in generating or ameliorating these global problems (Mansfield 2005).

5.2.3 Participation and community

Participation—who participates and how—is another central thread in definitions of integrated management. Bastien-Daigle et al. (2008) envision integrated management as a collaborative negotiation of public policy for sustainable development: "IM's objective is to instigate a voluntary collaborative process where actors negotiate public policies based on multi-criteria and participatory decision-making process … This consultative, negotiative and cooperative forum will inform on the consequences of human activities, limit environmental degradation and build consensus on how sustainable development should be achieved" (p. 97). By this definition, the ideal of regional actors negotiating public policy in a cooperative forum draws heavily from notions of community engagement as, in principle, these actors have an ongoing role in determining what and how activities are pursued in a given area, in relation to overarching ideas about what regional sustainable development looks like. In addition to the risk of generating an institutional void characterised by uneven power relations, or of the state redefining its power through decentralised governance, as discussed above, these definitions call attention to the need to consider how community is absorbed into thinking about integrated management.

In natural resource management thinking and practice, communities have been represented as small in scale and conservation oriented (but see Smith and Wishnie 2000; Li 2002) or community is erased from fisheries and oceans management altogether and replaced by the individual rational actor, as understood through the 'tragedy of the commons' model (Hardin 1968). Communities are often seen as the site of impact (Olson 2005) and the smallest scale in the hierarchical nest (Bulkeley 2005). Yet, critical geographers and activists reject this notion, as "localities or local practices can constitute multi-scalar system operating across scales' (Bulkeley 2005, p. 897). Bulkeley calls our attention to these networks, as does Escobar: "people are not only 'local'; we are all indissolubly linked to both local and extralocal places through what might be called networks" (Escobar 2001, p. 44). In response to Hardin (1968) anthropologists and others provided illustrations of small scale, long-enduring institutions for managing access to and allocation of common property resources (see, for example, McCay 1995; Feeny et al. 1990; Ostrom 1990). Communities are thus more complex than perceived in fisheries and integrated management policy which represents them as small scale, lacking agency, land-based and needing integration into the market economy.

This review has touched on the human dimensions of dominant approaches to marine and coastal governance. Reorganisation of spatial and scalar relations are at the heart of integrated approaches; the case study of the Annapolis Basin explores how power relationships are negotiated in uncodified spaces, how scalar politics are employed to capture power or exclude livelihoods, and how scale can also be used for resistance. Integrated management is part of a shift away from state control to decentralised governance; but is that process complete in the Annapolis Basin? And how does the Canadian government maintain power at a distance, under these new arrangements? Whose knowledge dominates this shift? Finally, the concept of community has been edged out of modern marine and coastal governance. New governance can work by instilling a sense of responsibility and citizenship on individual and collectives of fishermen. The neocommunitarian argument in support of devolved governance can further marginalise and lock communities into prescribed notions of conflict (excluding access) and conservation. How are different visions of community included or excluded from integrated management processes in the Annapolis Basin? This chapter next details how a discursive approach to policy helps to address these questions.

5.3 Interdisciplinary Discourse Analysis

Jentoft argues that "we need to understand how power is expressed in fisheries and coastal management discourse—how management institutions frame, legalize and validate discourse—who argues what, from what positions of power and with what impact?" (2007, p. 433). Discursive, or narrative, forms of policy analysis aim to identify dominant policy narratives and uncover how policy narratives developed, by placing them in a broader social-political-economic framework. This type of analysis sets out to "identify the grounds for contentions that arise from theoretical assumptions, conceptual orientations, methodological commitments, disciplinary practices, and rhetorical approaches closely intertwined in policy disputes" (Fischer 2003, p. 14). Discursive approaches encourage a more democratic policy analysis by examining in particular the dominant perspectives that typically go unchallenged and by engaging communities that are often excluded (Fischer 2003).

To examine these processes in light of changes to marine and coastal governance, the next section considers how policy discourses are used by different actors to frame social/policy problems, to shape the range of possible solutions, and to permit or constrain participation of people and forms of knowledge in the policy process.

Dominant discourses close down "reference to questions they cannot address", specifically political-economic questions, or those "that might cast doubt upon the completeness of their diagnoses or the feasibility of their solutions" (Li 2007, p. 11). Among opportunities for resistance, like scale framing, or discursive deliberation (Dryzek 2001; Parkins and Mitchell 2005), are switch points, or "conditions under which expert discourse is punctured by a challenge it cannot contain; moments where the targets of expert schemes reveal, in word or deed, their own critical analysis of the problems that confront them" (Li 2007, p. 11). In light of these proposals, the next section also asks whether alternative discourses are successful in altering the distribution of discursive power by opening switch points.

5.3.1 Methods

The texts for policy analysis are written policy documents and the like (authored texts) but also what policy makers do (constructed texts) (Yanow 2000). The institutional context in which things are said co-determines what can be said meaningfully (Hajer 1997). Therefore, data for the discourse analysis consist of policy texts relevant to fisheries and oceans governance in Canada, such as the Oceans Action Plan (DFO 2005) and Integrated Management Policy (DFO 2002), 45 interviews conducted in 2008–2010 with 36 key informants from Canadian government and community-based organisations in the Maritime region, and participant observation in multiple meetings over the study period. The discourse analysis consisted of organising and analysing this textual data. Using Atlas ti. (designed by Scientific Software Development), codes were developed based on a line of text, an individual word, or a part of an image in a document. Memos about codes track emerging connections and theoretical insight about the codes (Braun and Clarke 2006). Models called networks illustrate linkages between codes and code families. In Atlas ti.'s network building tool, the researcher specifies the relationships between codes, to help uncover underlying ideas and assumptions connecting the policy vocabularies. In this way, the conceptual and theoretical frameworks guide coding and model development while codes, themes and patterns are simultaneously allowed to emerge from the data. These network diagrams were then used as the basis for describing the discourses found in the case study. Following Teräväinen (2010), Venn diagrams were used to represent discourses to signal the terms (drawn from codes) that make up each discourse, as well as to highlight where discourses overlap.

5.4 Application: Annapolis Basin, Nova Scotia, Canada

In the Bay of Fundy's Annapolis Basin, changes to the Canadian national shellfish sanitation programme protocol led to the closure to clam digging of much of the basin's beaches during the summer of 2008. Previously, harvest rights to several of the basin's most productive clam beaches were transitioned from one year leases to ten year leases, all held by the same leaseholder. These two events dramatically diminished access of independent clam harvesters to the clam resource. This case centres around attempts to collaboratively address issues in the Annapolis Basin's clam fishery, which hinge on different perceptions of the problems and of possible solutions, including the role of integrated management, and different relative powers of discourses used to negotiate solutions. The data presented here illuminate how terms like health and food safety, conservation and restoration, privatisation, and integration have affected policy and programme implementation, thereby altering conditions of access for one group of harvesters in the Annapolis Basin.

The main positions in this case study are articulated along the lines of three discourses. Food safety is used by government to ensure compliance with export agreements, therefore to preserve the export-oriented clam industry and its trade relationships. The food safety discourse taps into fears about risk of human illness, and achieves discursive dominance by presenting risk as objective, and manageable in certain prescribed ways. The second is a variant of the tragedy of the community concerned with clamming as a last resort, and property rights as a way to achieve stewardship, used to promote sole ownership of access rights. Finally, the discourse of moral economies is used by clam harvesters and their advocates to try to ensure access rights and to restore a livelihoods-centred approach to the clam industry. Certain individuals within, and policy documents produced by, the government of Nova Scotia also participate in this discourse; despite this, the subsistence and moral economies discourse is the least powerful of the three in the context of the Annapolis Basin.

Fig. 5.1 The Bay of Fundy. (Chmura 2001)

5.4.1 Context

The Bay of Fundy is intersected by two provincial jurisdictions (New Brunswick and Nova Scotia) and one international border (Canada/US) (See Fig. 5.1). Due to a confluence of geomorphic features, the Bay of Fundy has one of the highest tidal ranges in the world, up to 15 m in the upper reaches. The Bay of Fundy is home to a host of both rich and highly exploited marine and coastal ecosystems, and fisheries range from handline and herring weirs to industrial vessels in excess of 20 m. Current development concerns also include tidal energy, liquid natural gas exploration and production, marine tourism, international shipping and finfish and bivalve aquaculture, to name a few. The first peoples to inhabit the area were Mi'kmaq, Maliseet and Passamaquoddy peoples, followed by French and British colonists, and later American loyalists. Community-based natural resource management took early root in the bay in the form of fishing cooperatives, such as the herring marketing cooperative, and today community groundfish quotas remain, albeit in a small segment of an otherwise quasi-privatised fleet (Bigney 2005; Kearney 2005). The bay is also the site of multiple large and small scale integrated management initiatives, both formal such as Atlantic Coastal Action Program (ACAP) sites in New Brunswick (NB) and Nova Scotia (NS), and informal, as described below. While the Bay of Fundy is not currently a Large Ocean Management Area (LOMA), Fisheries

Fig. 5.2 Clam harvesting in the Annapolis Basin. (Sullivan 2007)

and Oceans Canada (DFO) (Oceans Branch) had in 2009 assigned a staff member to explore how integrated management might be formally implemented in the region.

The Annapolis Basin is fed by the Annapolis, Bear and Moose Rivers, and measures approximately 24 km (south west to north east) by 6 km (south east to north west). The basin is bisected by the Digby-Annapolis county line with the two significant population centres, the towns of Digby (population 2,311) and of Annapolis Royal (population 411) in each county, respectively. While communities in the basin and along the Annapolis river, such as Granville Ferry, Bridgetown and Middleton, formerly boasted prosperous ship building industries, these diminished in importance and were definitively brought to a close by the construction of the Annapolis causeway in the 1960s. The region had close ties through trade and tourism with the Eastern Seaboard of the US and tourism remains a key industry. The provincial government is responsible for aquaculture, due to a Memorandum of Understanding with the federal government, which is formally responsible for activities at the high water mark and above.

5.4.2 *Soft Shelled Clam*

The soft shelled clam (*Mya arenaria*) has been harvested in the Bay of Fundy by First Nations people for thousands of years, as evidenced by shell middens found near aboriginal settlements (BoFEP 2003). Harvest is demanding and physically challenging work (Fig. 5.2). The harvest was plentiful and unregulated until a two inch size limit was established in the 1940s. The tidal barrage built in the 1960s is widely blamed for altering sediment flow and blanketing clam beaches. Nevertheless, after the tidal station was built, the clam industry's heyday resumed with processing plants opening through the 1970s. Soft shelled clams are susceptible to contamination, whether in the form of faecal matter from sewage, agricultural run-off, or other sources. Shellfish from contaminated waters are able to be processed via depuration, in which clean sea water is used to flush the contaminated animal. Faecal contamination first closed some of the basin's beaches in 1973 and at the time, an economic analysis was conducted to determine whether a depuration plant would be feasible. The first depuration facility was opened in 1991/92 and today there remains one depuration plant in the area.

In 1993, the first licences were required to harvest clams, and shellfish harvesting areas were designated in 1996. Today there are 279 clam harvesting licences in harvest area II, which includes the Annapolis basin, fewer than one hundred of which are being used to harvest clams. The Area II Clam Harvesters' Association (A2CA) represents clam harvesters on local industry-government groups like the Southwest NS Soft-Shell Clam Advisory Committee, and the NS Shellfish Working Group.

5.4.3 Beach Leases

In 1997, the depuration company was granted ten year licences to sites in nearby St. Mary's Bay (where quahog, or cherry stone clam, *Mercenaria mercenaria,* are harvested) and to most of the beaches in the Annapolis Basin that are closed to public, or non-depuration, clam harvesting. These licences were granted as part of a Memorandum of Understanding (MOU) between the NS DFA and the DFO that turned yearly DFO depuration licences into ten year aquaculture licences.[3] To secure these ten-year leases, the depuration company agreed to collaboratively fund research into the clam stocks.[4] The leaseholder was also granted the first right of refusal for beaches that would be subsequently closed to harvest and was delegated responsibility for water quality testing. Formerly a government responsibility, this also meant that previously accessible water quality data is now protected under Canadian privacy legislation.

The licences were granted without any apparent consultation, contrary to the requirements of fisheries and aquaculture legislation. Regional Aquaculture Development Advisory Committees (RADACs) are meant to work with communities on site selection, but none was formed in this case. It was only when community groups such as the MRC (Marine Resource Centre) heard of the proposal that public meetings were called. First Nations were not consulted about the leases, which, according to Frank Muese, then-Chief of Bear River First Nation, is in violation of their treaty rights[5]. Other locals also reacted vehemently. Other concerns included the length of the lease, first right of refusal, and that decisions about who can harvest clams now rest with the company.

[3] While aquaculture licences are typically 25 m from the mean low water level, the provincial Minister may issue licences up to the high water mark (Wiber and Bull 2009).

[4] This resulted in a 3-year project funded at almost CAN$ 200,000 in federal funding (Wiber and Bull 2009).

[5] All 13 Chiefs and Councils supported the letter. In it, Muese argued that the DFO must consult with the First Nations due to treaty rights including land title. The letter outlines the details of this process.

The rhetoric used to justify the length and security of the leases was that of stewardship, job security, security of business investment, and of particular interest, food safety. A DFO representative "confirmed that depuration licences are issued to a company to ensure a higher level of public safety and to maintain accountability and continuity."[6] Regulators argued the licences were simply 'migrating' from one year federal fisheries licences to provincial aquaculture licences for identical parcels of contaminated land, and that increased landings at the leaseholder's depuration beaches were a sign of good stewardship. DFO representatives also argued that the ten-year length was essential for long term planning into a costly venture and were, therefore, granted out of "fairness and assurance to the business community."[7] One interview participant described the several year-long process of acquiring the leases, which included the refusal of several reluctant Ministers, until one finally assented and the leases were granted. This Ministerial reluctance complicates claims about leases leading to stewardship, durable investment, and food security.

During the 2008 clamming season, only two beaches remained open to public clamming, meaning that most if not all licensed clam harvesters would harvest closed beaches for the leaseholder. The leaseholder's labour practices were a central topic of discussion during many interviews. The leaseholder is said to set lower prices than other buyers, which some feel forces pressure on open beaches (Wiber and Bull 2009). According to several interview participants, the company requires clam diggers to have harvested 2,500 kg from open beaches before they are employed for the season. One harvester complained that clams are sorted and weighed by the company, not in the presence of diggers. Clam diggers remain technically self-employed, yet the leaseholder controls the distribution of fishing licences. A clam harvester of more than 30 years was cited as saying, "you are telling me, as an independent self-employed clam licence holder that I have to dig for one company and one company alone in order to make living."[8]

Finally, there is a widespread belief that though the leaseholder is meant to be reseeding soft shelled clam (and indeed, the company's licences are for aquaculture, rather than for harvest and depuration of closed beaches), the company is not. The leaseholder's data for cherry stone clams in adjacent Saint Mary's Bay—landings and results of bacteriological testing of meat—have not been released due to data privacy. When asked for evidence of reseeding, or other signs of stewardship (as opposed to simple harvest), federal and provincial officials said they relied primarily on the landed value as an indicator of stewardship undertaken under this lease. According the MRC, "landed value could just as easily be an indicator of increased effort, as any kind of stewardship."[9] Many interview participants reflected that

[6] http://www.novanewsnow.com/article-72579-Stakeholders-question-10year-aquacultureleases, p. 2.

[7] Meeting minutes, Yarmouth NS, January 30 2007, recorded by DFO.

[8] Press release, MRC, April 2008.

[9] Press release, MRC, April 2008. See Wiber and Bull (2009) for more on research into quahog population dynamics post-privatisation in St. Mary's bay.

the long term nature of the leases actually removed incentives for environmental remediation or restoration: Digby-Annapolis politician, Harold Theriault, is quoted to have said that the proposed changes could remove any incentive to eliminate contamination on the beaches.

5.4.4 Wastewater Treatment and Food Safety

In the summer of 2008, beaches were closed for the better part of the summer (128 of 251 possible days) due to successive failures of the Digby town waste water treatment plant (WWTP). In this case, food safety, a well-known and well-justified concern with respect to seafood, was again cited as the rationale for increased closures of clam beaches to harvesting.

Canada exports most of its shellfish to the United States and since 1948 has agreed to harmonise its approach to 'sanitary practices'. This relationship allows the US Food and Drug Administration (FDA) to audit its suppliers of seafood. In 2004, the Canadian Shellfish Sanitation Program (CSSP, a joint programme administered by the Canadian Food Inspection Agency, Environment Canada and Fisheries and Oceans Canada) was audited by the US FDA and found to be lacking. The audit results were released in 2005, and the 2008 clam season saw the closure of multiple beaches for much of the season. Country-specific audit results are not available, nor is the Canadian response. It can be surmised that waste water treatment plant provisions and the overall risk management programmes were deficient. The intervening years presumably saw the Canadian regulators prepare their response to the FDA audit; indeed, a WWTP addition to the CSSP programme was officially added in March 2009, a few months before the FDA was due to revisit the Annapolis-Digby area. As part of these additions, a new risk based Hazard Analysis and Critical Control Points (HACCP) process was also put into place[10], in addition to area-specific management plans.

The CSSP revisions make no mention of an FDA audit or of export requirements in general; but a CSSP notice says that "it is critical that effective response measures are put in place to prevent affected shellfish from reaching domestic and international markets."[11] According to the CSSP 2009 Business plan, "any misalignment of Canadian inspection systems with international demands and standards could increase the risks associated with trade related delays and diminished market access for the Canadian agri-food industry" (CFIA n.d) These documents convey a message of concern about trade relationships first, while later messaging prioritise food safety. Depuration is also cited as a way to ensure food safety.

[10] The new approach adopted by the US FDA and required of its suppliers controls risk by identifying and managing 'critical control points', which is a shift from test-based (for some pathogen at some point in the system) or command and control to a mostly process-based assessment (Unnevehr and Jensen 1999).

[11] Ibid

5.4.5 *Institutional Improvisation: WWTP Meetings*

Clam harvesters, processors, First Nations representatives, and local government officials were angered when beaches were closed after WWTP failures. Many saw the post-rainfall closures as the result of American interference, and questioned why the WWTP in Digby was targeted when rainfall amounts had not been excessive and no structural change to the plant itself could be readily identified. No explanation about the FDA audit and subsequent changes to the CSSP was offered. In fact, regulators insisted the new response to the WWTP failures were internally driven. The Marine Resource Centre convened an ad hoc group of all parties with responsibility for or an interest in the clam fishery. The meetings were public and due to the urgency of the situation drew many clam harvesters, citizens, local politicians and media.

The conservation harvest plan was presented at these meetings and contained several changes. As these changes had already been implemented, the meetings were informational rather than consultative. Some reclassification of beaches post-overflow events meant that access to both open and closed beaches was altered. The new CHP presented changes to the terminology and to the boundaries of the areas. The changed harvest areas were presented, along with the previous harvest areas, as part of the new CHP. The scales, colouring and shading of the two maps (pre and post changes) were different, so that the maps were difficult to compare, and the slight changes in terminology also appeared to be confusing.

No data were presented in the CHP or at the meetings in support of the seven day closure period. One interview participant speculated that earlier testing would be prohibitively expensive. A CFIA representative insisted that such decisions were "based on science"; DFO representatives at the meeting made reference to hydrological studies, hypotheses and parameters, though these terms were not explained and no data were ever presented. Meanwhile non-government interview participants argued that the basin flushes every two to three days, which led them to question the scientific rationale for the seven day closure altogether.

Other than the timing of the closures, and the mandatory seven day closure post WWTP failure, controversies included: (i) the lack of willingness of Environment Canada to share results from water quality testing; (ii) communication of closures (one processor reported having heard of a recent closure on the radio); (iii) consultation around the development of the CHP (there had been none before the meeting) and; (iv) compensation for lost wages.

5.5 Discourses Operating in the Annapolis Basin

The allocation of clam leases and the closure of beaches are part of different yet inter-related policy processes. In both situations, key discourses are used, subtly or openly, as part of a process of altering power relations to grant one party increased access to and control of natural resource governance. These discourses

Fig. 5.3 Venn diagram displaying key components, relationships between the areas of overlap of dominant and counter-discourses identified in the Annapolis Basin case study

are represented in Fig. 5.3. The first is a variant of the economic prosperity discourse that is concerned with food safety. The second is a variant of the tragedy of the community concerned with clamming as a last resort, and property rights as a way to achieve stewardship. Finally, the discourse of moral economies is used by clam harvesters and their advocates to try to ensure access rights and to restore a livelihoods-centred approach to the clam industry.

5.5.1 Food Safety Discourse

The food safety discourse construes clams as risky, and prioritises consumer and trade relationships. Within this discourse, data are corporately owned or government controlled. Combined with changes to the CHP, poor labour relations and differential access to capital and other resources serve to maintain or even narrow access to clam grounds. The shellfish sanitation programme sees globalisation, health and changing markets as interrelated risks, and the switch to HACCP as the best way to address those risks. Depuration and risk management are both modern solutions. Food safety as a way to look at risk justifies withholding data and keeping harvesters out of clam beaches and of policy mechanisms. Food safety also allows the problem to be rendered technical by attribution to the WWTP, which removes a multi-stakeholder approach from the list of possible solutions.

This discourse hinges on an approach to risk that treats risk as real, objective and measurable. The new approach adopted by the US FDA and required of its suppliers controls risk from a command and control to a mostly process-based assessment

(Unnevehr and Jensen 1999). Both process and command and control approaches consider clams to be risky objects. Other risks include scarce resources, globalisation, loss of markets and public relations.

The rationale for the shift to HACCP was not explained by the Canadian regulators to the clam harvesters and their supporters, nor is there space within this discourse for other perspectives of risk. For example, harvesters and some managers asked were not clear on what was specifically unsafe about the previous system of inspections. Yet, the previous command and control system itself was inaccessible to clam harvesters as it was based on science in which they did not participate and data to which they did not have access to. Indeed, little epidemiological evidence of illness is presented in the policy documents, which seems at odds with the heavy focus on food safety and risk to public health. This is not to suggest that food-based illness does not exist, but rather that those illnesses are taken for granted rather than evidenced in the policy documents. In addition, control of raw data ensured the government and depuration company could maintain control of the narrative told through data interpretation. The power of the risk and food safety discourse is to make itself so dominant as to be unassailable when the clam harvest is, to paraphrase, made to be about safety, by decision-makers insisting on risk and science.

Omissions were central to this discourse, namely the role of the US FDA and the absence of water quality data. These omissions became central when other discourse coalitions focused their attention on them. Those espousing the risk discourse were then forced to explain these absences, in particular the missing data, explanations which were unsatisfying because clam harvesters and their allies suspected they were covers for the 'real' explanations—that the US FDA standards were in fact the real drivers for change and not new test results.

Data collection and dissemination nearly became switch points (Li 2007). But privacy laws protecting the private company along with the dominance of the scientific knowledge paradigm combined to make that point of entry impossible. The debate was shut down and became technical instead of political (Li 2007). The clammers and their allies were not able to harness their discursive power to rephrase the debate.

5.5.2 *Tragedy of the Community Discourse*

The government CSSP programme is confusing even to those involved and many meetings featured frustration at the perception of unwillingness of government participants to take responsibilities by statements such as "I'm not Environment" (meaning in this case an employee of the Province's Department of the Environment) or "I don't have the test results". The CSSP programme may function within government (though as internal documents indicate, there are coordination and leadership problems) but as a liaison to harvesters, communication and integration failed. The programme complexities and risk orientation (as well as a paucity of resources) served to reinforce a lack of integration and maintain solutions at the technical level. Internally, DFO Fisheries and Oceans branches were also at odds with regard to the Annapolis basin, one perceived to be working "in the weeds" on practical day-to-day matters of fisheries management (Fisheries) and one at "thirty thousand feet" of the policy world (Oceans).

The motivation for the depuration company to assist in ameliorating the conditions of the Annapolis Basin beaches is questionable when the leaseholder's profit depends on beaches being closed. With the number of open beaches declining, combined with difficult labour practices, clam diggers are squeezed into working for a company that many of them resent. In response to the suggestion that the depurator had a monopolistic control over the industry, government officials argued that another group could invest in a depuration facility and also apply for the leases. While regulators claim that any application for a depuration licence will be considered, according to Wiber and Bull (2009), "closed beaches are a resource … only for those with the capital to invest in depuration plants and other infrastructure that meet federal inspection guidelines for accessing, transporting, processing and marketing clams from contaminated areas" (p. 160). In fact, interview participants stated that the clam harvesters did not want to enter the depuration industry, nor did they wish to consider an aquaculture licence as neither conforms with their values, and due to concerns that they would lose the lease to the current depurator. By granting exclusive access to closed clam harvest beaches, the federal and provincial governments altered power relations such that harvesters are forced to work for an employer that has little apparent incentive to remediate a polluted ecosystem. It is unclear how overall food safety standards are improved without this long-term incentive. In addition to an apparent lack of interest in reseeding or other programmes that might ameliorate the clam stocks, the company does not facilitate clam harvesters to do this work independently by, for example, providing spat or under-sized clams.

As is highlighted by Wiber and Bull (2009), aquaculture is associated with progress, while clam digging with low-technology manual labour; this helps to "privilege a corporate actor over pre-existing resource users" (p. 160). In the policy imagination, communities can be construed as less integrated in the market or less industrialised, distant both geographically and temporally. Further, in a northern context, community can be seen by the dominant paradigm as part of culture and, therefore, not intrinsic to the economy or the policy sector (Olson 2005; St Martin 2006). In this way, resource-based communities can be construed as under-developed and policy interventions are designed to increase modernisation (aquaculture, integration with markets and professional specialisation). In this case the depuration company is the more modern of the local players, with capacity to navigate complex regulatory environments and engage in the political system. Enacted through meeting dynamics and in discussions around poverty, crime, migration and requests for compensation, the clamming community is constructed and understood to be less modern than other players, and less sophisticated, and possessing less agency in negotiating policy change. While the clam industry was encouraged to adopt the "industry restructuring" perspective in order to gain favour for their proposals, this shift was either insufficient by itself, in light of the dominance of other discourses, or was insufficiently completed, maintaining too much of the social and moral economies discourse. Here, integrated management alters scalar relationships and practices within government: the scalar politics related to complicated jurisdictional issues at the coast allowed questions of responsibility and authority to either drive action or excuse inaction.

5.5.3 Social and Moral Economies Discourse

Finally the subsistence and moral economies discourse connects place, ecology and people through the concepts of restoration, subsistence and livelihoods. From within this discourse come video, song and concepts like *Nutukulimk*—a Mi'kmaq concept for the connection of people and the natural world that includes rights, responsibilities, inter-generational equity, sustainability and spirituality[12]—as alternative ways to communicate and think about natural resource management. This discourse also links scales, defying the perception that communities operate only locally. Within this discourse, there can be a strategic benefit of opting out of governance processes, in particular when integrated management or multi-stakeholder processes are seen as a way to neutralise community practices and resistance by bringing parties together on an unequal playing field. Integrated management is, however, recognised as an essential part of coastal development planning and strategy. Certain policies and branches of government make use of this discourse, namely within the NS government, such as the community development strategy and the NS Voluntary Planning Agency. The clam harvesters and their advocates attempted to expand this discourse, or to 'hitch on' to the dominant discourses (Hajer 2003, p. 107) by including 'industry restructuring' as one of their goals.

This discourse is expressed by clam harvesters and their supporters (local NGOs such as the MRC) at meetings and other public fora; one poignant expression is through the songs of clam harvester Terry Wilkinson who sings of poverty that accompanies the hard physical labour of clam harvesting: "Pocket fashion dictates Frenchy's[13] Clothes, cause a poor man's life the only one I've known/ With calloused hands and the sweat upon my brow, I work the salty water earthen plow."[14] Prosperity is a reward from "salt water earth" for demanding physical labour. His identity as a fisherman is tied to working for himself: "Much more than a lifestyle to me/I stand in the life that is free"[15] These lyrics display a tension between pride

[12] According to the website of the Unama'ki Institute of Natural Resources, in Cape Breton, Nova Scotia, "Netukulimk is the use of the natural bounty provided by the Creator for the self-support and well-being of the individual and the community. Netukulimk is achieving adequate standards of community nutrition and economic well-being without jeopardizing the integrity, diversity, or productivity of our environment. As Mi'kmaq we have an inherent right to access and use our resources and we have have a responsibility to use those resources in a sustainable way. The Mi'kmaq way of resource management includes a spiritual element that ties together people, plants, animals, and the environment." Accessible at http://www.uinr.ca/2009/01/netukulimk/

[13] A regional chain of second hand clothing stores.

[14] From T. Wilkinson, "Blue Fishin'".

[15] From T. Wilkinson, "Clear Waters".

in one's livelihood, including sharing in a family occupation, and relative powerlessness that can accompany being poorly compensated for one's labour. Still, clam harvesting is portrayed as a choice, and importantly, as a fishery, albeit unique, but linked with the traditions and some of the privilege of more powerful fishing actors. Dignified employment (conceived of as decently compensated and independent) connects people to the natural system and forges a sense of place.

Within this discourse, communities are tied to use of and connection with a specific place. They are connected to nearby communities, those in other provinces and nationally and also internationally through the work of community leaders who represent their communities at fora such as the World Forum of Fisher Peoples. This counter-discourse construes place as multi-scale and as essential to livelihood. While regulators up-scale or down-scale problems, rendering them accidental as opposed to systemic, to justify a limited state response, local actors up-scale by connecting to state policies or actions, to insist on state responsibility, thus intervention (Harrison 2006).

Participants operating within this discourse are suspicious of 'integrated management' and with many other government interventions. They either participate in integrated management strategically, due to fear of being left out of a discussion that will impact them, instead of commitment to the process, or choose to opt out altogether. If and when clam harvesters do participate in integrated management fora, including the ad hoc meetings in the Annapolis Basin, the tools they use, including video and song, and the language they use do not correspond to what is expected in the settings of integrated management. Torgerson (2003) relates similar findings from the MacKenzie Valley pipeline public process. There, aboriginal people, invited to testify, "did so in a way that was not limited to rational argument, but included their own stories, poetry and songs. These bore witness to an experience of the north not as a frontier to conquer, but as a loved place shaping the lives and identities of people who called it home" (Torgerson 2003, p. 119).

Government agencies and departments do have policies that draw on features of this discourse. The Canadian Food Inspection Agency (CFIA) for its part recognises "that traditional industries such as agriculture and forestry have long been anchors of our nation's economic, environmental and social well-being. The vigour of these industries depends, in part, on the health and sustainability of the resource base on which these industries rely."(CFIA n.d) The Nova Scotia government has also written policies, which if enacted more thoroughly would enable citizen participation. Yet, this discourse is not heard, in part because there is no space for it, and it in part because those using it seem to speak another language. This results in fishers' practical knowledge, and the connection between policy interventions like privatising access and poverty, being ignored. In coastal and marine governance, certain spaces and scales are categorised, or framed, as capitalist and part of the modern economy while others are excluded (St Martin 2001, 2005). So for example, pre-existing management regimes are displaced: "socially important non-modern livelihoods (e.g. artisanal fishing) may be regulated out of existence to create space for state and internationally sponsored projects such as aquaculture development" (Nichols 1999, p. 390). The more modern partner is favoured, and the less modern clam harvest is construed as needing to restructure itself.

5.6 Conclusion

It is clear that competing models of integrated management operate in the Annapolis Basin. The model represented in the counter discourse is driven by concerns over conditions of access to and health of the clam resources, preservation of local livelihoods and indigenous rights. This model is open to multiple participants, to shifting institutional structures, and to dialogue between harvesters, regulators, the broader local and international communities. The principles of integrated management espoused in the *Oceans Act* would appear to be embodied in this process, which would also support coastal communities as they struggle to sustain their ways of life. Yet, in this case, the more powerful actors, via dominant discourses, frame the problems in the clam harvest as technical rather than political. Knowledge relevant to solve these problems comes from experts and scientists and is about risk, while relevant problem-solving tools and technologies include Hazard Analysis and Critical Control Point (HACCP) audits, conservation harvest plans and adversarial meetings. Food safety also scaled the discussion away from being a local problem with local solutions, albeit in a clandestine manner, as the international context of clam exports to the United States was downplayed by decision-makers yet understood by all to be the critical one. Within the state scale, despite the inter-department and agency CSSP, blame was shifted due to the complexity of regulation surrounding shellfish harvest, further undermining integrated solutions. Dominant discourses also help form identities or subjectivities for clam harvesters, oriented around poverty and migration (Agrawal 2005). These notions are taken up by clam harvesters as well, albeit in different discursive ways. Ultimately, the dominant discourses determine that the modern, industrial fishery is the depuration fishery with secure property rights, orderly business model, control of its workforce, and addresses risk in a way that is responsive to the dominant discourse and institutional needs. Fishers and the broader community are expected to benefit from employment, though fishers dispute the dignity of that employment.

Integrated management, while proposing to encompass a host of approaches that hold promise for community empowerment and sustainability, is vulnerable. The communities in the Coastal Community-University Research Alliance knew this, which in part explains their resistance to the language of the policy. The literature on integrated management as a governance concept hints at power dynamics behind these vulnerabilities, such as the possibility of creating an institutional void if new stakeholder bodies are not constructed with attention to the power relations between those in charge and those at the receiving end of policy changes, and the danger of flattening scalar relationships without attending to how certain livelihoods and communities become vulnerable in this new context. As the case study presented here demonstrates, instead of advancing collaborative management of a fishery essential for the sustainability of community livelihoods, institutional improvisation for integrated management was undermined by long-established relationships whereby government rather than governance dominates.

References

Abrahamsen, R. (2004). The power of partnerships in global governance. *Third World Quarterly, 25*(8), 1453–1467.

Agardy, T. (2005). Global marine conservation policy versus site-level implementation: The mismatch of scale and its implications. *MEPS, 300,* 242–248.

Agrawal, A. (2005). *Environmentality: Technologies of government and the making of subjects*. Duke UP: Durham, Michigan.

Bastien-Daigle, S., Vanderlinden, J. P., & Chouinard, O. (2008). Learning the ropes: Lessons in integrated management of coastal resources in Canada's maritime provinces. *Ocean and Coastal Management, 51,* 96–125.

Bay of Fundy Ecosystem Partnership (BoFEP) (2003). Living lightly on the land and water: Native people and the Bay of Fundy. Fundy Issues 23, Autumn 2003. http://www.bofep.org/native_resource.htm

Bigney, K. (2005). *Navigating troubled waters: Community quota in the Scotia-Fundy groundfishery*. MES Thesis. Dalhousie University: Halifax NS.

Braun, V., & Clarke, V. (2006). Using thematic analysis in psychology. *Qualitative Research in Psychology, 3*(2), 77–101.

Bulkeley, H. (2005). Reconfiguring environmental governance: Towards a politics of scales and networks. *Political Geography, 24,* 875–902.

CFIA (Canadian Food Inspection Agency) (no date). 2009-2010 Estimates. Part III- Report on Plans and Priorities. Archived at http://www.tbs-sct.gc.ca/rpp/2009-2010/inst/ica/ica-eng.pdf.

Cicin-Sain, B., & Knecht, R. (1998). *Integrated coastal and ocean management: Concepts and practices*. Washington DC: Island Press.

Chmura, G. L., Coffey, A., & Crago, R. (2001). Variation in surface sediment deposition on salt marshes in the Bay of Fundy. *Journal of Coastal Research, 17,* 221–227.

DFO (Fisheries and Oceans Canada). (2002). Canada's Ocean Strategy: Policy and Operational Framework for Integrated Management of Estuarine, Coastal and Marine Environments in Canada. DFO: Ottawa. http://www.dfo-mpo.gc.ca/oceans/publications/cosframework-cadresoc/pdf/im-gi-eng.pdf. Accessed 1 Nov 2012.

DFO (Fisheries and Oceans Canada). (2005). Canada's Ocean Acction Plan: For present and future generations. DFO: Ottawa. http://www.dfo-mpo.gc.ca/oceans/publications/oap-pao/pdf/oap-eng.pdf. Accessed 1 Nov 2012

Digeser, P. (1992). The fourth face of power. *The Journal of Politics, 54*(4), 977–1007.

Dryzek, J. S. (2001). Legitimacy and economy in deliberative democracy. *Political Theory, 29*(5), 651–669.

Escobar, A. (2001). Culture sits in places: Reflections on globalism and subaltern strategies of localization. *Political Geography, 20,* 139–174.

Feeny, D., Berkes, F., McCay, B.J., and Acheson., J.M. (1990). The tragedy of the commons: Twenty-two years later. *Human Ecology,* 18(1), 1-19.

Fischer, F. (2003). *Reframing public policy: Discursive politics and deliberative practices*. Oxford: Oxford University Press.

GESAMP. (1996). *The contributions of science to Coastal Zone Management*. Rep. Stud. GESAMP, IMO/FAO/UNESCO-IOC/WMO/WHO/IAEA/UN/UNEP Joint Group of Experts on the Scientific Aspects of Marine Environmental Protection, 61, p. 66.

Griffin, L. (2010). The limits to good governance and the state of exception: A case study of north sea fisheries. *Geoforum, 41,* 282–292.

Hajer, M. A. (1997). *The politics of environmental discourse*. Oxford: Oxford University Press.

Hajer, M. (2003). A frame in the fields: Policymaking and the reinvention of politics. In M. Hajer & H. Wagenaar (Eds.), *Deliberative policy analysis: Understanding governance in the network society* (pp. 88–113). Cambridge University Press: Cambridge.

Hajer, M., & Wagenaar, H. (2003). Introduction. In M. Hajer & H. Wagenaar (Eds.), *Deliberative policy analysis: Understanding governance in the network society* (pp. 1–33). Cambridge: Cambridge University Press.

Hardin, G. (1968). The tragedy of the commons. *Science, 162*(3859), 1243–1248.

Harrison, J. L. (2006). Accidents and invisibilities: Scaled discourse and the naturalization of regulatory neglect in California's pesticide drift conflict. *Political Geography, 25,* 506–529.

ICES. (2007). *Report of the working group on Integrated Coastal Zone Management* (WGICZM), 17–20 April 2007.

Jentoft, S. (2007). In the power of power: The understated aspect of fisheries and coastal management. *Human Organization, 66*(4), 426–434.

Kearney, J. (2005). Community-based fisheries management on the Bay of Fundy: Sustaining communities through resistance and hope. In B. Child & M. W. Lyman (Eds.), *Natural resources as community assets: Lessons from two continents* (pp. 83–100). Washington, D. C.: Sand County Foundation and Aspen Institute.

Li, T. M. (2002). Engaging simplifications: Community-based resource management market processes and state agendas in upland southeast Asia. *World development, 30*(2), 265–283.

Li, T. M. (2007). *The will to improve: governmentality, development and the practice of politics.* Duke University Press: Durham.

Mansfield, B. (2005). Beyond rescaling: Reintegration of the national as a dimension of scalar relations. *Progress in Human Geography, 29*(4), 458–473.

McCay, B. J. (1995). Social and ecological implications of ITQs: An overview. *Ocean & Coastal Management, 28*(2), 3–22.

Nichols, K. (1999). Coming to terms with "integrated coastal management": Problems of meaning and method in a new arena of resource regulation. *Professional Geographer, 51*(3), 388–399.

Olson, J. (2005). Re-placing the space of community: A story of cultural politics, policies and fisheries management. *Anthropological Quarterly, 78,* 247–269.

Ostrom, E. (1990). *Governing the commons: The evolution of institutions for collective action.* Cambridge: Cambridge University Press.

Parkins, J. R., & Mitchell, R. E. (2005). Public participation as public debate: A deliberative turn in natural resource management. *Society and Natural Resources, 18,* 529–540.

Rhodes, R. A. W. (1996). The new governance: governing without government. *Political Studies, XLIV,* 652–667.

Ricketts, P., & Harrison, P. (2007). Coastal and ocean management in Canada: Moving into the 21st century. *Coastal Management, 35*(1), 5–22.

Smith, E. A., & Wishnie, M. (2000). Conservation and subsistence in small scale societies. *Annual Review of Anthropology, 29,* 493–524.

St Martin, K. (2001). Making space for community resource management in fisheries. *Annals of the Association of American Geographers, 91*(1), 122–142.

St Martin, K. (2005). Mapping economic diversity in the First World: The case of fisheries. *Environment and Planning A, 37,* 959–979.

St Martin, K. (2006). The impact of "community" on fisheries management in the US Northeast. *Geoforum, 37,* 169–184.

Sullivan, D. (2007). *Digging into the past: Clam harvesting in the Annapolis Basin.* Prepared for the Marine Resource Centre as part of the Coastal CURA research project. Published by the Clean Annapolis River Project.

Swyngedouw, E. (2005). Governance innovation and the citizen: The Janus face of governance-beyond-the-state. *Urban Studies, 42*(11), 1991–2006.

Teräväinen, T. (2010). Political opportunities and storylines in Finnish climate policy negotiations. *Environmental Politics, 19*(2), 196–216.

Torgerson, D. (2003). Democracy through policy discourse. In M. Hajer & H. Wagenaar (Eds.), *Deliberative policy analysis: Understanding governance in the network society* (pp. 113–139). Cambridge: Cambridge University Press.

Unnevehr, J., & Jensen, H. H. (1999). The economic implications of using HACCP as a food safety regulatory standard. *Food Policy, 24,* 625–635.

Wiber, M., & Bull, A. (2009). Re-scaling governance for better resource management? In von F. Benda-Beckmann, von K. Benda-Beckmann, & J. Eckert (Eds.), *Rules of law and laws of ruling. Surrey* (pp. 151–170). UK: Ashgate.

Yanow, D. (2000). *Conducting interpretive policy analysis. Qualitative Methods series 47*. Sage Publications: Thousand Oaks.

Zimmerer, K. S. (2006). Cultural ecology: At the interface with political ecology—the new geographies of environmental conservation and globalization. *Progress in Human Geography, 30*(1), 63–78.

Part II
Issues in Local Ecological Knowledge and Scientific Evidence

Chapter 6
Stewardship and the Recovery of Threatened Wolffish in Eastern Canadian Waters

Kathleen Blanchard, Jennifer Dawe and Russell Wall

Keywords Eastern Canada • Stewardship • Local ecological knowledge • Wolffish • Intervale Associates

6.1 Introduction

In the past decade, the application of stewardship to the work of conservation has increased dramatically, encompassing a broad spectrum of species, habitats and landscapes. In Canada alone, there are numerous public policy applications and more than 1,000 groups and nongovernmental organisations involved in stewardship (Wolthausen et al. 2010). While the term 'stewardship' possesses no commonly accepted definition (Roach et al. 2006), in the context of nature conservation, stewardship means, simply, people caring for the earth. It encompasses both an ethic and action, similar to the philosophy and practical management described by the American ecologist Aldo Leopold during the first half of the 20th century (Leopold 1949). It is based upon the recognition of shared responsibility for the care and management of natural capital-the resources of water, land, air and biodiversity, and the benefits they provide. The goal is to conserve, protect or restore those resources and their environmental, social and economic values for future generations (Mitchell and Brown 1999; Roach et al. 2006; Wolthausen et al. 2010). Usually the care is directed locally, where the impact is more meaningful, but its reach can extend to global environmental issues. Some conservation leaders apply stewardship to a broader landscape/seascape approach in which people care for the natural, cultural and historic heritage of their local environment (Brown et al. 2005). Since

K. Blanchard (✉) · J. Dawe · R. Wall
Intervale Associates, PO Box 172, Doyles, NL A0N 1JO, Canada
e-mail: kblanchard@intervale.ca

J. Dawe
e-mail: jennifer.dawe@gmail.com

R. Wall
e-mail: russell_j_wall@hotmail.com

J. Urquhart et al. (eds.), *Social Issues in Sustainable Fisheries Management*, MARE Publication Series 9, DOI 10.1007/978-94-007-7911-2_6,

stewardship concerns both ethic and action, it carries two broad objectives: to cultivate in people a sense of responsibility and to engage them in the management of resources. Government programmes often view stewardship as a partnership arrangement with organisations and individuals who have the potential to impact the conservation outcome.

The theoretical basis for stewardship's application in the secular world of conservation is somewhat vague (Worrell and Appleby 2000), with researchers lacking consensus over whether its focus should be on encouraging specific conservation behaviours, through techniques such as social marketing (McKenzie-Mohr 2011), or cultivating a stewardship ethic, through appropriate education and engagement (Jacobson 1995; Martin and James 2005). Although both approaches have proven effective at achieving an increase in environmental behaviours (see Monroe 2003 for an overview), few documented studies have demonstrated a cause-and-effect relationship between stewardship programmes and the ultimate goal of species recovery, in part due to the complexity of the issues, the lengthy time required and the short-term timetables of many funding programmes (Kapos et al. 2008). Meanwhile, many practitioners are faced with selecting strategies and designing activities without knowledge of which ones work best. While programme planning ought to draw on relevant scientific and local ecological knowledge, very often the impetus depends on such factors as legislated mandates, financial resources, familiarity with conventional delivery mechanisms, and the need for quick turnaround of measurable results.

Although some stewardship programmes are implemented through regulation, the majority in nature conservation operate through voluntary means. Authentic stewardship should not be motivated by the threat of punitive measures (Roach et al. 2006; McLaren et al. 2008) but by intrinsic incentives such as personal ethics, social norms, genuine interest or empowerment, which may be more enduring and possibly transferable to other resources or ecosystems (Blanchard and Monroe 1990; Martin and James 2005). In the stewardship of marine resources, often it is important to understand the incentives of fish harvesters, since their actions are key to the recovery of several marine species at risk (Richardson et al. 2005).

Stewardship is centred around positive relationship between the species, resource or habitat, and the caregiver. The starting point should be to unleash and encourage the positive values people may already possess about nature, rather than force the immediate adoption of new ones (Blanchard 2000). However, it is difficult to foster positive relationship with a species if people are being forced to alter their behaviour without convincing knowledge of how they and the species may benefit from the change. It is especially difficult if they fear that helping the species may cause them personal harm or interfere with their livelihood. In such cases it may be necessary to spend more time understanding the audience. In the stewardship of marine species, it is often fish harvesters who are most directly affected. For them, positive relationship relies upon several things, not the least of which include: trust in the science that declared the species threatened, respect for local knowledge and values, consultation, targeted goals and timetables for recovery, consistent release of information about progress being made, and appreciation for their contributions.

The case involving three species of wolffish in Newfoundland and Labrador waters demonstrates stewardship's effectiveness in encouraging conservation behaviours aimed at reducing mortality of at-risk species. The case also reveals the importance of cultivating a stewardship ethic as a potentially enduring strategy independent of regulation. The need for building a trustworthy partnership relationship with the primary stakeholder group cannot be overstated—one which is built upon shared objectives for ecological sustainability as well as cultural and economic values.

The knowledge, opinions and actions of fish harvesters in the province of Newfoundland and Labrador, Canada, was the subject of a five-year study from 2004 to 2008 by the nonprofit organisation, Intervale, and of research in 2009 by a graduate student at Memorial University of Newfoundland. We conducted semi-structured interviews with active harvesters to uncover local ecological knowledge about wolffish and opinions about their management, specifically relating to new regulations affecting wolffish bycatch. We used that information to inform our approach to stewardship and to create a baseline from which to track changes in harvesters over a five-year period.

With funding from Canada's Habitat Stewardship Program for Species at Risk, the Intervale team was required to follow the recommended actions of a wolffish recovery strategy pertaining to stewardship: promoting safe handling and live release of wolffish, educating resource users in identification and biology of wolffish, raising awareness about the importance of new laws that promote species recovery, and enhancing consultative activities (Kulka et al. 2008). The graduate student research was funded by the Community-University Research for Recovery Alliance (CURRA) at Memorial University of Newfoundland. CURRA, of which Intervale is a partner, is a five-year research programme designed to help communities and organisations along Newfoundland's west coast develop strategies for the recovery of fish stocks and fishing communities.

6.2 COSEWIC, the Species at Risk Act, and the Status of Wolffish in Canadian Waters

The Committee on the Status of Endangered Wildlife in Canada (COSEWIC) was created in 1977 with the mandate to designate the conservation status of indigenous wildlife species at risk in Canada (Shank 1999; Freedman et al. 2001). Based on status reports that are prepared for species, COSEWIC may designate species as *extinct* or *extirpated, endangered, threatened* or *special concern*. The term *threatened* refers to a species likely to become endangered if nothing is done to reverse the factors leading to its extirpation or extinction; the term *special concern* refers to a species that may become a threatened or endangered species because of a combination of biological characteristics and identified threats (SARA 2009). The COSEWIC has no legal authority and there are no regulatory consequences associated with its listings.

Canada's *Species at Risk Act* (SARA) was enacted in 2002 for the purpose of preventing wildlife species from becoming extinct; to provide for recovery of species that are extirpated, endangered, or threatened as a result of human activity; and to manage species of special concern to prevent them from becoming at risk. Sections 10 and 11 of the Act specifically address stewardship (SARA 2009). For a species to be listed under Schedule 1 of the Act, it must be recommended by COSEWIC and approved by the federal minister of authority. Before the federal government makes a final decision on the status of a recommended species, the species status report is given to stakeholders and posted on the SARA public registry for consultation (Bourdages and Labelle 2003). For marine species, this may include consultations with fish harvesters and possibly an examination of both scientific and local ecological knowledge.

Three species of wolffish (Family *Anarhichadidae*), northern *Anarhichas denticulatus*, spotted *A. minor*, and Atlantic *A. lupus*, are demersal fish that inhabit the Atlantic and Arctic oceans. Their distribution in the northwestern Atlantic includes much of the coasts of Labrador and insular Newfoundland. The centre of distribution for the northern and spotted wolffish is the region encompassing the Labrador Shelf, northeastern Newfoundland and the Grand Banks. The distribution for Atlantic wolffish is slightly more southern and includes the Gulf of St. Lawrence, Scotian Shelf, Bay of Fundy and Gulf of Maine (Simpson and Kulka 1998). Wolffish in the northwest Atlantic do not form dense aggregations and do not exhibit strong migratory behaviour. Collectively they are characterised as 'low productivity' fishes, based on data from studies of Atlantic wolffish in US waters.

In 2001, COSEWIC designated northern and spotted wolffish as *threatened* based on declines greater than 90 % over three generations and significant reduction in distribution. In addition, COSEWIC designated Atlantic wolffish as *special concern*, since its numbers had declined at a similar rate over the portion of its range extending from the Labrador Shelf to northeastern Newfoundland. The trends in biomass and abundance that had been reported in the status reports were based on limited scientific data collected between 1978 and 1994 from offshore trawl surveys taken by the federal Department of Fisheries and Oceans (DFO 2004), covering the Labrador Shelf to Grand Banks region (Kulka et al. 2008). When the *Species at Risk Act* was enacted, northern, spotted and Atlantic wolffish as well as several terrestrial, aquatic and marine mammal species were grandfathered into the Act, before consultation was fully incorporated into the listing process (Dawe and Neis 2012). Wolffish became the first marine fish species to be listed under Schedule 1 of the Act.

Fish harvesters in eastern Canada refer to wolffish generally as catfish. Common names include jelly cat (northern wolffish), spotted catfish or leopard fish (spotted wolffish) and striped catfish (Atlantic wolffish). Although wolffish in Canadian waters have never been exploited commercially as a directed fishery, they are caught as bycatch, i.e. incidental catch, in nearly all fisheries of Newfoundland and Labrador using a wide variety of gear types. Bycatch occurs especially in directed fisheries for Atlantic cod *Gadus morhua*, Greenland halibut *Reinhardtius hippoglossides*, haddock *Meanogrammus aeglefinus* and yellowtail flounder *Limanda ferruginea*.

The COSEWIC assessment stated that specific threats to wolffish included bycatch mortality in commercial fisheries and habitat alteration by trawling gear. The strategy emphasised as a high priority working with harvesters and other ocean users to mitigate harm to wolffish and to promote stewardship initiatives. Recommended activities included consulting with resource users over gear modifications, promoting quick and safe release of incidentally caught wolffish to the site of capture and increasing awareness among resource users. The strategy stressed the involvement and support of the resource user community as critical to the success of wolffish recovery (Kulka et al. 2008).

Mandatory release of northern and spotted wolffish was instituted in Newfoundland and Labrador through a licence amendment. Harvesters were issued an Incidental Harm Permit, in accordance with subsection 83 (4) of SARA, which allowed for wolffish incidental catch. Techniques for live release were developed through onboard observations and in consultation with harvesters, after which they were promoted in professional training courses and conveyed through information materials by Fisheries and Oceans Canada (DFO). Due to Atlantic wolffish status as *special concern*, live release of that species, although not required, was recommended as a voluntary measure and at the discretion of the DFO regions (Simpson and Kulka 2002; DFO 2004; Kulka et al. 2008).

6.3 Methods

6.3.1 Stewardship Survey and Dockside Dialogue

A team from Intervale conducted a face-to-face survey of fish harvesters each year during a five-year period, 2004–2008. The purpose of the survey was to gather information on three general topics: (1) harvester encounters with wolffish; (2) harvester knowledge, opinions and actions regarding wolffish, wolffish listings and the SARA regulations; and (3) progress with the implementation of recovery actions. A total of 329 harvesters were interviewed independently and the yearly sample size varied as follows:79 (2004), 76 (2005), 74 (2006), 56 (2007) and 44 (2008). Harvesters interviewed represented a wide range of fisheries, vessel sizes, and gear types, the majority of which were inshore fleets targetting cod, crab, lobster and halibut with vessels 65 feet (20 m) or less in length. In conjunction with the interviews, the team engaged an additional 255 harvesters in what Intervale termed 'dockside dialogue', aimed at promoting consultation and stewardship in marine species at risk, for a total of 584 harvesters overall.

Our sampling strategy and geographic coverage consisted of intercepting harvesters at their vessels in ports from the north coast of Labrador to the southeast coast of insular Newfoundland (Fig. 6.1). In a total of 101 ports, we interviewed harvesters who were actively fishing waters of NAFO regions 2H, 2J, 3K, 3L, 3Ps, 3Pn and 4R. Because fishing quotas and season durations varied from year to year,

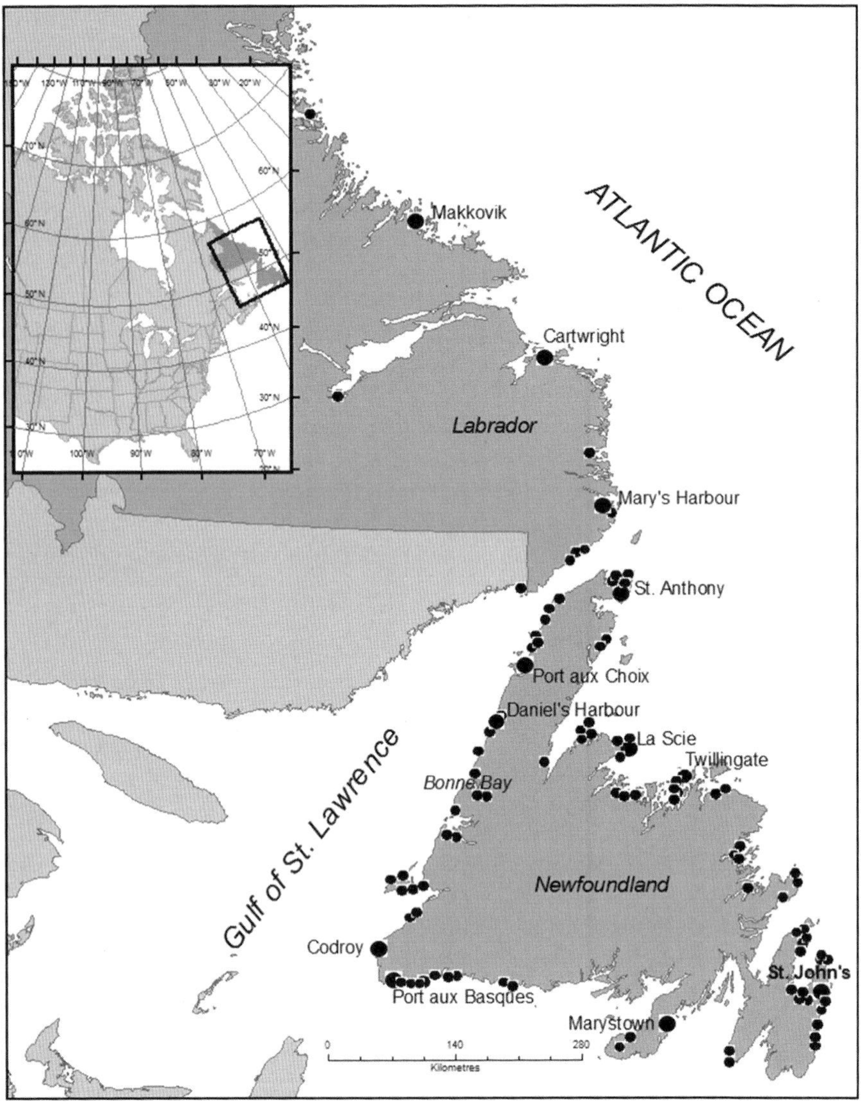

Fig. 6.1 The study area, including coastal communities where interviews were conducted

we achieved our targeted sample size by returning to some of the busier ports in multiple years.

Following a brief introduction, we requested consent for an interview with the skipper or a senior member of crew. Non-response bias was not an issue, as nearly all consented to the interview. The interview duration was 10–20 min and was followed by a post-interview dockside dialogue with additional crew members lasting between 5 and 30 min. The purpose of this post-interview dialogue was to gain further insight into social issues affecting recovery of species at risk and to encourage stewardship actions. During this dialogue, we offered harvesters information materials such as identification charts and we expressed our appreciation for their cooperation in the recommended recovery actions.

6.3.2 *In-depth Interviews*

In 2009, semi-structured face-to-face interviews were conducted with 21 fish harvesters along the southwest and west coasts of insular Newfoundland (Fig. 6.1). In a quiet setting of a home or other location, harvesters were asked open-ended questions about their fishing history as well as their opinions about conservation, stewardship and the listing of species. Interviews were recorded and transcribed.

6.4 Results

6.4.1 *Stewardship Survey and Dockside Dialogue*

All harvesters demonstrated familiarity with wolffish, although descriptions of the three species overlapped occasionally. Most harvesters referred to the wolffish group as catfish and used a variety of common names to distinguish the species. For the five-year period, an average of 89 % of harvesters had either experienced bycatch of wolffish that same year or knew someone nearby who had. The species encountered by the highest proportion of harvesters was spotted wolffish (83.4 % average, all years), followed by northern wolffish (65.6 %, all years) and Atlantic wolffish (39.2 %, all years). The targeted fishery in which wolffish were encountered most often was Atlantic cod. In 2004, the year in which we recorded the highest number of wolffish caught as bycatch, we recorded total season estimates of 10,218 fish and an additional 92,750 lbs. of wolffish that had been caught among the 79 harvesters interviewed. Harvesters expressed their bycatch using either measurement.

Harvesters demonstrated a high level of awareness about regulations concerning wolffish. The proportion that had heard of the *Species at Risk Act* was 98.5 % in 2004 ($n=73$), 98.6 % in 2005 ($n=73$), 98.5 % in 2006 ($n=65$) and 88.9 % in 2007 ($n=56$). All respondents knew that northern and spotted wolffish were protected under SARA and that they were required to release them live and in the place where they were caught (Table 6.1). In 2004, most harvesters recalled having learned about the status of wolffish from information distributed by Fisheries and Oceans Canada (DFO), including identification cards created by Intervale and DFO.

Even though harvesters had heard about SARA and understood the regulations, they did not agree with the listings for each of the three species of wolffish. In 2004, 96 % of harvesters knew of the listings for northern, spotted and Atlantic wolffish but 81.9 % did not agree that northern and spotted wolffish should be listed as

Table 6.1 Percentage of harvesters that knew the regulations under SARA for northern and spotted wolffish

	2004	2005	2006	2007	2008
Knew the regulations	100.0	98.6	100.0	100.0	100.0

Table 6.2 Percentage of harvesters that knew the listing status for northern and spotted wolffish and percentage of harvesters that did not agree with the listings for wolffish

	2004	2005	2006	2007	2008
Knew that northern and spotted wolffish are listed as threatened	95.6	100.0	98.6	97.8	n/a
Did not agree with listing for northern, spotted wolffish	81.9	88.7	87.3	64.1	n/a
Did not agree with listing for Atlantic wolffish	95.5	94.9	91.5	64.1	n/a

Table 6.3 Percentage of harvesters that stated they release wolffish live

	2004	2005	2006	2007	2008
Release wolffish live	93.3	98.5	95.4	96.4	97.3
Release in absence of authority on-site	78.5	77.7	91.5	95.7	n/a

threatened and 95.5 % did not agree that Atlantic wolffish should be listed as *special concern* (Table 6.2). We were concerned that such disconnect between scientific knowledge, which formed the basis for the assessment of all three species, and the opinions of harvesters would work against the cultivation of authentic stewardship. In fact, our fears worsened by 2007, when many harvesters appeared to being resigned to the regulations out of what they described as 'wolffish fatigue' —a condition reportedly caused by too much attention on relatively unimportant species.

The stewardship programme's goal of encouraging responsible behaviour, specifically live release of wolffish, seemed to be working. In spite of harvesters disagreeing with the listing of wolffish, a high percentage claimed to be complying with the new regulations, a claim that was verified by Fishery Observers working onboard approximately one-third of the inshore fleet. The vast majority of harvesters—an average of 96 % over all five years—stated that they were releasing wolffish live and in the place where they were caught (Table 6.3). Moreover, after the first few years, the adoption of the new recovery actions grew in strength, as the percentage of harvesters that stated they would release wolffish in the absence of an authority figure such as government fishery officer or onboard Fishery Observer increased from 78.5 % in 2004 to 95.7 % in 2007 (Table 6.3). As a further indication of the degree of compliance with the new regulations, the vast majority believed that other harvesters in their area were releasing wolffish live, with the least harm, in the place where they were caught (Table 6.4). We recognise the potential for harvesters not to reveal actions that run counter to the law; nonetheless, we believe that the consistency of

Table 6.4 Responses to the question of whether harvesters believed that others in their area were releasing wolffish live

Believed other harvesters in their area were releasing wolffish live	%		
	Yes	No	Did not know
2004	88.0	4.0	8.0
2005	84.1	5.8	10.1
2006	98.4	1.6	0.0
2007	90.4	3.8	5.8
2008	94.5	0.0	5.5

Table 6.5 Reasons harvesters gave for releasing wolffish live

	%			
Reasons for releasing	2004	2005	2006	2007
No good to us	10.9	52.9	51.7	76.3
It's the law	15.1	26.5	20.0	44.7
There is a fine	5.5	2.9	13.3	10.5
Importance to food web	0.0	0.0	10.0	15.8
Might become a commercial resource	1.4	19.1	13.3	2.6
Might become a food fishery	1.4	5.9	0.0	13.2
Right to live	49.3	14.7	23.3	0.0

response over broad spatial and temporal scales, combined with the relatively non-threatening nature of the dockside interview, argue in favour of these results.

We heard strong opinions from harvesters who believed wolffish were a 'nuisance' or 'destructive' or that they negatively impacted fisheries, either by eating bait or by competing for the directed species. Given these opinions and the fact that most harvesters disagreed with the listings for wolffish, we explored whether their actions in releasing wolffish were being motivated by a stewardship ethic or some other influence such as fear of penalties for being noncompliant with the law. From 2004 through 2007 we asked harvesters their reasons for releasing wolffish live. In 2004, the highest percentage of harvesters stated that "they have a right to live," which is an indicator of intrinsic value (Table 6.5). However, beginning in 2005, the emphasis shifted: the highest percentage of response was "they are no good to us" and the second highest was "it's the law." These results suggested that, over time, harvesters' actions were being motivated more by the absence of economic incentive and the influence of the law and less so by a stewardship ethic.

6.4.2 In-depth Interviews

The 21 harvesters who participated in the in-depth interviews in 2009 had fished between 13 and 39 years (mean=28). Eighteen of the 21 harvesters were skippers of their own boats. Interviews lasted between 23 and 66 min. Interview length was influenced by the number of years harvesters had been catching wolffish, the size of the catches, the number of fisheries in which they participated and the extent of the discussion evoked by the questions.

Results confirmed that wolffish were caught as bycatch in the major fisheries of the southwest and west coasts of Newfoundland, the most important of which were Atlantic cod (90 % of harvesters interviewed) and American lobster *Homarus americanus* (95 %). All harvesters interviewed (100 %) reported catches containing *A. lupus* and *A. minor*, 76 % reported catches of *A. denticulatus*.

All harvesters knew about SARA and that wolffish were listed. Seventy percent of harvesters found out about the listing through DFO and information materials that accompanied the mail-out of licences. The other 30 % of harvesters found out through meetings organised by the Fish, Food, and Allied Workers union and other meetings. Fifty-seven percent agreed with the listing of northern and spotted wolffish and 43 % were neutral, i.e. had no opinion or did not care. In contrast, 67 % disagreed with the listing for Atlantic wolffish and 33 % were neutral. Further results from the in-depth interviews are presented in a thesis by Dawe (2010).

6.5 Discussion

Preliminary fieldwork by Blanchard in 2003 revealed that many harvesters worried about the new regulations under SARA and their possible implications for fisheries. As prior studies in fishing communities of the Gulf of St. Lawrence had demonstrated, it is important to consider the local knowledge and opinions of resource users before introducing conservation initiatives that they do not support (Blanchard 1994, 2005). Intervale launched its dockside dialogue in 2004 in order to better understand the perspectives of harvesters while, at the same time, encouraging the recommended recovery actions. Crucial to that dialogue was the ability to listen intently to harvesters' complaints and to express appreciation for their efforts in recovery. Many harvesters told us that they were seldom thanked for their efforts in general and never for recovery actions targeting wolffish. These findings suggested the need to build stronger partnership relations involving harvesters.

The prohibitions under SARA involved penalties of up to $50,000 in fines for causing harm to endangered or threatened species. This served as an effective 'wake-up' call to harvesters who previously had been utilising or discarding wolffish caught as bycatch. The federal Department of Fisheries and Oceans, in partnership with the FFAW union and other organisations, launched a rapid information and training campaign for industry, which focused on what harvesters need to know about SARA and techniques for live release. Meanwhile, funding that was made available from the federal government enabled nongovernmental organisations such as Intervale to develop programmes promoting stewardship. Many information materials were produced by DFO and its partners, including identification cards (Fig. 6.2), instructions in the methods of live release, posters for schools and public areas, and industry training videos. Results of the stewardship survey suggest that this information and training strategy contributed substantially to increased

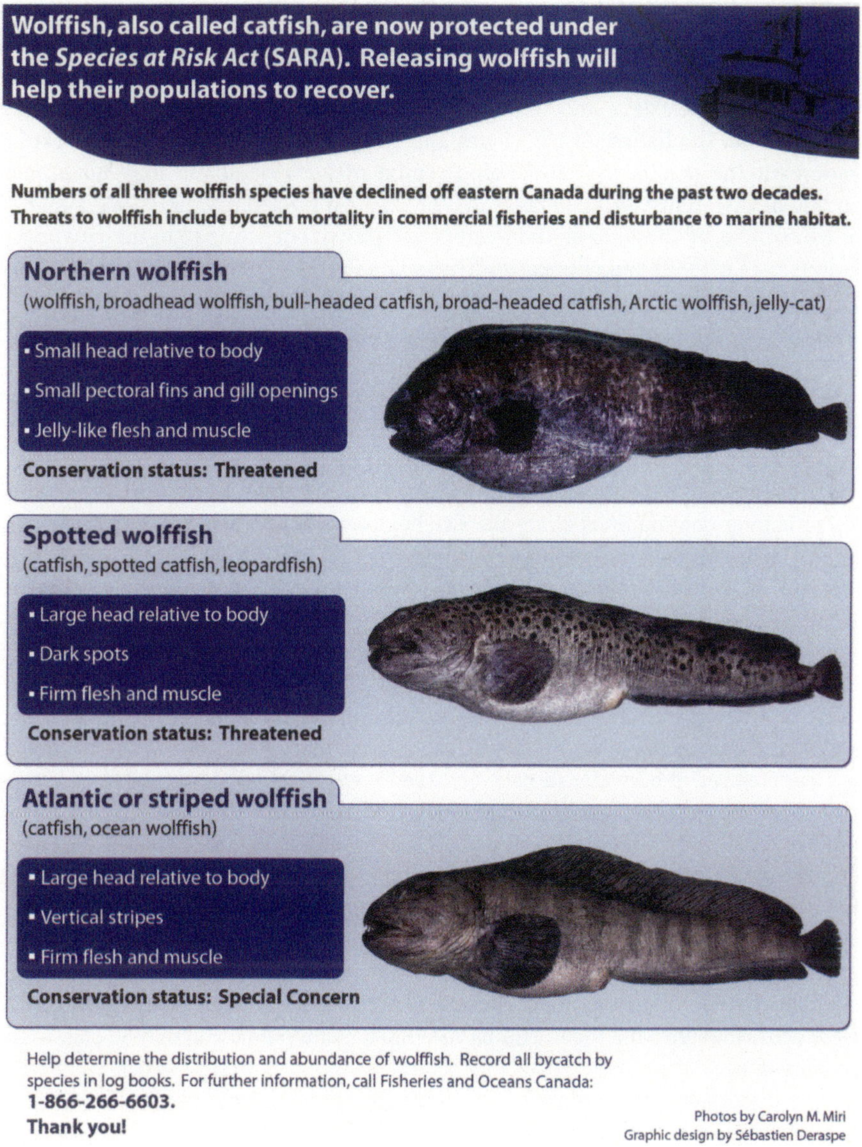

Fig. 6.2 A wolffish identification card, produced by Intervale with the support of DFO and other partners. Approximately 20,000 cards were issued in English, French and Inuktitut and distributed to harvesters and industry groups in eastern Canada

harvester knowledge and the adoption of new behaviour, i.e. live release of wolffish. However, since the recovery actions had been mandated, the recovery actions were not voluntary and, therefore, were less likely to be a reflection of authentic stewardship at work.

The stewardship survey revealed that even though harvesters were compliant with regulations for live release, the majority disagreed with the SARA listings during the early years of 2004–2006. The in-depth interviews showed that harvesters in southwestern and western Newfoundland either agreed with or were neutral about the listings for northern and spotted wolffish. One possible explanation for the results in that specific region of Newfoundland may have been an actual difference in abundance of wolffish in adjacent waters. We suggest, however, that by 2009, many harvesters had grown accustomed to the new regulations for northern and spotted wolffish and may have reached a point at which they simply no longer cared to disagree with the listings. Harvesters did not feel empowered when it came to decision-making about the recovery and management of wolffish.

Of concern to us were statements by harvesters that their actions were motivated by the need to be compliant with new regulations under the *Species at Risk Act*, as well as the fact that wolffish were no longer of economic or utilitarian value to them, as opposed to being motivated by a stewardship ethic. In contrast, other studies in fishing communities along the Quebec North Shore had reported beneficial effects of a similar conservation initiative for seabirds, in which the encouragement of stewardship appealed to the culture's own ecological or nature-oriented values (Blanchard 2000). For this reason, in 2007 we launched a more positive approach to engaging harvesters in wolffish recovery. We developed an educational tool for the purpose of describing the ecological role of wolffish, thus attempting to strike a chord with harvesters' inherent belief in the value of marine ecosystems and their common passion for the sea. The documentary DVD, *Wolffish: A Balance of Life*, featured underwater footage of Atlantic wolffish and described the species' role as predator of sea urchins, *Strongylocentrotus droebachiensis*, in helping to protect kelp beds as vital nursery habitat for fisheries (Fig. 6.3). Harvesters were the sole spokespersons in the DVD. Their comments revealed both fear and respect for wolffish while portraying themselves as passionate defenders of the marine ecosystem. Investigations into the video's effectiveness among harvesters six-months after viewing revealed noticeable change in that harvesters voluntarily expressed the ecological value of wolffish and their need for recovery. Other positive outcomes were confirmed through focus groups with the general public and in nine classroom presentations.

The importance of introducing stewardship at the start of a conservation initiative cannot be overstated. Because wolffish had been grandfathered into the *Species at Risk Act* without consultation, the early stage of the recovery process for wolffish had passed by the time stewardship was called into play. Therefore it was difficult to realise some of the broader benefits of stewardship, such as its capacity for building positive relationships, its attention to the cultural context, the appeal to personal ethics and social norms, and its empowerment value. The last item is especially important in fostering a sense of responsibility and ownership for the stewardship initiative among the stakeholders who are closest to the resource of concern. This is a key strategy for cultivating sustained stewardship.

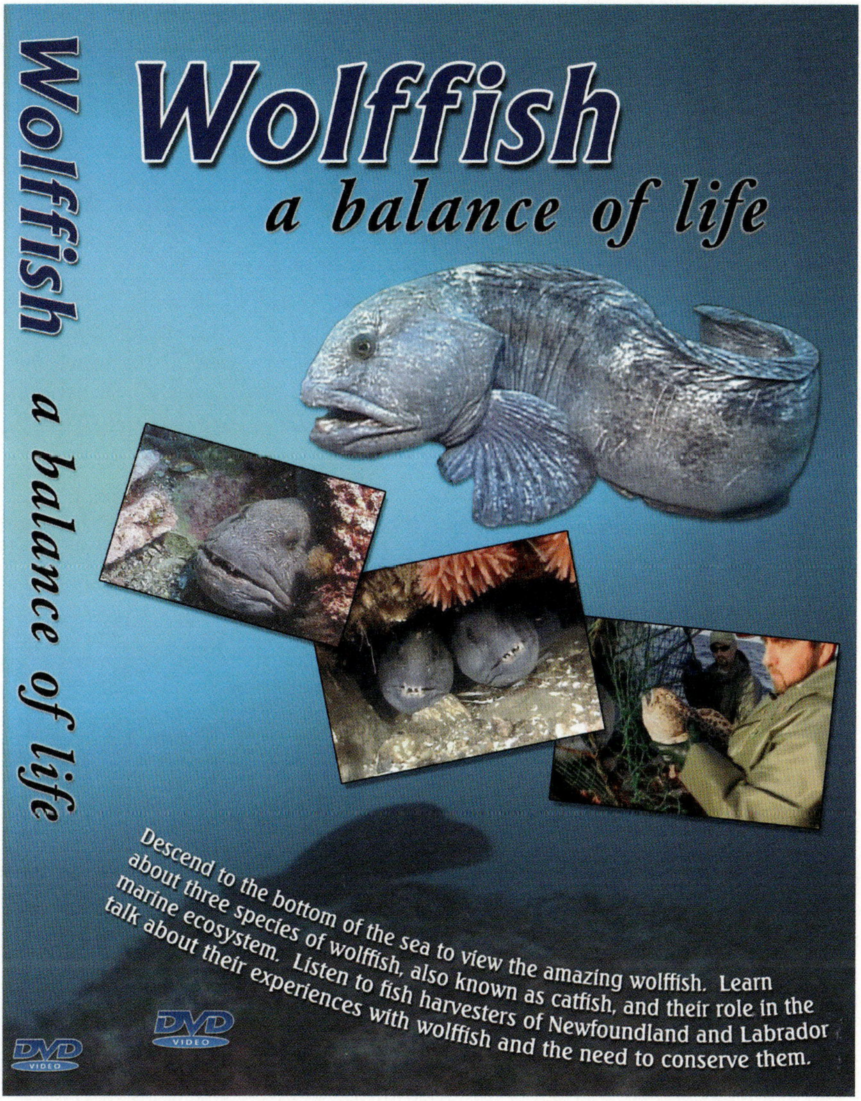

Fig. 6.3 The documentary DVD, *Wolffish: A balance of life*, produced by Intervale

6.6 Conclusions

The implementation of a recovery strategy for wolffish as called for under Canada's *Species at Risk Act* influenced significantly the actions and awareness of fish harvesters in Newfoundland and Labrador, the vast majority of whom interacted directly with wolffish at sea. Three species of wolffish had been grandfathered into the

Act without the incorporation of local ecological knowledge and without prior consultation with the fishing industry. The recovery strategy and information campaign by DFO, in collaboration with the FFAW union and other partnering organisations, were effective in rapidly disseminating information to harvesters and achieving a high compliance rate with the SARA regulations, as harvesters quickly adopted the new actions for releasing wolffish live. Stewardship's objective of changing the specific behaviour of a target audience was achieved. The strategy was successful as a crucial first test of the strength and effectiveness of SARA within the commercial fisheries sector. However, we believe that the strategy could have applied more effort towards stewardship's other objective of encouraging an ethic of care and responsibility. A stronger partnership relationship with harvesters might have been accomplished by offering more consultation at an early stage, utilising harvesters' local ecological knowledge and providing harvesters with regular progress reports on meeting targets for recovery. By encouraging both ethic and action, the strategy could prove more durable in the long term and potentially lead the way to effective stewardship of other at-risk species.

Fish harvesters in Newfoundland and Labrador, on the whole, are passionate defenders of the marine ecosystem and possess a practical and well-developed understanding of what constitutes good stewardship. While it is not yet known what long term impact the recovery actions of tens of thousands of harvesters in Newfoundland and Labrador may be making on the status of wolffish populations in eastern Canadian waters, the substantial number of wolffish that harvesters are releasing live each year is a remarkable demonstration of industry collaboration in marine species recovery.

Stewardship at its core involves a positive, beneficial relationship between the resource, species or habitat, and the caregiver. It is encouraged by many factors independent from the use of force. In order for stewardship to work in the recovery of marine species, ecological values and feelings of responsibility need to be strengthened at both the individual and community levels, where personal ethics and social norms take effect. Educational tools such as the documentary DVD, when used in conjunction with group dialogue, are proving useful in helping to strengthen those values.

As stewardship's potential role in recovery is better realised, it ought to be made an integral part of the conservation initiative from the very beginning and utilised throughout the recovery process. Stewardship helps make the work of recovery meaningful and beneficial to the people most affected by the conservation initiative. It assists in building partnerships based on trust, common goals and frequent face-to-face dialogue. It strengthens commitment and empowers stakeholder groups to play a more active role as partners in conservation action. It appeals to personal ethics and social norms. Stewardship takes time to build, but it is more sustainable than alternatives that are 'quick fix'. We believe that the benefits of stewardship will be realised more as we move from compliance to stewardship in marine species recovery.

Acknowledgements We are grateful to the many fish harvesters who allowed us to interview them and who shared their knowledge about wolffish. We thank staff of the FFAW for their assistance and the many university students who assisted with the interviews and dockside dialogue. We extend a special appreciation to Dr Catherine Hood for giving Blanchard the opportunity to participate in wolffish recovery and for comments on this manuscript. We especially thank Dr Barbara Neis, principal investigator of the Community-University Research for Recovery Alliance (CURRA), for many project-related opportunities and for advisory support to Jennifer Dawe. Major funding for this project was provided by the Government of Canada's Habitat Stewardship Program and from the CURRA at Memorial University of Newfoundland, which is funded by the Social Sciences and Humanities Research Council of Canada, Memorial University, and multiple community partners.

References

Blanchard, K. A. (1994). Culture and seabird conservation: The north shore of the Gulf of St. Lawrence, Canada. In *Seabirds on Islands: Threats, Case Studies and Action Plans* (pp. 294-310). Bird Life Conservation Series, No. 1, Cambridge, England.

Blanchard, K. A. (2000). Rachel Carson and the human dimension of fish and wildlife management. *Human Dimensions of Wildlife*, *5*(1), 52–66.

Blanchard, K. A. (2005). Seabird populations of the North Shore of the Gulf of St. Lawrence: Culture and conservation. In M. Lyman (Ed.), *Natural resources as community assets: Lessons from two continents* (pp. 211–237). Washington: Aspen Institute.

Blanchard, K. A., & Monroe, M. C. (1990). Effective educational strategies for reversing population declines in seabirds. *Trans. N. American Wildlife and Natural Resources Conf, 55*, 108–117.

Brown, J., Mitchell, N., & Beresford, M. (2005). Protected landscapes: A conservation approach that links nature, culture and community. In J. Brown, N. Mitchell & M. Beresford (Eds.), *The protected landscape approach: Linking nature, culture and community* (p. 268). Gland: IUCN.

Bourdages, J. L., & Labelle, C. (2003). *Protecting wild species at risk in Canada* (p. 29). Ottawa: Library of Parliament.

Dawe, J. (2010). Marine fish, local ecological knowledge, and the *Species at Risk Act* in Canada: Lessons from the case study of three species of wolffish. M.Sc. thesis. Memorial University of Newfoundland, St. John's, NL.

Dawe, J. L., & Neis, B. (2012). Species at risk in Canada: Lessons learned from the listing of three species of wolffish. *Marine Policy*, *36*, 405–413.

Department of Fisheries and Oceans. (2004). Allowable harm assessment for spotted and northern wolffish. *DFO Can. Sci. Advis. Sec. Stock Status Report 2004/031.*

Freedman, B., Roger, L., Ewins, P., & Green, D. M. (2001). Species at risk in Canada. In K. Beazley & R. Boardman (Eds.), *Politics of the wild: Canada and endangered species*. Toronto: Oxford University Press.

Intervale Associates. (2007). *Wolffish: A balance of life*. Educational DVD. 20 minutes. Doyles, NL.

Jacobson, S. (Ed.). (1995). *Conserving wildlife: International education and communication approaches* (p. 302). New York: Columbia University Press.

Kapos, V., Balmford, A., Aveling, R., Bubb, P., Carey, P., Entwistle, A., & Manica, A. (2008). Calibrating conservation: New tools for measuring success. *Conservation Letters*, *1*, 155–164.

Kulka, D., Hood, C., & Huntington, J. (2008). *Recovery strategy for Northern wolffish (Anarhichas denticulatus) and Spotted Wolffish (Anarhichas minor), and Management Plan for Atlantic Wolffish (Anarhichas lupus) in Canada* (pp. x+103). St. John's: Fisheries and Oceans Canada: Newfoundland and Labrador Region.

Leopold, A. (1949). *Sand county almanac* (p. 240). UK: Oxford University Press.

Martin, K., & James, M. (2005). The need for altruism: Engendering a stewardship ethic amongst fishers for the conservation of sea turtles in Canada. *MAST (Maritime Studies)*, *3*(2) and *4*(1),105–118.

McLaren, B., Hollis, T., Roach, C., Blanchard, K., Chaurette, E., & Bavington, D. (2008). *Knowledge flows, conservation values and municipal wetlands stewardship. How knowledge moves.* Montreal: McGill-Queens University Press.

Mckenzie-Mohr, D. (2011). *Fostering sustainable behaviour: An introduction to community-based social marketing* (p. 192). Gabriola Island: New Society Pub.

Mitchell, B., & Brown, J. (1999). Stewardship: A working definition. In J. Brown & B. Mitchell (Eds.), *Stewardship: An international perspective. Environments: A Journal of Interdisciplinary Studies*, *26*(1) 1998. University of Waterloo, Waterloo, Ontario.

Monroe, M. C. (2003). Two avenues for encouraging conservation behaviors. *Human Ecology Review*, *10*(2), 113–125.

Richardson, E. A., Kaiser, M. J., & Edwards-Jones, G. (2005). Variation in fishers' attitudes within an inshore fishery: Implications for management. *Environmental Conservation*, *32*(3), 213–225.

Roach, C., Hollis, T., McLaren, B., & Bavington, D. (2006). Ducks, bogs, and guns: A case study of stewardship ethics in Newfoundland. *Ethics & the Environment*, *11*(1), 43–70.

SARA. (2009). *Species at Risk Act* public registry. www.sararegistry.gc.ca.

Shank, C. C. (1999). The committee on the status of endangered wildlife in Canada (COSEWIC): A 21-year retrospective. *Canadian Field-Naturalist*, *113*(2), 318–341.

Simpson, M. R., & Kulka, D. W. (2002). *Formulation of an incidental harm permit strategy for wolffish species (Anarhichadidae).* DFO Can. Sci. Advis. Secretariat Research Document 2003/047.

Wolthausen, D., McIntyre, L., Neave, E., & Rubec, C. (2010). *An ecological recovery plan for Canada: Creating a healthier environment and economy through effective stewardship.* Contract report by the Centre for Environmental Stewardship and Conservation for the Canadian Stewardship Coalition. Edmonton, Alberta.

Worrell, R., & Appleby, M. C. (2000). Stewardship of natural resources: Definition, ethical and practical aspects. *Journal of Agricultural and Environmental Ethics*, *12*(3), 263–277.

Chapter 7
Building bridges among scientists and fishermen with participatory action research

Steven Mackinson and Douglas Clyde Kongshøj Wilson

Keywords Fishermen's knowledge • Stakeholder engagement • Participatory Research • Governance • Co-management

7.1 Introduction

7.1.1 The Challenge of Action Research in European Fisheries

Among developed countries, the European Union (EU) has made comparatively little progress in finding and applying solutions to the crisis of sustainability facing fisheries on a global scale. EU fisheries are managed by an agreement known as the Common Fisheries Policy (CFP). It is a large and unwieldy attempt to manage fisheries that is often more focused on solving political problems around dividing fish than it is on sustainability (Wilson 2009). It is one of the few policy arenas where Member States have ceded decision making power to the EU, giving it political influence beyond that which its economic and environmental importance would suggest.

With its top-down approach, the CFP is unresponsive to local conditions and lacks support from both the communities reliant on fish resources for a living, and other stakeholders interested in the long-term wellbeing of the ecosystem. Fisheries stakeholders in particular view the governance system as top-down controlled, characterised by a history of negative incentives. At the same time, management has failed to meet its own resource-related objectives, with many fish stocks being

S. Mackinson (✉)
Centre for Environment, Fisheries and Aquaculture Science (Cefas), Pakefield Road, Lowestoft, Suffolk, UK
e-mail: steve.mackinson@cefas.co.uk, www.cefas.co.uk

D. C. K. Wilson
Innovative Fisheries Management, Aalborg University Research CentreFibigerstræde 13, 66c, 9220, Aalborg Ø, Denmark
e-mail: dw@ifm.aau.dk

J. Urquhart et al. (eds.), *Social Issues in Sustainable Fisheries Management,* MARE Publication Series 9, DOI 10.1007/978-94-007-7911-2_7,

in a poor state (ICES 2011). The European Commission itself tells us that 88% of EU stocks are being fished beyond agreed targets, and 30% of these stocks are so depleted they may not be able to replenish (CEC 2009).

7.1.2 The Need for Participatory Action Research

Despite being perhaps the most science-driven policy arena in Europe, the CFP suffers, more than comparable fisheries regimes, from a legitimacy crisis focused on the knowledge base for management decision making (Schwach et al. 2007). It requires a large and constant stream of scientific advice—around 1,600 pages per year—just to make its routine administrative decisions (Wilson 2009). This advice is mainly quantitative assessments of the size of fish stocks from which sustainable catches can be derived and these assessments are often highly uncertain. The foundation of the legitimacy crisis rests with how the science-policy system makes informed management decisions when it is known (or believed) that the underpinning science is uncertain. Understandably, these questions of legitimacy undermine credibility of the institutions responsible for assessment and advice.

Box 1. Regional Advisory Councils (RACs) in brief

Under the auspices of the Common Fisheries Policy, RACs were established by Council Decision (EC) 256/2004 with the intention of increasing the participation of those affected by the CFP in the fisheries management decision-making process. They are the main body for engaging with stakeholders on issues that directly (fisheries management and research) and indirectly (e.g. wind farms, aggregate extraction, conservation planning) affect fisheries, although stakeholders also have the opportunity to provide input independent of the RACs. Two thirds of the seats are allotted to the fisheries sector and one third to other interest groups. Either directly or at the request of the Commission or a Member State, RACs submit recommendations and suggestions to the Commission on matters relating to fisheries.

The command and control management paradigm of the CFP has meant that until recently, conditions have not been fertile for catalysing stakeholder-led or cooperative research initiatives necessary to rebuild trust and credibility. But things are changing. Reflection on the failings of the CFP have led to a tangible change in attitude, both in the policy and the scientific arena, with particular momentum being gained since the inception of the Regional Advisory Councils (RACs)(see Box 1) in 2003. Conditioned by this backdrop, efforts to bring together the knowledge and know-how of scientists and fishermen in Europe are finding more favour.

Fisheries management is a science-policy arena in which interested lay people have a great deal of experience-based knowledge (EBK) to supplement the research-based knowledge (RBK) of scientists. Over the past 20 years an extensive literature on the importance and usefulness of EBK in management has arisen (see, for example, Felt 1994; Pálsson 1995; Mackinson & Nottestad 1998; Neis & Felt; 2001). Fishermens' EBK may include detailed and long term information on fish behaviour, patterns in distribution and abundance, knowledge of habitats, responses to environment and more (Pederson & Hall-Arber 1999; Mackinson 2001). Many believe that it should be further incorporated in management (Grafton & Silva-Echenique 1997) to increase the credibility of information (Pinkerton 1989), provide additional indices for stock assessments (Rochet et al. 2008), increase knowledge about poorly understood species, and suggest novel hypotheses (Neis & Felt 2001). Indeed, research seeking to integrate the experiences of stakeholders in the knowledge base for management is a rapidly developing field (Bergmann et al. 2004;Murray et al. 2006; Hoefnagel et al. 2006; Ommer et al. 2007; Shackeroff & Campbell 2007; Prigent et al. 2008; Moreno-Baez et al. 2010; Feinholtz 2011). One common driver is using cooperative research funding as an indirect mechanism for financial support to fishermen needing to make drastic cuts in catches (Johnson 2007).The objectives can include legitimate and equitable management, cost-efficient research, and more efficient enforcement due to higher legitimacy among stakeholders.

The incorporation of EBK in management is not easy. Enabling stakeholder participation in research at the European level requires connection and alignment of European management policies, research policies, structure of the funding system and funding instruments. This does not usually occur. When combined with stakeholders' limited capacity for engagement, real or perceived barriers may prevent them from collaborating (Mackinson et al. 2010). Moreover, several studies have shown high variation among fishers' own observations (Felt 1994; Wilson et al. 2006) making direct use of EBK in management decision making difficult. Part of the difficulty lies in the fact that fishermen tend to view the resource on smaller scales than managers and as much more complex systems than stock assessment models can capture (Berkes 1993; Pinkerton 1989). Differences in perception of resources can arise simply as a consequence of the alternative 'windows' that fishermen and scientists use to view the resource (Mackinson & van der Kooij 2006). Many fishermen are also reluctant to share knowledge, fearful that it might be used as a rationale for reducing fishing opportunities (Pederson & Hall-Arber 1999).

As Pálsson (1995) has argued, the metaphor of knowledge as a sort of mental script or 'container' is not accurate. Fishermens' knowledge is part of their overall fishing skill and the knowledge that underlies a skill is intuitive and not easily articulated or even necessarily understood well by the scientist. To make knowledge useful for management requires taking this tacit knowledge out of the local context in which it is embedded and creating more explicit, discursive knowledge (Wilson 2003). This transformation is more than just the translation and transcription (Latour 1987) of the EBK. As Holm (2003) emphasises, it is also a process of 'purification' in which many kinds of beliefs, speculations, hopes and exaggerations are stripped from the EBK, transforming it in to a discourse that can 'hold its own' in scientific debates. Agrawal (1995) argues that this process can change the EBK so much that it becomes unrecognisable to the resource users. Hence, the intention of empowering fishermen and other resource users by mobilising their experience-based knowledge can actually disempower them, as their knowledge is removed from its context, transformed, alienated or even distorted (Maurstad 2001).

Well-designed participatory action research (PAR)[1] is one strategy that has been shown to be effective in addressing these complex issues of knowledge, participation and management decision making (Reid & Hartley 2006; Johnson & van Densen 2007). What we mean by 'well designed' will be discussed in detail below, but it boils down to an ongoing interchange based on genuine respect for participants' perspectives and contributions. PAR creates not just a set of new knowledge, but a social network of learning; the action research aspect then seeks to link this network to the decision processes of marine management.

There is a broad literature on PAR and, for the most part, the academic lessons about experiences in PAR cut across disciplines and have many similarities that can be used to help design and develop successful approaches. Recent studies have shown that PAR in fisheries can be a learning platform (Leeuwis & Pyburn 2002) that can produce useful science-policy 'boundary objects' for getting to grips with complex issues like the ecosystem approach to fisheries management. Following Cash et al. (2002), boundary objects are knowledge products produced jointly by scientists and others in a policy arena that exhibit high legitimacy and policy saliency as well as scientific credibility. Boundary objects in fisheries management include, for example 'the precautionary approach', 'sustainable fishing', 'long term management plans', 'ecosystem approach', 'Good Environmental Status', 'maximum sustainable yield' and 'biodiversity', but they can also include more specific products such as indicators, models and action plans as long as they have been jointly produced. These objects enable communication and collaboration across a wide diversity of actor groups, while still maintaining local interpretation of their meaning to each of the actor groups. The importance of such concepts increases when the shared meanings become stronger compared to the local interpretations. Joint production of such 'boundary objects' can help steer the relationship between science, managers and fishermen away from the impasses that have been common in the past (Johnson & van Densen 2007; Reid & Hartly 2006). Within the vast science-policy machine of the CFP, PAR remains marginal, but it is growing in frequency and creating stories that give new ways of talking about effective reform that provide common ground among divergent interests.

Embracing these challenges has been the stimulus for the GAP programme (Bridging the gap between science and stakeholders: phase 1- common ground; www.gap1.eu), where participatory action research is at the heart of efforts to demonstrate how by combining their knowledge and know-how, fishermen and scientists can make a difference to achieve sustainable fisheries.

[1] 'Participatory action research'is a type of collaborative or cooperative research, and thus about processes as well as scientific outcomes. It involves stakeholders and scientists working (and learning) together in the planning and delivery of research. The common aim is to improve the knowledge base and quality of scientific information for management advice and legislation.

7.1.3 Casting the Net Wider

While the focus of this chapter is mainly on the interaction among scientists and fishermen, it is not our intention to suggest that the scope of participatory action research should be restricted to this group of stakeholders alone. In general, two types of stakeholders have the most influence on the management of fisheries and the marine environment. The first type we refer to as fisheries stakeholders—individuals and organisations representing direct interests in fisheries, including fishermen, shore side businesses/workers, crew or fishing-reliant families and communities. The second type are environmental non-governmental organisations (ENGOs), and other citizens whose interest and concern is with the wellbeing of the marine environment (and dependent fisheries).

While fishermen have a reputation for having a very independent mind set (see, for example, Creative Research Ltd 2009), the fact that they have to share a common resource means that they derive a number of benefits from being members of fishermens' organisations. Such organisations or associations link the catching sector with processors, marketers, distributors and management in a structured way (Jentoft & Davis 1993; Nielsen et al. 2004). Within Europe, all Member States have national fishermens' associations that have local representatives, and there are many other regional and community-based associations with various levels of formality and organisation. Working with the fishing industry almost always means working with and through these kinds of organisations. At a higher level, they are represented on the RACs. Lessons learned from around the world about the sustainable management of common pool resources, such as fisheries, are that the support and participation of those whose livelihoods are made by exploiting the resource is critical to its success (Ostrom 1990). Obvious tensions arising from the tradeoffs between the desire for short-term economic benefits and long-term societal well-being require that participation is balanced by various interests. Tapping into the potential of PAR to help establish a sustainable future for EU fisheries requires balancing these tensions.

ENGOs have played an increasingly important role in the management of fisheries in the last two decades, lobbying to place fisheries within the context of broader environmental considerations. While the fishing industry places high importance on maintaining sustainable fish stocks, they are under constant short-term economic pressure and often lobby for exploitation levels based on the most optimistic resource assessments. It was not until the 1980s when ENGOs began their own lobbying campaigns that industry lobbying began to be balanced and pressure began to build for a more precautionary approach. This role for ENGOs has now become institutionalised in several forms, most critically in Europe with permanent ENGO seats on the RACs. A problem with this institutionalisation is that responsibility has been placed on the ENGOs to maintain and use their limited funds to play this balancing role and there is a serious question regarding the sustainability of this approach. The role of ENGOs in PAR in fisheries is similar to their role overall, their participation is not as intensely active as that of the industry, but their help in formulating questions and reviewing results is critical for maintaining the saliency and legitimacy of results.

7.2 Learning by Doing

7.2.1 The Gap Programme: Bridging the Gap Between Scientists, Stakeholders and Policy Makers

Phase 1 of GAP (GAP1) was a cooperative planning process funded by the EU's 7th Framework Programme. Its goal was to prepare for a series of PAR efforts by: a) bringing scientists, fisheries organisations, ENGOs and managers together to plan specific PAR case studies focused on issues of shared concern; and, b) designing GAP2, a much larger project that would carry out and monitor the actual PAR efforts and promote a deeper, systematic engagement among fishermen and scientists at the European level. GAP1 consisted of partners in 11 countries, working on regional case studies that focused on addressing science and sustainability issues in the marine environment. These are the PAR studies that are now being undertaken in GAP2 (www.gap2.eu). The future of the GAP programme depends to a large extent on the outcomes of phase 2, but the initial programme laid the conceptual foundation for establishing structures and processes that enable a systematic engagement of stakeholders in research and the governance of EU fisheries. Most of this chapter focuses on what was learned during GAP1 through a 'Good Practice' workshop, joint planning of PAR case studies, and a sociological study (Jacobsen et al. 2011) of the process of initiating participatory research.

GAP1 understood participation in research (i.e. PAR) and participation in management decision making in the following way. While there are common features of the two processes, and the persons involved may be the same, the key distinguishing feature in PAR is that there is an attempt to discuss and reduce the influence of any policy agenda associated with research. Participatory research aims at improving the knowledge and evidence for informed management decision-making. Paradoxically, the way to achieve this, we found, was to link the research to questions relevant to management policy so that the issues on the table and incentive to engage were clear to everyone.

GAP1 involved workshops at both the European and individual case study level. Research plans were developed at the case study level and then reflections on these experiences were made in European level workshops where fishermen, fisheries scientists and ENGOs were in attendance. Examples of the eleven case studies for PAR planning include: the behaviour and migration of brown crabs in the United Kingdom; the behaviour and spatial population dynamics of the spider crab (*Majasquinado*) in Spain; evaluating management objectives for spring spawning herring in the Skagerrak, Kattegat and Western Baltic in Denmark; identifying essential habitat for demersal fish in the Northern Adriatic Sea in Italy; and, investigating the implications of the proposed 25 nm Maltese fisheries management zone in Malta.

A sociological study consisting of interviews with participants and other parties interested in the relevant fisheries was carried out in three of the case studies. In Marsaxlokk, Malta, fishermen perceive that trawling efforts should increase in the fishery management zone. Scientists propose to share knowledge, perspectives and survey data with them so that they may jointly consider the effectiveness of the management regime of the demersal trawl fishery. In Lake Vättern, Sweden, scientists, regional stakeholders and a fisheries co-management initiative are starting to work together on developing selective gear for whitefish fishing. In South Devon in England, fishermen and scientists are sharing knowledge on the behaviour and migration of brown crab, and using it to assess the sustainability of the crab fishery. The case studies were visited by an anthropologist, whose aim was to follow the sociological aspects of the participatory process. Using an open-question qualitative approach, 19 interviews were undertaken, 11 with fishermen and 8 with scientists. As reflections of the PAR process, it was through these interviews that we hoped to learn to do better. The interviews were transcribed and analysed using a grounded theory approach (Glaser and Strauss 1967) and this analysis was subsequently supplemented by feedback from other GAP1 cases at special plenary session meetings(see Jacobsen et al. 2011, for full details).

7.2.2 Understanding Incentives

Our experiences highlight that where research involves outcomes targeted to benefit society as a whole, these must still translate into tangible benefits for the participants, since this is the basis of their individual incentive to participate. Because of the differences among stakeholders, it is important to clearly identify the benefits and who might be expected to receive them. During GAP1, we held a 'Good Practice' workshop where different stakeholders (fishermen, ENGOs, scientists) described their experiences of PAR and discussed differences in incentives and benefits (Box 2). We found that there is a diversity of incentives among stakeholders and many of these are shared. Generally speaking though, incentives for fishermen tend to focus on both short-term and long-term interests in the factors that influence the success of fisheries. Incentives for scientists and ENGOs are more aligned with the generation and accessibility of knowledge, the shift in attitudes and behaviours that this brings about and the long-term benefits that society receives from sustainable resource use.

This learning was a valuable aid to establishing the PAR case studies and was used to establish a good practice guide and code of conduct for cooperative research (see Mackinson et al. 2008; Mackinson & Neville 2009). However, although a Good Practice Guide and general rules of thumb can be a useful starting point for PAR, they should not be thought of as a recipe. Specific planning of PAR requires much more detailed understanding of the incentives for individuals to get involved.

The question that runs through everyone's mind is "What's in it for me?", so it is important to make an effort to understand this.

To foster exchange of knowledge and know-how among scientists and fishermen, we sought to focus on research issues that were less likely to get subsumed by political/sectoral arguments relating to management. However, we found that the incentive to engage was strong when the research questions were clearly linked to management policy issues. The sociological study identified mainly benefits for fishermen. Among these was that through their involvement they obtain more ownership of the project and that this ownership translates into greater confidence in the results. They stated that their involvement in PAR would also help clarify the reasons behind a management rule when it originates from the research. Less intuitive was the benefit identified by some fishermen that engaging in PAR allows them to clear their name when they are wrongly accused of damaging the resource or its habitat.

Box 2. Incentives and benefits of participatory research

Being recognised and valued

- An opportunity to express opinions.
- An opportunity to get a better reputation.
- Improving relationships with other stakeholders.

Improving sustainability

- Greater compliance with management decisions as fishermen have a feeling of ownership over the data provided to decision makers.
- Longer-term agreements can be reached due to improved communication, trust and respect between fisheries' stakeholders, researchers and decision makers.
- Development of co-management arrangements catalysed by successful and mature participatory research processes.

Making better use of available information

- Identification of research priorities of direct relevance to resource management.
- Research that is more focused on finding solutions that lead to more sustainable management of the marine environment.
- Including fishermen's knowledge for improving research design and data accuracy
- More efficient use of available knowledge by partnering with existing activities.

Improving knowledge and understanding

- Improved knowledge and understanding of issues of common concern.
- Catalyst for new ideas and innovative research methods.
- Co-education of fisheries stakeholders and researchers.

- Changing perceptions and attitudes.
- Builds trust between fishermen and public research institutions.
- Mutual respect gained through shared understanding of challenges, expectations and views.
- Fosters long-term shifts in attitudes, helping to engage wider society

The sociological study also explored reasons why fishermen and scientists may not want to be involved in PAR. Three reasons well known in the literature were also recognised in GAP1: fishermen do not have time to do extra tasks; they have negative opinions about research projects; and they are afraid the results will be used against them. Two new reasons were also identified: fishermen have other priorities and there are areas that some fishermen would prefer not to be examined in research. From the scientists' perspective, the extra time that participatory research takes was the most common reason for not wanting to be involved in participatory research. A particularly de-motivating situation experienced by scientists was when fishermen agree to participate and then do not show up at meetings.

On many occasions, we learned firsthand that it is not so much what is written about PAR that counts, but how attitudes and decisions change as a result of engagement. By its nature the value of action research is in the doing.

7.2.3 *Recognising and Respecting Differences*

Planning PAR requires not only recognition of individual differences, but also how these can be embedded in different social and cultural contexts. These can be subtle issues to understand, but go a long way in preventing many small but potentially significant problems. In GAP1 we found differences that related to alternative belief systems, and different professional and cultural aspects relating to ways of working (Box 3).

Box 3. What needs to be understood and taken into account when working with each other?

About fishermen	About researchers
Way of life	
Love of the sea	Love of the sea
Fishing is a way of life	Driven by curiosity and academic moti-vation
Livelihood – money is important, but not all financially motivated	Not all motivated by academic 'fame'
Want to be involved, feel use-ful/ important	Want to be involved, feel use-ful/important

About fishermen	About researchers
Not just short-term vision (but some do)	Try to provide knowledge produced for better stewardship
Education and authority	
Education levels variable	May lack skills for collaborative work
Scientists can be perceived as the 'authority' because of links to government and policy	May need
Perceptions of fish stocks and sustainability	
Embedded in experience and observations of how fish stocks and environments change	Based on scientific understanding of mechanisms that influence population dynamics
Assessment of sustainability deter-mined mainly by their experience of changes over time, catching pat-terns. Less likely to believe models. Views of other fishermen also very influential. Agree on sustainability as the key goal for all stakeholders	Assessment of sustainability mainly guided by scientific surveys and output from models, for which scientists are more trusting. Views of other scientists also very influential. Agree on sustainability as the key goal for all stakeholders
Ways of working	
Time rhythms (tide and seasons) guide work patterns but unpredicta-ble weather can lead to changes in plans at short notice	Constrained by available time of research vessels and weather
Need to fish efficiently as possible to maximise income	Need to design surveys that provide robust scientific information

Some of the fundamental elements necessary for establishing common ground for participatory research among fishermen and scientists are neatly captured in statements made during the workshop:

> Message from the fishermen
> "The hardest thing for scientists is to explain to fishermen the long term benefits"…. "it's important to learn about ways of working with fishermen and how to convince them of the value of science in helping them to conserve the resource"…. "they need to respect traditional/experienced-based knowledge and see that it can be used in a systematic way valuable to science"…. "Scientists should welcome fishermen to science events and come and talk with fishermen."

> Message from the scientists
> "Stakeholders need to respect the research process and results, even if it does not meet their expectations or provide the certainty they hoped for"…. "they need to understand that it may not change anything from a political point of view"… "On a practical note, fishermen need to try to welcome scientists on board their vessels, talk with them to understand the reasons for scientific sampling and appreciate the difficulties and time required for research."

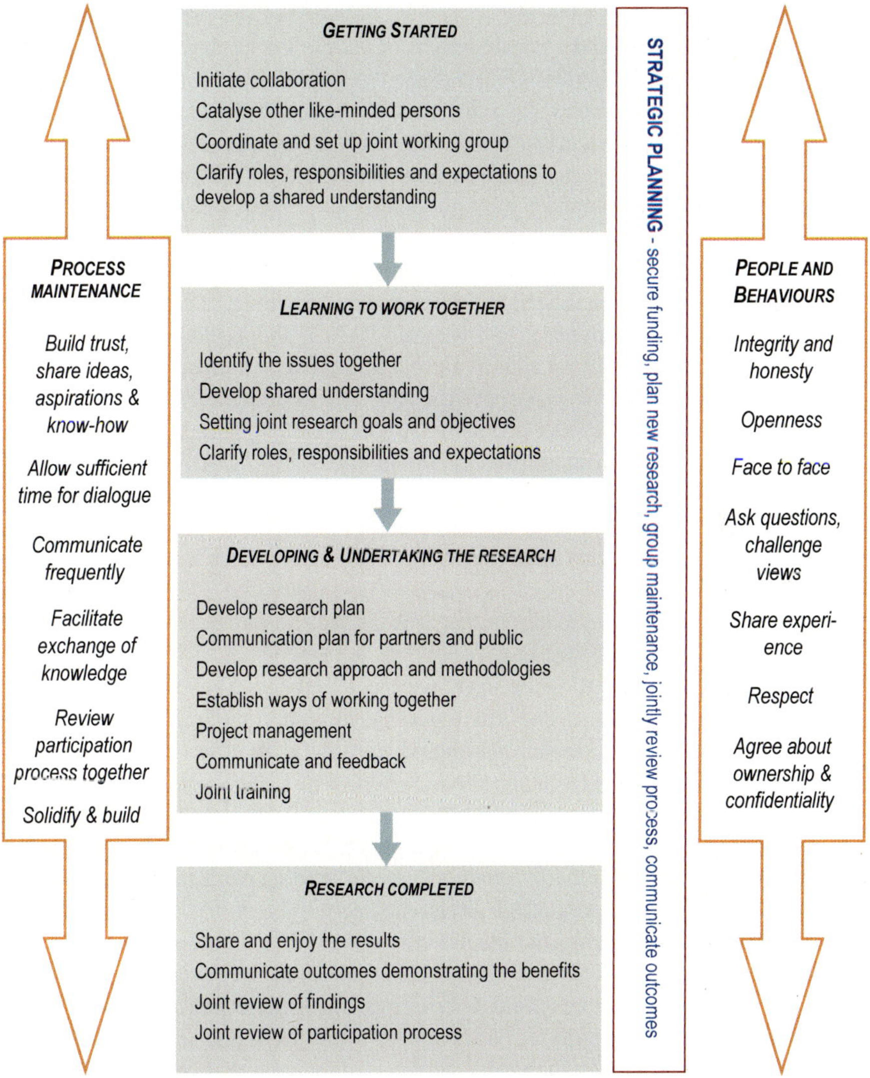

Fig. 12.1 Initiating collaborative research. (See Mackinson et al. 2008 for detailed version)

7.2.4 Balancing People, Process and Delivery

The practical aspects of developing participatory research should follow a logical sequence (Fig. 12.1) similar to that in any well-managed project. To be effective, we learned that (i) the cooperative process needs to be actively managed to work towards outcomes that make a real difference to informing management/ policy; and (ii) establishing and maintaining the participatory processes is arguably the most important aspect in ensuring successful delivery of the project. This means paying special attention to the consideration of people's roles and their behaviours.

While the project deliberately emphasised the importance of understanding, respecting and giving equal weight to different views and knowledge, we found that it was not necessary that all activities should involve everyone. Excessive and inappropriate involvement at times leads to poor focus and procrastination. It emphasised the importance of including the right people in the right actions at a time when they can have a real influence over the process. In general, we believe that deciding who to involve and when to involve them should be determined by their roles given the specific situational needs of the research. However, because it is difficult to know this clearly in advance, and because PAR is about getting involved, this aspect is very much part of an adaptive process. By this we mean a PAR planning process that is flexible so as to adapt to the particular individuals involved and how they work together to overcome problems. A good example here is the decision not to involve the GAP1 coordinator in regional meetings as originally intended. When we understood more clearly the specificities and interpersonal dynamics of the case study meetings, it was clear that the coordinators involvement might interrupt the natural flow of dialogue, by either requiring the meeting to be held in English (or translated) and because the coordinator would be an outsider. The benefit of being flexible in planning PAR studies is that it provides the opportunity for emergent leadership.

It is well known that getting the 'right people' together is a key ingredient in successful PAR. Experiences of PAR discussed by workshop participants tell that the 'right people' are tuned to working with and learning from others, having personal attributes that enable them to catalyse trust. Within a mix of participants, three types of attributes emerge as being important: 'facilitators'—that are able to listen and ask appropriate questions which help achieve understanding and respect for the knowledge and views of others; 'enablers'—who tend to work to enable effective participation by helping prevent or overcome seemingly insurmountable barriers; and 'leaders'—who motivate and inspire others toward a common goal.

The sociological study revealed that where fishermen were working with scientists prior to GAP1, fishermen were performing one or more of three roles: providing research platforms, acting as data collectors and providing ideas. While some fishermen liked to maintain such roles, others wished to have stronger involvement in generating research hypotheses, planning, data analysis (rarely) and as providers of suggestions to management. Some scientists found it challenging to broaden their view on the roles that fishermen might undertake in research. From a practical point of view there was a clear message that finding ways of working together had to ensure that it did not interfere with the daily activities of the fishermen.

Careful consideration was given to the planning of the regional and European workshops. For regional meetings, we adopted a directive yet sympathetic approach, whereby the local lead scientists were responsible for initiating engagement and making plans for meetings. Informal meetings held at times most convenient to fishermen and in their mother tongues were found to lead to greatest participation. In contrast, European workshops were led by the coordinator and held in English.

Building a personal, yet informal approach, plans for the workshops were communicated to stakeholders directly by the project coordinator. Early on we found that encouraging participation of non-scientists required us to spell out clearly the purpose, expectations and anticipated outcomes. For fishermen, where a day's meeting would mean a day's lost earnings, the offer of financial support was important in their decision whether to attend. Striving to ensure a 50:50 science-stakeholder balance among participants helped set the tone for the meeting, with participants expecting that activities would be focused on building partnerships and developing opportunities for shared learning. Because of possible tensions among some of the stakeholders, a facilitator was employed to help design and run the meeting in a neutral environment. Through a series of engaging activities, workshop participants discussed issues and how best to share knowledge and know-how towards achieving a common aim.

Much of what has been discussed already can be broadly described as the need to get the communication right; a centre stage issue in PAR. The sociological study revealed some specific insights in helping fishermen and scientists to get it right. The importance of one-on-one contact and using the native language of fishermen in communications was emphasised by all interviewees. Among other suggestions, the use of videos instead of written material was a common suggestion for improving communication. A preferred place for meetings with fishermen was on their vessels, or over lunch. Conversations at the fish market or requests to attend meetings were not always welcomed, and meetings when the weather conditions were good for fishing were welcomed even less. Scientists providing feedback were important for maintaining a good research environment:"[...] we offer the fishermen lots of rewards but ultimately, what they would like are some results or some information about the tags that they returned [...] they are interested in the knowledge".

Some of the key tactics that worked well in GAP1 are shown in Box 4.

Box 4. Top Tactics

Face to face is best: Throughout GAP1, emphasis was placed on face-to-face meetings, giving the opportunity to openly discuss expectation, fears, ideas and to resolve any concerns. This helped develop depth in understanding, which was beneficial because it enabled individuals to learn how best to help themselves.

Saying it with meaning: In several cases, initial reluctance of stakeholders to participate was overcome by making sure that written material was produced in their native language, even in cases where English was widely used. This demonstrated the genuine effort to connect with those whose involvement was paramount.

A comfortable 'atmosphere': Regional meetings were kept 'informal' using local language, and avoiding unnecessarily involving others for the tokenism of inclusivity.

Scientists and fishermen on board: When asked about what makes for productive cooperation, fishermen and scientists both suggested that scientists should go out on fishing vessels more. Both agreed that this leads to closer relationships and more productive interactions. However, some fishermen told us that they were not completely comfortable with that kind of exposure to outsiders and that they have reservations about inexperienced people on board.

7.2.5 Barriers and Recurring Dilemmas in PAR

Our experience from GAP1 and other work is that enabling stakeholder participation in fisheries research at the European level can be challenging. Some of the possible constraints that either make it difficult, or provide insufficient incentives for both stakeholders and scientists to get involved in participatory research on fisheries and the marine environment are elaborated below. During GAP1, overcoming such issues required that sufficient opportunity was given for the fears, motives and expectations of fisheries' stakeholders and scientists to be discussed openly.

Research policies focused on developing the science required to underpin the CFP has rarely involved collaborative research with stakeholders. Until very recently, research policies have not connected well with aspirations of the Commission to improve the basis of decision-making on the CFP by increasing participation of stakeholders. Even now steps in this direction are tentative both because information derived from the small geographical areas on which PAR is most meaningful are often insufficient to answer the questions the Commission needs answering and because of a reasonable fear of being seen favouring a commercial stakeholder.

Communication among the sections of DG Research and DG Mare that facilitates research on governance and science of fisheries and the marine environment could be improved. The structure of the EU Framework system for tendering for research projects is daunting for scientists experienced with the system, let alone stakeholders who may not be. For the most part, stakeholders simply do not have the capacity to instigate and lead proposals. Rarely are they official project partners. Funding for cooperative research processes is difficult to obtain, but needed for developing the capacity to engage. The newer programmes funding science and society linkages—such as the one that funded GAP1 and GAP2—are an important advance, but even these programmes tend to not fund a great deal of collaborate research as such.

Reluctance of stakeholders to participate in research can be a more significant problem when everything is going well in the fisheries and the corresponding political will diminishes. In times of hardship, innovation and collaboration become essential, with fishermen seeking improvements in economic performance/efficiency that might arise directly through development of new methods or as a result of management action based on outcomes of the research.

Reluctance of scientists to work with fishermen. As noted previously in the discussion on incentives, scientists too may be reluctant to collaborate. In our discussions with scientists it emerged that apublication-based reward system may deter scientists from getting involved. Two reasons for this were cited (i) the length of time it takes to yield publications from collaborative research, (ii) being put off by resistance to publications by those who consider PAR to be 'soft/second class science'.

Our sociological investigation named five ongoing dilemmas in PAR for which there are no clear answers or solutions, but which nevertheless require attention and sensitivity. These are sets of issues that remain difficult to resolve even with a great deal of good will. These five dilemmas are summarised from Jacobsen et al. (2011):

1. *Should research-management links be emphasised or deemphasised?* Fishermen often hope that research results will lead to a modification of regulations while scientists need their research to be useful for management. However, fishermen are also concerned that data may used to reduce fishing opportunities, while scientists may be trusted more if they are seen as distant from the management process.
2. *How close and frequent should scientist/fishermen/manager interactions be?* One goal of participatory action research is to have fishermen, scientists, and managers discussing the basis of regulations before they are implemented. Each of the three cases that were examined was different in this regard and each one had showed problems of its own. In one, there was almost no contact and the fishermen were very dissatisfied. In another fishermen were asked for input but saw no results emerging from that input. In the last case the fishermen found their discussions of management to be satisfactory in terms of substance, but found that they took up a great deal of their time.
3. *How widely should the data gathered in collaborative research be shared?* Can information collected by one project be shared among different scientists? Fishermen do not like to be asked the same questions by different scientists, but, sharing information too widely can lead to information being shared without the fishermen's consent.
4. *How to handle differences in work demands?* Both fishermen and scientists are busy professionals. However working conditions for fishermen can be considerably different to scientists' operating procedures for research. Fishermen are pressed for both time and the desire to be part of a project that has its premises in another working culture.
5. *How to communicate across professional cultures.* Scientists want to disseminate information to fishermen quickly and clearly and fishermen advise them to 'do it on our terms'. The direct presentation of the material, however, requires making use of local communication channels such as the fisheries association. If they rely on the association to communicate in their place this can result in unclear messages attributed to scientists.

7.3 Summary and Conclusions

Research seeking to integrate the experiences of stakeholders in the knowledge base for management is a rapidly developing field. GAP1 made apparent the disparity between the political desire to actively engage a broad range of stakeholders and the practical means by which to achieve it (Mackinson et al. 2010). It challenged the barriers and promoted ideas to better enable the participation of stakeholders in research.

One thing that emerges from the lessons of GAP1 is the multiplicity of the roles that stakeholders can play. ENGOs can provide not just a perspective that balances that of the industry; they provide skills that facilitate the development and acceptance of a useful knowledge base in many different ways. Fisheries stakeholders can do much more than just lobby for fishing interests, they can mobilise the support needed for change.

Box 5. Project 50 %[2]

Project 50% funded by Defra, UK is a recent example where an innovative partnership between scientists and Devon beam trawlermen was set up with the aim to help to protect fish stocks by reducing the amount of juvenile fish discarded overboard by over 50%. Time was spent to understand clearly fishermen's motivations, concerns and incentives for change. Together, fishermen and scientists identified barriers to reducing discards and identified measures to overcome them. Social marketing approaches were used to help motivate behavioural change by enhancing fishermen's innovation and responsibility. Fishermen and scientists contributed to modified net designs; they were inspired and motivated to participate as the measures were not imposed by the government. The benefits to the fishermen alone were remarkable: fewer discards meaningless work for crew, improved catch quality, reduced drag and lower fuel costs. In addition, the participation of the fishermen in the development of the modified fishing nets has had significant benefits for the sustainability of fish stocks, the environment, and therefore society as a whole. The fishermen's involvement was entirely voluntary and no charter payments, additional quota or extra days at sea were given, demonstrating how participation by stakeholders can lead to more cost-efficient research.

[2] (http://www.cefas.defra.gov.uk/our-science/fisheries-information/discards-and-fishing-gear-technology/project-50.aspx).

GAP1 also reinforced the understanding that a lot of PAR is about social behavioural change. When participatory processes are appropriately implemented, there are significant benefits for the environment, fish stocks, stakeholders and society. These benefits arise when people find the right incentive for choosing to change their attitudes and behaviours. Project 50% (see Box 5) is widely recognised as a great European example of the power of PAR in motivating behavioural change towards achieving sustainable fishing outcomes. But one example is not enough to bring about the required stimulusto change the institutional behaviour and structures required for lasting change. The CFP is a large, sluggish system and so far the various attempts at making it more participatory have been marginal. This includes the RACs, which while making a great deal of progress, are highly constrained, including having strong limits on both access to and participation in research. Reforms continue and scommitments from the Commission to move towards greater regionalisation and industry participation are hopeful (CEC 2012, CION 2012).

For many reasons outlined above, PAR has the potential for making an important contribution to the struggle for a more responsive, adaptive and sustainable European fisheries system. The shared experiences in GAP1 have shaped 13 PAR case studies now being carried out across Europe through GAP2. These PAR case studies and efforts facilitating a more systematic engagement of stakeholders is lending impetus to this change. Paradoxically, the success of PAR in making a recognisable difference to management hinges upon the creation of a governance structure where stakeholders have a central role in linking research with policy outcomes.

Acknowledgments We would like to thank all the participants in the GAP project, without which this work would not have been possible. A special thank you to Paulina Ramirez-Monsalve and Rikke Becker Jacobsen and participants of the interviews.This work was support by European Commission Grant Agreement 217639 (GAP1) 266544 (GAP2), Defra M1001.

References

Agrawal, A. (1995). Dismantling the divide between indigenous and scientific knowledge. *Development and Change, 26,* 413–439.

Bergmann, M., Hinz, H., Blyth, R. E., Kaiser, M. J., Rogers, S. I., & Armstrong, M. (2004). Using knowledge from fishers and fisheries scientists to identify possible groundfish 'Essential Fish Habitats'. *Fisheries Research, 66,* 373–379.

Berkes, F. (1993). Traditional ecological knowledge in perspective. Pp 1–9. In J.T. Inglis (ed.) *Traditional Ecological Knowledge: Concepts and Cases.* Ottawa: International Program on Traditional Ecological Knowledge, International Development Research Center.

CEC (Commission of the European Community). (2002). *Towards a reinforced culture of consultation and dialogue—General principles and minimum standards for consultation of interested parties by the Commission.* Brussels COM (2002) communication 704.

CEC (Commission of the European Community). (2009). *GREEN PAPER: Reform of the Common Fisheries Policy.* Brussels: DGMARE.

CEC (Commission of the European Community) (2012). Commission non-paper: key points on regionalisationhvttp://ec.europa.eu/fisheries/reform/docs/regionalisation_en.pdf

CION. (2012). Proposal for a regulation of the european parliament and of the council on the common fisheries Policy. Non paper from the Danish presidency. 12514/11 PECHE 187 CODEC 1166– COM (2011) 425 final. http://register.consilium.europa.eu/pdf/en/12/st09/st09016.en12.pdf

Cash, D., Clark, W., Alcock, F., Dickson, N., Eckley, N., & Jaager, J. (2002). Saliency, credibility, legitimacy and boundaries: linking research, assessment and decision making. John F. Kennedy school of government, Harvard university faculty research working paper series RWP02–046.

Clark, W. C., Mitchell, R. B., Cash, D. W., & Alcock, F. (2002). Information as Influence: How institutions mediate the impact of scientific assessments on global environmental affairs. John F. Kennedy School of Government, Harvard University, Faculty Research Working Paper Series RWP02–044.

Creative Research Ltd. (2009). A fishermen's tale: being a fishermen in England in 2009. December 2009/ Job No. 578/ Version 5. Creative Research Ltd: London. p. 146.

FAO (Food and Agriculture Organization of the United Nations). (2009). The state of world fisheries and aquaculture 2008 Rome: FAO.

Feinholtz, C. (2011). Engaging stakeholders in coastal management through participatory mapping. NOAA Digital Coast Webinar Series. Available at: http://www.csc.noaa.gov/digitalcoast/webinar/archive.html.

Felt, L. (1994). Two tales of a fish: the social bases of indigenous knowledge among commercial atlantic salmon fisheries. Pp 251–286. In C. L. Dyer & , J. R. McGoodwin (Eds) *Folk Management of the World's Fisheries*. Boulder: University of Colorado Press.

Grafton, R. Q., & Silva-Echenique, J. (1997). How to manage nature? strategies, predator-prey models and chaos. *Marine Resource Economics, 12,* 127–143.

Glaser, B. G., & Strauss, A. L. (1967) The Discovery of Grounded Theory: Strategies for Qualitative Research. Aldine Publishing Company, Chicago.

Hoefnagel, E., Burnett, A., & Wilson, D. C. (2006). The knowledge base of co-management. In: *The Knowledge Base for Fisheries Management. L. Motos and D.C. Wilson (eds.) Elsevier, Amsterdam. Developments in Aquaculture and Fisheries Science Series, 36,* 85–108.

Holm, P. (2003). Crossing the Border: On the relationship between science and fishermens's knowledge in a Resource Management Context. *MAST, 2*(1), pp. 5–33.

ICES. (2011). Advisory committee. http://www.ices.dk/advice/icesadvice.asp

Jacobsen, R., Ramirez, P., & Wilson, D. C. K. (2011). 'Empowerment and regulation—dilemmas in participatory fisheries science'. *Fish and Fisheries* DOI: 10.1111/j.1467–2979.2011.00434.x.

Jentoft, S., & Davis, A. (1993). Self and sacrifice: An investigation of small boat fisher individualism and its implication for producer cooperatives. *Human Organization, 52*(4), 356.

Johnson, T. R. (2007). Integrating fishermen and their knowledge in the science policy process: case studies of cooperative research in the Northeastern U.S. Dissertation: Rutgers University.

Johnson, T. R., & van Densen, W. L. T. (2007). The benefits and organization of cooperative research for fisheries management.—*ICES Journal of Marine Science*, *64*, 834–840.

Latour, B. (1987). *Science in Action.* Cambridge: Harvard University.

Leeuwis, C., & Pyburn, R. (eds.). (2002). *Wheelbarrows full of frogs: Social learning in rural resource management Assen.* The Netherlands: Koninklijke Van Gorcum.

Mackinson, S. (2001). Integrating local and scientific knowledge: An example in fisheries science. *Environmental Management, 27*(4), 533–545.

Mackinson, S., & Nøttestad, L. (1998). Combining local and scientific knowledge. *Reviews in Fish Biology and Fisheries, 8*(4), 481–490.

Mackinson, S., & van der Kooij, J. (2006). Perceptions of fish distribution, abundance and behaviour: Observations revealed by alternative survey strategies made by scientific and fishing vessels. *Fisheries Research, 81,* 306–315.

Mackinson, S., Neville, S., Raicevich, S., & Worsøe Clausen, L. (eds) (2008). Good practice guide to participatory research between fisheries stakeholders and scientists. GAP project deliverable 1, pp. 23.

Mackinson, S., & Neville, S. (eds). (2009). Bridging the GAP between science and Phase 1– Common Ground 1, Final Report. Pp. 46. Available on request steve.mackinson@cefas.co.uk

Mackinson, S., Wilson, D. C., Galiay, P., & Deas, P. (2010). Engaging stakeholders in marine research. *Marine Policy, 35,* 18–24.

Maurstad, A. (2001). Trapped in biology: an interdisciplinary attempt to integrate fish harvesters' knowledge into norwegian fisheries management. pp 135–152. In B. A Neis & L. Felt (Eds) *Finding Our Sea Legs: Linking Fishery People and Their Knowledge with Science and Management*. St. John's: ISER Books.

Moreno-Báez, M., Orr, B. J., Cudney-Bueno, R., & Shaw, W. W. (2010). Using fishers' local knowledge to aid management at regional scales: Spatial distribution of small-scale fisheries in the Northern Gulf of California, Mexico. *Bulletin of Marine Science, 86*(2), 339–353.

Murray, G., Neis, B., & Johnsen, P. (2006). Lessons learned from reconstructing interactions between local ecological knowledge, fisheries science, and fisheries management in the commercial fisheries of the Newfoundland and Labrador, Canada. *Human Ecology, 4,* 549–571.

Neis, B., & Felt, L. (2001). *Finding our sea legs: linking fishery people and their knowledge with science and management.* St. John's: ISER Books.

Nielsen, J. R., Degnbol, P., Viswanathan, K., Ahmed, M., Hara, M., & Abdullah, N. M. R. (2004). Fisheries co-management—an institutional innovation? Lessons from Sout East Asia and Southern Africa. *Marine Policy 28.*

Ommer, R. E., Perry, R. I., & Neis, B. (2007). Bridging the gap between social and natural fisheries science: why is this necessary and how can it be done? *American Fisheries Society Symposium 49,* 587–595.

Ostrom, E. (1990). Governing the commons New York: Cambridge University Press.

Pálsson, G. (1995). Learning by fishing: practical science and scientific practice. In: S. Hanna & M. Munasinghe (Eds.), *Property rights in a social and ecological context: case studies and design applications.* Stockholm: The Beijer Institute (1995) 85–97.

Pederson, J., & Hall-Arber, M. (1999). Fish habitat: a focus on new england fishermens's perspectives. *American Fisheries Society Symposium, 22,* 188–211.

Pinkerton, E. (1989). Introduction: attaining better fisheries management through co-management—prospects, problems and propositions. Pp. 3–36. In E. Pinkerton (ed.) *Cooperative Management of Local Fisheries.* Vancouver: University of British Columbia Press.

Prigent, M., Fontanelle, G., Rochet, M. J., & Trenkel, V. M. (2008). Using cognitive maps to investigate fishers' ecosystem objectives and knowledge.*Ocean and Coastal management, 51*(6), 450–462.

Reid, A. N., & Hartley, T. W. (eds.). (2006). *Partnership for a common purpose: Cooperative Fisheries Research and Management. American Fisheries Society Symposium 52, Proceedings of the Symposium Fisheries Society/Sea Grant Symposium Partnerships for a Common Purpose: Cooperative Fisheries Research and Management held in Anchorage, Alaska, U.S.A., 13–14 September 2005.* Bethesda: American Fisheries Society.

Rochet, M. J., Prigent, M., Bertrand, A., Carpentier, A., Coppin, F., Delpech, J. P., Fontenelle, G., Foucher, E., Mahé, K., Rostiaux, E., & Trenkel, V. M. (2008). Ecosystem trends: Evidence for agreement between fishers' perceptions and scientific information. *ICES Journal of Marine Science, 65,* 1057–1068.

Schwach, V., Bailey, D., Christensen, A.-S., Delaney, A. E., Degnbol, P., van Densen, W. L. T., Petter, P., McLay, A., Nielsen, K. N., Pastoors, M. A., Reeves, S. A., & Wilson, D. C. (2007). "Policy and knowledge in fisheries management: A policy brief. ICES. *Journal of Marine Science, 64*(4), 798–803.

Verweij, M. C., van Densen, W. L. T., & Mol., A., J. P. (2010). The tower of Babel: Different perceptions and controversies on change and status of North Sea fish stocks in multi-stakeholder settings. *Marine Policy, 34*(3), 522–533.

Wilson, D. C. (2003). "Fisheries Co-management and the Knowledge Base for Management Decisions" Chap. 5. In D. C. Wilson, J. R. Nielsen & P. Degnbol (eds). *The Fisheries Co-management Experience: Accomplishments, Challenges and Prospects.* Dordrecht, The Netherlands: Kluwer Acahdemic Publishers.

Wilson, D. C. (2009). *The Paradoxes of Transparency: Science and the Ecosystem Approach to Fisheries Management in Europe.* Amsterdam: Amsterdam University Press.

Wilson, D. C., Nielsen, J. R., & Degnbol, P. (2006). Local ecological knowledge and practical fisheries management in the tropics: A Policy Brief. *Marine Policy, 30,* 794–801.

Part III
Issues in Sustaining Fisher Livelihoods

Chapter 8
Ghost Boats and Human Freight: The Social Wellbeing Impacts of the Salmon Ban on Lough Foyle's Fishing Communities

Easkey Britton

Keywords Social wellbeing • Coastal livelihoods • Inshore fisheries • Atlantic salmon • Lough Foyle • Drift net ban • Fisheries management

8.1 Introduction

This chapter analyses the social wellbeing impacts of changing access to fish in a specific context. Its focus is the conflict between EU-endorsed fisheries policy with a pro-conservation rationale and sustaining local livelihoods and cultural heritage. Lough Foyle was one of the last bodies of water to be influenced by the EU Habitats Directive[1] through the imposition of a moratorium on the commercial at-sea driftnet fishery at the end of 2006 and restrictions on estuarine driftnetting. Implementation was delayed in response to a request by the Irish government to allow more time to adapt to the proposed closure. The reasons for the closure were biological, concerned with protecting and restoring salmon stocks following a decline in wild North Atlantic Salmon stocks and a decline of Irish Atlantic salmon (*Salmo salar. L.)* stocks of more than two thirds in the past 30 years (Brennan and Rodwell 2008). There is a general lack of understanding of the impacts of fisheries conservation policy from a user-perspective within EU fisheries management and according to Jentoft (2000, p. 143): "User's interpretations on issues such as rationality, equity and fairness pertaining to fisheries management are rarely investigated (…) either prior or after the design and implementation." The majority of studies of Lough Foyle fisheries are biological. To date there has been no published assessment of social impacts following closure of the fishery in 2010.

This chapter explores how the wellbeing of individual resource users and their community are affected by reduced access to fish through a multi-dimensional

[1] Council Directive 92/43/EEC on the Conservation of Natural Habitats and Wild Flora and Fauna.

E. Britton (✉)
School of Environmental Sciences, University of Ulster, Cromore Road, BT52 1SA, Coleraine, Co. Londonderry, Ireland
e-mail: easkey.britton@gmail.com

J. Urquhart et al. (eds.), *Social Issues in Sustainable Fisheries Management*, MARE Publication Series 9, DOI 10.1007/978-94-007-7911-2_8,

social wellbeing lens. A number of articles have recently highlighted how the need for a more holistic and multidisciplinary approach to fisheries managment can be addressed through a social wellbeing approach (McGregor 2009; Coulthard et al. 2011; Coulthard 2012a; Britton and Coulthard 2013). They advocate the application of a three-dimensional wellbeing framework (cf. Gough and McGregor 2007), which considers both objective and subjective aspects of wellbeing from a person-centered perspective. A social conception of wellbeing attempts to bring together resources, relations and subjective reflections on life satisfaction, that is, the material, relational and cognitive dimensions of wellbeing (McGregor 2007). This approach broadens the scope of analysis emphasising the importance of engaging with the meanings that people have in their lives, their 'goals and aspirations', recognising that wellbeing is pursued through relationships with others (Gough and McGregor 2007). As the social wellbeing approach allows for a more 'local view' it can help identify management indicators that resonate with local contexts (WeD 2008). Armitage et al. (2012) also use a social conception of wellbeing for its promise in dynamically linking human interests and ecological systems, adding 'environment' to the definition of social wellbeing viz a state of being with others and the natural environment which arises where human needs are met, where individuals and groups can act meaningfully to pursue their goals, and where they are satisfied with their way of life (adapted from McGregor 2008).

Salmon has been fished for millenia on the island of Ireland and has become an important cultural identifier. The maritime culture and heritage of the island of Ireland has been greatly influenced and shaped by this species celebrated in legends, folk songs and the distinct ecological knowledge of those who pursue them (Gregory 1994). There is a strong sense of identity, place and belonging associated with the salmon driftnet fishery (Carraher 2006). The Foyle fishery provides a good micro-context for examining the social wellbeing impacts of a policy instrument at an individual and community level.

8.2 Lough Foyle and the Lost Salmon Fishery

Lough Foyle, covering about 186 km^2, is a shallow estuarine sea lough straddling the border of County Donegal in the Republic of Ireland (ROI) and Northern Ireland (NI) (Fig. 8.1). Three main freshwater rivers flow into it and there is a diversity of important coastal habitats and species. The main fishing ports, Greencastle (2,174 inhabitants) and Moville (807 inhabitants), are located in County Donegal.

8.2.1 Fishing Fleet

Lough Foyle has supported fisheries for hundreds of years, predominantly for oysters, mussels and salmon. Greencastle in particular is highly dependent on its

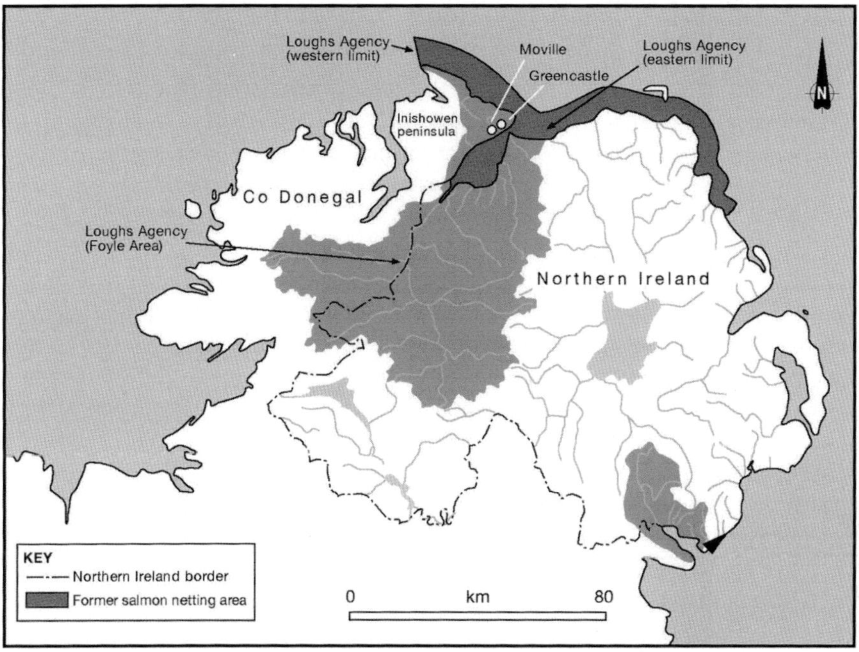

Fig. 8.1 An overview of coastal commercial salmon fisheries and rivers where angling takes place in the Foyle catchment area and the location of the main case study sites. It should be noted that the extent of the commercial fishery (netting) reflects the position in 2006. (Adapted from: DCAL 2008)

fisheries with 56 % of employment directly related to fisheries (e.g. catching, processing, ancillary) (CEFAS 2007). Ranked 8th among Irish fishing ports, in 2002, it was the major whitefish port for the region. However, the fishing fleet is in decline with a 64 % decrease in the number of fishing vessels from 2003 to 2006 (CEFAS 2007), and this trend is continuing. The majority of the 40 or so vessels are small-scale 'half-deckers', under 10 m in length, though a number of these are inactive. The decline has been variously attributed to fuel price rises (Tingley 2006; Abernethy et al. 2010), decommissioning (White 2005), declining salmon stocks (NASCO 2008; Loughs Agency 2009) and regulation aimed at limiting fishing effort and conserving stocks, which has encouraged diversification into shellfisheries, particularly farmed mussels with an estimated 14 mussel dredgers fishing the Lough. An important native oyster fishery supports about 20–30 boats. Conflicts have arisen between fishers and local management authorities with a recent move to formalise control of shellfisheries following increasing effort in the mussel fishery and perceived threats of over-exploitation of the native oyster stock (Allen 2010).

8.2.2 The 'Lost' Traditional Salmon Fishery

Salmon drift netting carried out by small-scale fishers during June and July[2] was an important seasonal source of income (NASCO 2008). Access to the traditional fishery had been maintained informally for centuries in Ireland, legitimising territoriality, access rights and traditional institutions the loss of which may further affect the social wellbeing of the community. The fishery was targeted by young and old, encouraging the intergenerational transfer of knowledge and skills as well as providing a source of social interaction for elderly members of the community who would otherwise have remained isolated in a rural area and wholly dependent on farming. Since 2002 access has been managed by the local governing body (Loughs Agency), which introduced a shorter season and reduced quota. In 2007, the Agency prohibited driftnetting seaward of Lough Foyle and achieved a reduction of 84 % in the drift net fishery within the Lough (Loughs Agency 2009b). As conservation limits were being met within the catchment, a small number (N= 18) of drift net licences were retained after the initial ban inside the Lough was imposed. Measures to restrict the commercial fishery in the Republic of Ireland in 2007 complemented the above approach in Lough Foyle and provided a precautionary response to stock status around the island of Ireland (*Ad Hoc* Review Group 2008). Following evidence of continuing decline in the stock the remaining licences were not renewed in 2010 and no commercial fishing has taken place since.

There was widespread belief amongst commercial drift netters that the ban was unjust, in part due to the continuation of angling licences after the drift-net ban (Carraher 2006). In 2007, 365 salmon rod licences were issued for the Foyle area, whereas the number of commercial fishing licences fell to 18 (NASCO 2008). The estimated number of salmon caught by rods actually rose while net caught salmon dramatically declined following the initial 2007 ban. In 2009, the last year of the commercial drift net fishery, a total of 1,937 salmon were caught and, in 2010, an estimated 4,730 salmon were caught by anglers (DCAL 2012). The ban failed to have a positive impact on salmon stocks and in 2012 Foyle salmon stocks were declared to be 'in serious decline' by the Agency (BBC 2012). A system of catch and release was introduced for anglers in 2011 (DCAL 2012).

8.2.3 Policy and Management

> The government is choking them with regulations to squeeze them out (small-scale fishers)…It's death by 1000 cuts. (L.Foyle 'gatekeeper')

As the above implies, the relationship between government and local fishers is a strained one. There has been a long history of management by different agencies

[2] Shortened from a May-August season.

in the Lough, which has been made difficult due to the cross-border nature of its location. Since at least the 17th century the Foyle salmon fishery has been at the centre of political conflict over ownership of maritime resources along Ireland and Northern Ireland's border (Healy 1913; Kennedy 2000). Following the Partition of Ireland in 1921, the dispute over the maritime border and legal jurisdiction of the Lough began between the UK and Ireland (Kennedy 2000), which continues to this day. The lack of regulatory control led to a rise in illegal fishing of salmon. Poaching became a 'Robin Hood' pursuit for rural Irish to assert their traditional rights and antagonise British rule (J.Hamilton, pers.comm.). Following the outbreak of World War II in 1939 the price of salmon soared, further encouraging illegal fishing (Hansard 1995). Viewed as 'lawbreakers and poachers' by the British government the 'poachers' claimed they were legally entitled to fish under licences issued by the Republic of Ireland. In response to the increasing need to regulate the natural resources of the Lough, the Irish and UK governments set-up the Foyle Fisheries Commission in 1952, one of the very first cross border bodies to be established, to manage and protect the fisheries in the Foyle area and regulate the salmon fishery in particular. In 1998, the Commission became the Loughs Agency, following the Good Friday/Belfast Agreement. The Loughs Agency now manage the fisheries of the Foyle however, political ownership has still not been resolved.

As well as the ban on salmon drift netting, regulations aimed at the recovery of cod stocks resulting in the closure of fishing areas north-west of Ireland in 2008 (ICES areas VIa and VII) were considered 'the final nail in the coffin' for many local fishers. These closures have had a significant impact on fishing behaviour and geographical mobility of the fleet with bigger whitefish and pelagic trawlers fishing further away from home for longer periods. However, small boats (under 10 m) are not able to travel the distance to escape the closure and are forced to stay and fish significantly reduced days at sea.

8.2.4 Socio-economic Factors

Evidence of the importance attached to local fisheries is provided in a letter from the Donegal Salmon and Inshore Fishermen's Association (Doherty 2006) to the Irish government, stating that direct revenue from fishing and the ancillary employment it provides acted as an important buffer in hard economic times and helped to keep the basic social and economic fabric of the coastal region intact. County Donegal has some of the highest levels of unemployment and dependency on commercial fishing in Ireland and is one of the poorest regions in terms of economic output (Morrissey and O'Donoghue 2012). The safety-net or welfare function of fisheries has been cited as grounds for greater investment in fisheries, however, the steps necessary to achieve this have yet to be outlined by government.

8.3 Methodology

In order to understand how closure of the Foyle drift net fishery in 2010 has impacted on the wellbeing of fishers and their families, semi-structured interviews were conducted with a sample of recent participants. Initial reluctance to being interviewed can be explained by the failure of previous investigations to exert any positive effects on the conduct of the fishery. Key contacts, or gatekeepers, and the use of a snowballing technique provided access to a sufficiently wide range of respondents. Interview questions covered views on the important aspects of salmon fishing, perceptions for decline and closure, perceived impacts of the ban and aspirations for the future. A mix of other qualitative research methods complimented the interviews such informal interviews and participant observation at local harbours and events such as the annual Greencastle Regatta and blessing of the fleet.

The resulting sample was small ($n=14$) and skewed in favour of those based in Co. Donegal (12) but provided a broad range of age groups (25–68 with a mean age of 48) and a mix of licence holders (8) and crew members (6)—this is important given the different entitlements to compensation. The majority (10) had finished school at the age of 15 or 16, with only one continuing into third level education. At the time of the interview, 11 were still engaged in fishing (5 full-time and 6 part-time): all but one were operating inshore mainly for shellfish—the exception being engaged full-time in the offshore pelagic sector. The three part-time fishers supplemented their fishing incomes with shore-based activities including the Rural Social Scheme (see below), trade or tourism.

The aim of the study is to understand how the drift net salmon fishery was valued from the perspective of the resource users and how the impact of the ban on the individual, family and community, were perceived. Campbell and Cattermoul (2008, p. 41) note that "the path of policies from intended to actual impacts travel through a set of contextual filters", resulting in differing outcomes across sub-sectors and from place to place. Within social wellbeing literature there is a strong emphasis on the importance of context and how people's perceptions of wellbeing differs depending on their cultural, social and economic settings (White and Ellison 2006; McGregor 2007; White 2011). This has implications for EU fisheries policy with its top-down approach and 'one size fits all' management, which fails to take account of local realities and has consistently failed to meet its own objectives (Jensen 1999; Symes and Phillipson 2009). By focusing on a broad range of social relationships, competing interests in fisheries, such as the mismatch between policy and local realities are highlighted. As a result, this can generate conflict, which undermines existing policy regimes (Coulthard et al. 2011).

Figure 8.2 illustrates the pathway of a policy instrument through a set of contextual filters. This framework is built around the concept of social wellbeing, which situates the individual within the wider social structures of household, community and society that shape how wellbeing is pursued and achieved (McGregor 2007; Coulthard et al. 2011). It allows for the complexity of the local to be better accounted for and seeks to bridge micro-level, local scales with macro-level global scales.

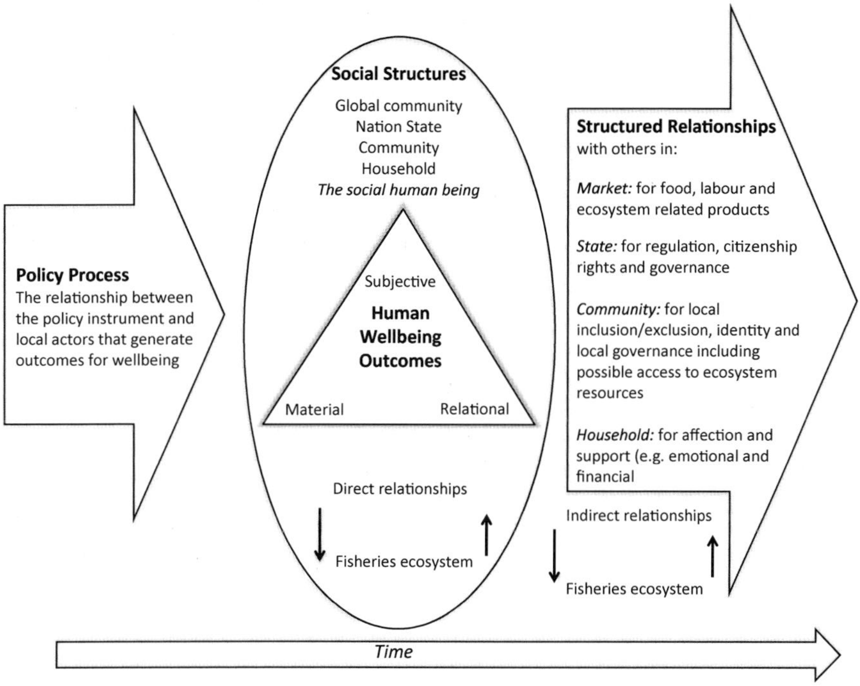

Fig. 8.2 Conceptual framework of the study. (Adapted from White 2010 and Coulthard et al. 2011)

8.3.1 Material Wellbeing Impacts: Why Compensation Will Never be Enough

In 2007, in response to the EU drift-net ban, a nation-wide hardship package worth € 25 million and funded by the EU was proposed by the Irish Government as compensation for individuals exiting the salmon drift-net fishery. Those eligible for compensation were licence holders only (Collins et al. 2006), with some forced out (licence-holders seaward the Lough) and others offered a voluntary buy-out (licence-holders within the Lough). A report by the Independent Salmon Group (Collins et al. 2006) stated that although the commercial salmon fishery may be locally important it made little significant economic contribution. The report failed to account for the lack of alternative employment opportunities in a rural, isolated area, especially at a time of economic recession and the safety net function of the fishery. The following comment from a participant in this study illustrates how the commercial salmon fishing acted as a safety net, or seasonal buffer, providing a cash flow in the community and, at times, a great source of wealth:

> When the fishing was good it definitely would have accounted for 50–60 % of annual income. A young lad aged 22, was able to build a house after three salmon seasons. (Ex-drifter and former fisher)

The salmon fishery also instilled individual characteristics for business success from an early age:

> I was skipper of the salmonboat when I was 15 years of age, the average age of drifters then was 17. The salmonfishing gave you a taste for running your own business from an early age. It gave me responsibility, accountability. It was a stepping stone for a lot of young lads. (Ex-drifter and mussel fisher)

As well as direct seasonal income generation for individual fishers and their families there was evidence of value for the wider economy. The salmon fishery created what Jentoft (2000) describes as a 'domino effect' or a multiplier effect (Béné 2006):

> Even if it wasn't a lot of money it was money being generated and circulated in the area and that all has a knock-on effect. (Ex-drifter and offshore fisher).

Béné (2006) critiques socio-economic analysis that focuses only on household income contribution, arguing that it misses the role income plays in other areas of life. The extra income from salmon fishing acted as a 'needs satisfier' (see Gartaula 2011), enabling people in local fishing society to fulfill their needs by, for example, achieving a good education or building a good house. Béné (2006) gives the example of the important role even low level supplementary income can play in supporting household needs, if it is generated at a crucial time of year when a household needs cash in order to purchase other household inputs. One respondent described how, due to the unpredictable nature of fishing, it was not possible to budget for a whole year so that, "A housewife of a fisherman, once she got this (salmon) money she went and bought everything in the first week—the school clothes for the wee'ans, food, everything." Cavanagh (2010) also reported similar findings where 'drifters' were able to pay their bills and debts only, 'after the salmon'.

When asked what they felt were the greatest impacts of the ban for their communities, responses from ex-drifters were overwhelmingly negative but also highlighted how impacts were felt differently, with people having differing capacities to cope with the change:

> Full-time fishermen are at other things now, other fisheries…For others it was a transition phase and they've moved on to other things, jobs outside the industry. And there are those who are in long-term unemployment with no options. (Ex-drifter and mussel fisher)

Evidence of decline was clearly visible during the former salmon season. On landing day there was little or no activity and it was difficult to find any active fishers during what would typically have been the busiest day of the week at the harbour, as this ex-drifter emphasised:

> If you drive over Greencastle Harbour there's 40–50 boats and they're tied up idle because there's no salmon, there's nothing to do. It wiped them out. It's like ghosts. It's kinda like a museum. (Ex-drifter and former fisher)

All respondents perceived the 2007 buy-out scheme as unfair. It failed to meet the needs of their household such as providing sufficient start-up capital for re-investment in another fishery, or for alternative enterprises like aquaculture, which is a rapidly growing, highly profitable industry but dominated by foreign companies.

Another issue was the creation of 'winners and losers' caused by the failure to recognise the contribution of crew members to the fishery, or family members, with only the licence holder receiving compensation. One respondent felt that the salmon fishery was a community asset and did not belong to individuals:

> The compensation wasn't adequate. The money went to individuals when it's a community asset. It wasn't a gift for the fishermen to give or the policy-makers to take. (Ex-drifter and former fisher)

The buy-out scheme was aimed at permanently removing fishing effort with no opportunity of return if the fishery ever improved in the future, which meant the loss of a traditional way of life, family heritage and all the knowledge and experience that went with the fishery:

> You had to hand in your net and you had to sign that you would never ever fish salmon again. So even if a change of government or something brought salmon back I couldn't because I have that signed. (Ex-drifter and inshore fisher)

The importance of small-scale fisheries for maintaining the social as well as economic fabric of coastal communities globally has been well documented in the literature (Binkley 2000, 2002; Allison and Ellis 2001; McGoodwin 2001; van Ginkel 2007; Jentoft and Eide 2011; Chuenpagdee 2011). According to a whitefish skipper, during Ireland's previous economic recession in the 1980s, fishing acted as a buffer, creating job opportunities in an area where little other opportunity for employment existed: 'I'm old enough to remember the last recession [early 1980s]. When we came out of school the only ones that were making any money were from fishing. This was a thriving place.' McClanahan et al. (2009, p. 43) state that "change and the existence of options are part of the resilience of households and many coastal resource users." However, Coulthard (2012b) argues that where 'the existence of options' are lacking or undesirable household resilience can be affected. Today, the Inishowen peninsula still lacks any major development and following the combined collapse of the local construction industry and fisheries, as this quote illustrates, there is a lack of alternative employment opportunities:

> Well, there's no Microsoft around here. There's no multi-nationals. That's what the government got wrong, they don't understand it. If you let the fishing go in the fishing communities—let them have it—it builds up like. (Ex-drifter and part-time fisher)

For those not able to continue fishing the alternatives are often either unemployment or leave to find work elsewhere. The following example highlights the impact of the loss of salmon, a source of upward mobility for youth in a small coastal community, and the emigration of youth:

> I have 5 children, 4 in New York and one in Scotland, 3 sons and 2 daughters. They'd still be here if the fishing was good but they'll never be back. There's nothing here to keep them. The salmon used to be a great way to get the kids into fishing. Human freight that's all we've got here now. I sent my children away as human freight. That's the main export now, emigration. (Ex-drifter and farmer)

The buy-out scheme failed to assess costs other than monetary value, such as the loss of cultural heritage and the human right to have a job of one's choosing, recognised in the FAO Code of Conduct for Responsible Fisheries (1995, article 6.18) as the 'right to a secure and just livelihood'. However, there is some social support for those who are not able to maintain an adequate living from fishing. The Rural Social Scheme (RSS) is a LEADER funded initiative aimed at low income farmers and fishers who rely on welfare payments. It provides additional income support through employment in community programmes whilst still allowing fishers the flexibility to fish part-time. However, up-take of fishers ($n=4$) has been very low compared to farmers ($n=26$). This is possibly due to the fact that the scheme is aimed at boat owners only, who must also be licence holders and in receipt of means tested social support. Therefore, crew members cannot participate and the scheme is now closed to new applicants as the total number of participants is limited to 30. A lack of willingness, and not just access to the grants, may be another reason for the low uptake. Farmers are more used to applying for and receiving grants as this is quite common in agriculture, whereas fishers are more independent and therefore may be less willing to accept grants (Pita et al. 2010). Social injustice and inequalities can exist and be reinforced by policy change within fishing communities, such as the failure to recognise access rights for crew, increasing vulnerability of some households. Intergenerational loss in families with a long history of fishing was also identified by respondents:

> It's sad for the area. I have a young fella and he'll never know what it is to catch a salmon and to me that's morally wrong. But…that's the way they left it. That's how it is. The powers that be. (Ex-drifter and mussel fisher)

It is this sense of powerlessness and relationship with the 'powers that be' that the chapter turns to next.

8.3.2 Relational Wellbeing Impacts of the Ban

Even more than employment and supplementary income generation, it was the intangible or 'invisible values' (see Turner et al. 2008) that were identified as most important, in particular 'the buzz' and 'good social life' generated during the height of the salmon season:

> It was more important as a means for social interaction, the glue, social fabric. (Ex-drifter and former fisher)

Respondents went so far as to make connections between youth employment in the salmon fishery and reduced anti-social behavior:

> Mostly summer jobs for young lads. It gave them income and made them tired! You could be working from 4am to 9pm, so you're only fit to fall into your bed at the end of the day. It cut down on vandalism and things like that. And a lot used the money for 3rd level education to get degrees who were mostly from working-class families. It was a great enabler. (Ex-drifter and mussel fisher)

When asked what was most important about the salmon fishery, an ex-drifter replied that it wasn't the money, but communication. He said it was, 'four men standing on a boat, talking to each other, sharing information and local news (…) even a form of 'neighbourhood watch.' That traditional system of communication has been lost. As a result, men have 'gone back into themselves' (N.Gallgher, pers. comm. representative from a local 'men's shed' project, see below).

A key factor determining how people respond to a particular policy instrument is the degree to which they perceive it as legitimate. Several authors emphasise that compliance is determined by perceived legitimacy (Jentoft 2000b; Raakjær Nielsen and Mathiesen 2003; Coulthard 2012b). Abernethy (2010, p. 35) states that "failure of further restrictions of effort controls undermines the science underpinning management and gives weight to fishers' arguments of lack of legitimacy of regulations." Confusion and conflict is rife among stakeholders within the Lough and the management of its natural resources. It was evident that the policy implementation process began without an understanding of what Chuenpagdee and Jentoft (2007) call 'step zero' for fisheries management. 'Step zero' is the pre-implementation stage that requires an understanding of the local context including the relations between those who favour or oppose implementation of the policy, to aid in evaluating the likelihood of success.

> Fishermen were given no chance to discuss the decision, consider the implications, and absolutely no consultation. (Ex-drifter and oyster fisher)

The right to be involved in decision making is referred to in the Code of Conduct for Responsible Fisheries (FAO 1995). However, as the above statement highlights, this is not always the reality. Furthermore, Chuenpagdee and Jentoft (2007) argue that the process of implementation of policies, not just the outcomes, must be considered. The top-down nature of the ban, under EU legislation and with Member States facing fines for failure to implement, prevented the establishment of forums where diverse opinions could be expressed and where a process for integrating the needs of the resource users and the ecosystem could be developed (McClanahan et al. 2009). It also removed responsibility for the natural resource and instilled a sense of disempowerment amongst those at the most local levels (Bavinck and Jentoft 2011). Fishers and communities were deprived of the opportunity to be involved in a decision making process to determine the how and why of resource allocation (see also Smith et al. 2003). This relationship between government and fishers was highlighted by Britton and Coulthard (2013) as the greatest influence (power over) on fishing behaviour but also the greatest source of dissatisfaction for fishers.

> They can't figure it out but they blamed us. (Ex-drifter)

This observation illustrates a 'sticky issue' for governance of the fishery, suggesting that the decision making process and policy implementation is grounded in uncertain and poorly communicated evidence. This lack of acknowledgement of uncertainty in science as an issue for governance is well documented in the literature (Majone 1989; Sweeting and Polunin 2005). Differing reasons for decline highlighted contestation over the extent and cause of decline. Management and media reports (NASCO 2008; BBC 2012) highlighted an over dependence on 'expert' driven decision making with a lack of appreciation for complexity. Often decisions were based on evidence from a single scientific discipline. Fishers felt blamed for the decline of salmon (*Donegal News, 28 February* 2012) and, except for coverage in local newspapers voicing fishermen's concerns, fishing families had to cope with the negative public perception of their way of life. The NI Department for Culture Arts and Leisure (DCAL) stated concern that fishers' were unfairly stigmatised: "We are disturbed that salmon netsmen are being portrayed to DCAL officials and the media as a major factor in the current shortage of fish" (NI Assembly 2012, p. 9).

A review of recent newspaper articles highlights the continuing conflict between fishers and management with serious implications for governance of the Lough. In 2006, after the ban was first announced a headline in the local newspaper read, '"We'll Defy the Salmon Ban" - angry driftnet fishermen' (Cullen 2006). Fishers adopt different strategies to cope with change and these strategies reflect different forms of agency (see Coulthard 2012b). In response to feeling 'left out' of decision making, there was a growing sense of political agency (see Lister 2004) following the ban, with fishermen 'getting organised' and establishing and mobilising fisher organisations (e.g. the Donegal Salmon and Inshore Fishermen's Association). However, as the following comment illustrates, fishers felt marginalised against the powerful lobby and strong financial base of the angling associations, "Everybody blamed us because we're the smallest. There was only 114 of us whereas maybe 1000s of anglers." There was also evidence of another expression of political agency, 'getting back at' (Lister 2004; Coulthard 2012b), or rule breaking by fishermen who felt frustrated that their voices were not being heard and that the rules lacked legitimacy and impinged on their rights (*Irish Times,* 15 February 2011, *Derry Journal,* 25 January 2011, 27 September 2012). More recently, media reports have begun to agree with the fishermen, criticising management and the failure of the ban to protect salmon stocks (BBC 2012).

A lack of voice or, rather, the ability to be heard and influence decisions, can erode willingness or ability to participate. Lister (2004) states that to act politically, one first requires a sense of personal agency or a 'belief that one can act', and, as the following quote illustrates, when this belief is eroded people feel disempowered:

> I was a fishing rep for the Foyle Fisheries Commission (former Loughs Agency). And the same issues 20 years later are still not being addressed. It's so futile no matter what effort you put into it. So I left. (Ex-drifter and mussel fisher)

A lack of participation in the decision making process and the isolated nature of the rural fishing community led to feelings of betrayal and mistrust and a high level of tension between those to be 'managed' and the 'managers' (see Marshall 2007).

8.3.3 Subjective Wellbeing Impacts: Cultural Identity and 'the Salmonof Knowledge'

There was a lack of consideration for potential social and cultural impacts of the ban for fishers and their communities in the impact report by Collins et al. (2006), with only one reference to non-economic values, despite the North Atlantic Salmon Conservation Organisation (NASCO) outlining guidelines for the integration of socio-economic factors into any salmon management plan. The NASCO Socio-Economic Working Group (2008, p. 3) highlighted major gaps in available information, particularly with regard to non-consumptive uses and existence of values stating that, "The complex and very hard to quantify values of subsistence fisheries to dependent communities were also lacking." It was these subjective values that emerged most strongly when respondents were asked what was important about the salmon drift-net fishery:

> It was part of society, part of culture. It was tradition here for many hundreds of years, it's not just a case of one generation gone, everyone's out. (Ex-drfiter and offshore fisher)

Coppens (2011) identified individual experience, family tradition and other fishermen as the main sources of knowledge transfer amongst fishers in her study in Belgium. Similar to Coppens, frequent comments from respondents highlighted that inter-generational knowledge sharing was one of the more valued, and threatened, aspects of salmon fishing:

> I'm 25 now but my father's in his late 60s but after his generation there'll be half the knowledge. I'm young—I'd be one of the youngest—and after me, nobody knows about salmon. It's all just from a tourist perspective. (Ex-drifter and inshore fisher)

Ex-drifters emphasised the 'added-value' of the fishery. Contrary to the popular economic rationalisation or rent maximisation theory (World Bank and FAO 2011), for respondents, it was more than a source of material or economic wellbeing. Salmon fishing offered greater challenge and excitement than other types of fishing or work, it was simply '*fishing for enjoyment*':

> It's not even the part of fishing for profit—it was fishing for enjoyment [...] drifting for salmon is nets, it's different. You had to know the coast. You had to know the rocks. You had to know the tides. And in another 10 years time nobody will know the coast as well as a salmon man did. You've lost all the information too. (Ex-drifter and 'pot' fisher)

This emphasis on knowledge is interesting given the mythology that surrounds the salmon, in particular the 'salmon of knowledge' or *bradán feasa*, a famous creature in Irish mythology said to possess all the knowledge of the world. The first person to taste it would gain this wisdom from the salmon (see Gregory 1994). The cultural significance of salmon has implications for the outcomes of the ban. It is clear that personal, subjective values permeate every dimension of wellbeing. For example, White (2009) argues that material wants are nested in cultural values and Appadurai (2004) notes that aspirations are embedded in culture and shaped through social interaction.

In 2010, the remaining licence holders (n=18) did not have their licences renewed, nor did they receive any compensation. These were the last of the drifters and this marked the final closure of the drift-net salmon fishery in Lough Foyle. Fishers' described the loss, in a local newspaper (*Derry Journal,* 18 June 2010), as heart-breaking, experiencing mental and psychological impacts as well as financial. Studies of the consequences of job loss or income decline in other primary sectors (e.g. farming and forestry) highlight the major negative impacts on individual and family health and decreases in social wellbeing. As Smith et al. (2003) highlight, this is especially the case when producers feel blamed for environmental destruction and betrayed by the government. The loss of these 'invisible values' (Turner et al. 2008) such as self-worth and social cohesion, particularly for older members of the fishery, is captured in the following:

> I think the greatest loss, which is something people don't realise, and that's the psychological or mental effect. For the older people it was the only thing they did (within the community). It gave them a sense of value and self-worth. A real interest in life (for those few months). Life is work, rest and play and it gave them the work so they could enjoy the rest and play. (Ex-drifter and mussel fisher)

8.3.4 Unintended Ecosystem Impacts

The initial aim of the ban was a temporary conservation measure to prevent further decline and to help restore salmon stocks in the Foyle. Although the ecological impact of the ban could not be fully assessed for 4 years (the typical lifecycle of the Atlantic salmon), an EU report on the salmon management plan for the area stated that it anticipated "compliance with conservation limits in UK-NI rivers will reflect an improvement in 2007 and thereafter" (2008, p. 15).

Reasons for decline identified by ex-drifters highlight close social-ecological interactions and linkages, which were poorly considered in the implementation of the ban. Similar to Brennan and Rodwell's (2008) study on sustainable management of wild Irish Atlantic salmon with various stakeholders, a combination of factors were identified by Lough Foyle fishers including the impact of seal predation, pollution, fish farms, overfishing offshore and in rivers (angling) and climate change. Despite some differing opinions as to the cause of decline the over-riding consensus was that drifters were not the primary cause. Evidence from the international Salmon at Sea (SALSEA) research programme suggest that the current phase of low survival may be due, in part, to changing sea conditions, such as surface temperature and the availability of food organisms, as a result of climate change. What is clear is that salmon are failing to return from the sea to the Foyle system.

There can be unintended knock-on effects for the natural resource arising from environmental policy such as the issue of over capacity in the inshore sector (Tingley 2006; Béné 2006). Despite this growing trend, effort displacement is poorly considered or documented (Degnbol and McCay 2006). In the case of the Foyle drift net ban pressure may have increased on the marine environment as fishing families remained in fishing but targeted specific species, such as mussels. A majority continued to fish full-time or part-time and three respondents remained highly dependent on fishing. Traditionally, in the inshore sector, a diversity of species could be targeted during the summer season (e.g. dogfish, salmon, skates), but following successive EU legislation banning the targeted capture of these species, the remaining effort has been concentrated almost entirely in the potting and shellfish sectors:

"The result[of the ban] was displacement with increased pressure on potting [...] everything is under severe strain" (ex-drifter and oyster fisher). Small, local inshore vessels are limited in their range and space is limited with competition and conflict over the best areas to fish increasing. These stocks would traditionally have had a seven week rest during the salmon season, now there is no informal rest period. Several respondents mentioned there was evidence of declining biodiversity, "I remember my grandfather fishing for mackerel, cod, pollock and even herring and haddock. I haven't seen a haddock for years" (ex-drifter, seasonal inshore fisher). A possible unintended consequence of the ban (combined with the promotion of selective fishing through a restrictive licensing that reduced the ability to switch between fisheries and/or gears) is a decline in ecosystem biodiversity as well as increased vulnerability of fishers (see Garcia 2011).

8.4 Future Aspirations

This section explores responses to the question; *what are your hopes for the future and what would have to happen that would help you realise these aspirations?* It seeks not only to identify aspirations but also explore how people plan to fulfill these aspirations. Appadurai's (2004, p. 63) discussion of the 'capacity to aspire' relates to Albert Hirshman's concept of 'voice' and 'exit', the decision to voice dissatisfaction and desire for change or the capacity to leave an unsatisfactory situation for opportunity elsewhere. This of course, raises the question of power. Ibrahim (2011) argues that poverty, in the broadest sense (not just material, but social and cultural) is associated with the failure of aspirations, as well as powerlessness and vulnerability. People may not have the power to change their position, lacking the capacity to exit or an adequate platform to give voice to their aspirations but that does not mean they don't have aspirations. Appadurai (2004) highlights the importance of understanding differences in capacity to aspire using the concept of the 'horizon of aspiration' to illustrate; the closer the horizon the more concerned it is with short-term strategies and more immediate needs or day-to-day survival. A more 'expansive horizon' is concerned with longer-term visions for the future.

As referred to in the previous section, 'capacity to aspire' is a cultural capacity, ideas for the future as well as ideas about the past are embedded in culture (Appadurai 2004). Béné (2006, p. 35) states that culture is also important at an individual, as well as collective level, for identity creation, sense of self-esteem and belonging. As Kurien (2011, p. 326) notes, identity creation or, 'how one defines oneself, how one sees oneself, has a strong impact on the making and working of governance arrangements.' Lough Foyle drifters' identity is connected to their past, to a traditional 'heritage' fishery perceived as their 'historical right' to fish, passed down through generations for centuries. Their identity is also influenced by the dynamic geopolitical seascape, situated in a politically contentious border region.

Table 8.1 Most frequent[a] aspirations for the future mentioned by respondents ($N=14$)

Aspiration	Pathway to aspiration
Ecological *A return of the salmon, healthy ecosystems*	Better research and understanding of the state of the Lough's ecosystem, in particular the need to assess the impact of the increasing seal population A broadening of knowledge sources is necessary with the involvement of fishers' knowledge as local experts e.g. fisheries-science partnerships (see Johannes et al. (2000); CEFAS (2012))
Material Sustainable livelihoods; *Support for the most vulnerable (e.g. crew, youth, older fishers)*	Improve capacity to sustain coastal livelihoods (see Allison et al. (2009)) e.g. supplement fisheries income with seasonal tourism activities
Relational *Successful management, improved participation*	Greater regionalisation of fisheries management; putting local resources 'back in the hands of the people.' Stewardship, encouraged through local ownership and in-depth knowledge of local ecosystems
Good governance; *Creation of 'common ground', an essential component of 'good governance'* (Kooiman and Bavinck (2005))	Improved mechanisms for inclusive participation and co-management. Better sharing of knowledge and understanding amongst all stakeholders

[a] Responses from qualitative interviews were grouped under these 4 main themes; some respondents gave more than one 'aspiration.'

The most frequently mentioned aspirations identified in Table 8.1. highlight an expansive, longer-term vision for the future of fishing society in Lough Foyle, with opportunities as well as problems. According to Bavinck and Jentoft (2011) understanding how the future is *imagined* is essential for good governance of fisheries. The four aspirational 'themes' for the future overlap and are interrelated such as the relationship between 'healthy ecosystem and sustaining livelihoods'. The idea expressed by one respondent that, '*fish are the rock in the foundation*' points to a shared value or common ground and possible starting point for governance. Respondents recognised not just socio-economic needs but also a sense of ownership, and a need for more ecosystem-based thinking that considers the system as a whole. The responses highlight the importance of taking local conditions into account, a point which has been stressed in the literature for some time now (Doeringer et al. 1986; Pollnac et al. 2001; Wilen 2004; Pita et al. 2010), combined with the effect of wider economic factors. Furthermore, the need to broaden the knowledge base for fisheries management, including local fisher knowledge was emphasised by respondents, similar to what Raakjær and Hegland (2012) refer to as moving knowledge 'down and out'.

8.5 Conclusion

The findings have highlighted the unequal impacts felt by individuals within fishing communities and between different groups of resource users (e.g. drifters and anglers) and the need to identify those who are most vulnerable (e.g. crew members). Improving capacity for multi-livelihood strategies may be another way to reduce vulnerability, such as supplementing income from fishing with seasonal income from tourism, and having the flexibility to switch between occupations seasonally which would require changes to current licensing legislation.

Improved mechansims for co-management (Jentoft 2000a; McClanahan and Cinner 2012) and the process of how (and why) are very important (Jentoft 2007). Given the current lack of participation and the contentious political process, it is perhaps unsurprising that there is a great deal of stress and anger directed at the decision to ban salmon drift-netting. In this context of distrust between resource users and resource managers, identifying who the stakeholders are and what level of power and influence they have will need careful consideration for any co-management arrangement and greater regionalisation of fisheries (Jentoft 2000a; Raakjær and Hegland 2012). Small-scale fishers are frequently marginalised from decision making processes as a result of their generally low level of representation and lack of membership in producer organisations. In line with recommendations for the reform of the CFP, greater regionalisation is a desired way forward and was strongly emphasised by respondents in their image of the future:

> Every country should have control of their own waters. In every county every area should have an organisation running it answerable to the main office in Dublin. Different species, different fish, different everything over a 400 mile area. The same law can't work in both. We're getting one law from Europe that suits France and Spain and we've to go by that law. So we have to have it regional. (Ex-drifter and part-time inshore fisher

However, one respondent cautioned against the promotion of 'community-based' management:

> You have to be careful what you mean by community? Who is the community? How do you decide that?(…) And what about the fella from Islay [Scotland] who has a small boat but fast and powerful engine and can zoom over here and fish in a couple of hours? (Young inshore fisher, north coast)

It is clear that localising is not always a good thing and can be double edged, especially given the cross border context. The above respondent argued that management could become too piece meal as opposed to the equally unsuccessful 'one size fits all' approach of the CFP, and instead suggested, 'what's needed is a middle way.' Gasalla (2011, p. 187) also emphasises 'the shared way' where, "the state would have fundamental responsibility to ensure that access to resources is fair and equitable, local communities would also have a commitment to sustainable use and protection of marine and fisheries resources." Small-scale fisheries, however, are diverse, and consideration for who does what, such as people who fish seasonally, as well as full-time, family members involved in the business and the numerous multiplier effects (Béné 2006) is important when deciding who to involve, and how. A multi-stakeholder approach, and 'independently faciliated workshops' between scientific, fishing and policy groups is advocated by Brennan and Rodwell (2008, p. 1079) as a way to help overcome "entrenched value judgements and deep-seated emotions" by 'actively listening' and learning from the views of others instead of the current one-way flow of information. Their study suggests that a major area of common ground between stakeholders is the desire to see the return of salmon, a key aspiration identified in this study also (Table 8.1). However, four years on from Brennan and Rodwell's study little has been achieved in developing a more constructive co-management of wild Irish Atlantic Salmon stocks and time is running out for what is left of remaining stocks. This sense of urgency may hinder the patience and time that is needed to develop successful co-management. It is argued that a social wellbeing approach goes some way to discerning 'who wants what and why', which is a major issue for governance (Chuenpagdee and Jentoft 2007)—what resources people have (material wellbeing), how their aspirations are negotiated and communicated with others (relational wellbeing) and how cultural values, etc. underpin these actions (subjective wellbeing).

Given the loss and unlikely return of the salmon fishery in the near future, there is a very poor platform for small-scale fisheries because of mistakes already made. A possible way to address these 'hard choices' may lie in exploring new ways of how qualities attached to the fishing way of life (e.g. social cohesion and communication) can be harnessed to embrace change, and not just put into resisting change. These could be new ways of 'doing' that integrate old ways of 'being;. For example, the success of 'men's sheds' in the fishing port of Killybegs, also in Co. Donegal (member of men's shed pers. comm. 26 November 2012; *Marine Times* August 2012), a community-based initiative that brings men together from the community (from fishing and non-fishing backgrounds) to learn new skills or apply existing skills to traditional maritime crafts such as boat building, as well as building on the succes of other community-based initatives such as the Rural Social Scheme.

The combined need for macro-economic policies and fine-tuned policies at the local level is necessary and strongly associated with empowerment (Bavinck and Jentoft 2011). Trimble and Johnson (2012, p. 2) argue that "governance processes in fisheries should engage artisanal fishers in ways that are sensitive to material, relational and subjective considerations." In Lough Foyle, as in Trimble and Johnson's (2012, p. 2) example from South American fisheries, fisheries management has failed to implement policies that consider all of these factors together which has resulted in a "legacy of mistrust and scepticism among (artisanal) fishers." This was evident by fishers' response to the salmon ban that they will not willingly hang up their nets in return for compensation that does not meet their needs. A key lesson is the likelihood of failure where policies ignore local realities. A social wellbeing approach can better take account of the wider social benefits and impacts, aiding assessment of the planning and implementation of various policy tools, as well as the outcomes.

References

Abernethy, K. E. (2010). Fishing for what? Understanding fisher decision-making in southwest England. PhD thesis. University of East Anglia, UK.

Abernethy, K. E., Trebilcock, P., Kebede, B., Allison, E. H., & Dulvy, N. K. (2010). Fuelling the decline of UK fishing communities? *Ices Journal of Marine Science*. [Advanced electronic access published January 8, 2010].

Ad hoc review group. (2008). *Atlantic salmon management strategy for Northern Ireland and the cross border Foyle and Carlingford catchments to meet the objectives of NASCO resolutions and agreements 2008–2012*. IP(07)11 Final Implementation plan European union—uk (northern ireland).

Allen, M. (2010). Lough Foyle—ownership, licensing, and levy issues relating to shellfish. Northern Ireland Assembly, Research and Library Service Briefing Paper. Paper 32/11.

Allison, E. H., & Ellis, F. (2001). The livelihoods approach and management of small-scale fisheries. *Marine Policy, 25*(5), 377–388.

Allison, E., Perry, A. L., Badjeck, M. C., Adger, W. N., Brown, K., Halls, A. S., Pilling, G. M., Reynolds, J. D., Andrew, N. L., & Dulvy, N. K. (2009). Vulnerability of national economies to the impacts of climate change on fisheries. *Fish and Fisheries, 10*(2), 173–196.

Appadurai, A. (2004). The capacity to aspire: Culture and the terms of recognition. In V. Rao & M. Walton (Eds.), *Culture and public action* (pp. 59–84). Palo Alto: Stanford University Press.

Armitage, D., Béné, C., Charles, A. T., Johnson, D., & Allison, E. (2012). The interplay of wellbeing and resilience concepts: Towards a transdisciplinary social-ecological perspective. *Ecology and Society, 17*(4), 15.

Bavinck, M., & Jentoft, S. (2011). Subsidiarity as a guiding principle for small-scale fisheries, Chapter 20. In R. Chuenpagdee (Ed.), *World small-scale fisheries: Contemporary visions*. Deflt: Eburon Academic Publishers.

BBC. (2012). Slamon stocks in the River Foyle continue to decline. http//www.bbc.co.uk/news/uk-northern-ireland-foyle-west-17176712. Accessed 27 Feb 2012.

Béné, C. (2006). Small-scale fisheries: Assessing their contribution to rural livelihoods in developing countries. FAO Fisheries Circular No. 1008. Rome: Food and Agriculture Organisation of the United Nations.

Binkley, M. (2000). 'Getting by' in tough times: Coping with the fisheries crisis. *Women's Studies International Forum, 23*(3), 323–332.

Binkley, M. (2002). *Set adrift: Fishing families*. Toronto: University of Toronto Press.

Brennan, R. E., & Rodwell, L. D. (2008). Sustainable management of wild Irish Atlantic salmon: Keys found through the looking glass. *Marine Policy, 32*(6), 1072–1079.

Britton, E., & Coulthard, S. (2013). Assessing the social wellbeing of Northern Ireland's fishing society using a three-dimensional approach. *Marine Policy, 37*, 28–36.

Campbell, J., & Cattermoul, B. (2008). *The socio-economic impacts of Fisheries management and policy designed to achieve biodiversity conservation*. Phase 1 report. Prepared for: The Tubney Charitable Trust, UK.

Carraher, B. (2006). '*The Drifters*'. Produced by Mind the Gap Films, 11 Dec 200, 22.35 BBC One NI.

Cavanagh, M. (2010). *Socio-economic impacts of the 2007 drift net fishery ban on communities in the Lough Foyle area*. Dissertation submitted for the degree of Marine Science BSc (Hons). Department of Environmental Sciences, Ulster of University.

CEFAS. (2007). *Baseline survey of shellfish resources in Lough Foyle*. Report C2697 for Loughs Agency. The Centre for Environment, Fisheries and Aquaculture Science, Lowestoft, UK.

CEFAS. (2012). *The fisheries science partnership (FSP)*. [online] URL: http://www.cefas.defra.gov.uk/industry-information/fisheries-science-partnership.aspx.

Chuenpagdee, R. (2011). *World small-scale fisheries: Contemporary visions*. Delft: Eburon Academic Publishers.

Chuenpagdee, R., & Jentoft, S. (2007). Step-zero for fisheries co-management: What precedes implementation. *Marine Policy, 31,* 657–668.

Collins, T., Malone, J., & White, P. (2006). *Report of the Independent Salmon Group—Established to Examine the Implications of Alignment with the Scientific Advice for the Commercial Salmon Fishing Sector in 2007 and Beyond.* Inland Fisheries Division, Dublin.

Coppens, J. (2011). *A social and cultural framework for local ecological knowledge: Case study of the Belgian fishery.* Paper presented at the Conference, 'It's not just about the fish': Social and cultural perspective of sustainable fisheries, University of Greenwich 4–5 April, 2011.

Coulthard, S. (2012a). What does the debate around social wellbeing have to offer sustainable fisheries? *Current Opinion in Environmental Sustainability, 4*(3), 358–363.

Coulthard, S. (2012b). Can we be both resilient and well, and what choices do people have? Incorporating agency into the resilience debate from a fisheries perspective. *Ecology and Society, 17*(1), 4.

Coulthard, S., Johnson, D., & McGregor, J. A. (2011). Poverty, sustainability and human wellbeing: A social wellbeing approach to the global fisheries crisis. *Global Environmental Change, 21*(2), 453–463.

Cullen, I. (2006). 'We'll defy the salmon ban'—Say angry driftnet fishermen. *Derry Journal*, 3 November 2006.

DCAL. (2008). *Focus Area Report on management of salmon fisheries in UK-NI. Fishereries Management Focus Area Report IP(08)4.* Department of Culture Arts and Leisure, Northern Ireland.

DCAL. (2012). Scientific and technical input to salmon conservation consultation. Department of Culture Arts and Leisure, Northern Ireland. Document available online: www.dcalni.gov.uk/salmon_conservation_-public_consultation_-_s.

Degnbol, P., & McCay, B. (2006). Unintended and perverse consequences of Ignoring linkages in fisheries systems. *ICES Journal of Marine Science, 64,* 793–797.

Doeringer, P., Moss, P., & Terkla, D. (1986). *The New England fishing economy: Jobs, income, and kinship.* Massachusetts: University of Massachusetts Press.

Doherty. (2006). Donegal Salmon and Inshore Fishermen's Association with Donegal Traditional Fishermen's Association, Submission—Part 1 (Letter) to: Joint Committee on Communication, Marine and Natural Resources, Dublin.

FAO. (1995). *Code of conduct for responsible fisheries.* Rome: Food and Agriculture Organisation

Garcia, S. M. (Ed.). (2011). *Selective fishing, balanced harvesting and sustainability of fisheries and ecosystems.* Report of a scientific workshop organized by the IUCN-CEM Fisheries Expert Group (FEG) and the European Bureau for Conservation and Development (EBCD) in Nagoya (Japan), 14–16 October 2010.

Gartaula, H. N. (2011). *The other side of migration in rural Nepal: Sociocultural transformation and the women left behind.* PhD Thesis, Waginen School of Social Sciences, Waginen University.

Gasalla, M. A. (2011). Do all answers lie within (the community)? Fishing rights and marine conservation, Chapter 11. In R. Chuenpagdee (Ed.), *World small-scale fisheries: Contemporary visions.* Deflt: Eburon Academic Publishers.

Gough, I. R., & McGregor, J. A. (Ed.). (2007). *Wellbeing in developing countries: New approaches and research strategies.* Cambridge: Cambridge University Press.

Gregory, A. (1994). *The coming of finn. Lady Gregory's complete Irish mythology* (pp. 117–123). London: Octopus Publishing Group Ltd.

Hansard. (1995). *Foyle fisheries commission.* HC Dec 08 February 1995 vol. 254 cc303-11. Available online at: http://hansard.millbanksystems.com/commons/1995/feb/08/foyle-fisheries-commission.

Healy, T. M. (1913). *Stolen waters.* London: Longman, Green & Co.

Ibrahim, S. (2011). Poverty, aspirations and wellbeing: Afraid to aspire and unable to reach a better life—Voices from Egypt. *BWPI Working Paper 141*, University of Manchester, UK.

Jentoft, S. (2000a). Legitimacy and disappointment in fisheries management. *Marine Policy, 24,* 141–148.

Jentoft, S. (2000b). The community: A missing link of fisheries management. *Marine Policy, 24,* 53–59.

Jentoft, S. (2007). In the power of power: The understated aspect of fisheries and coastal management. *Human Organisation, 66*(4), 426–437.

Jentoft, S., & Eide, A. (Eds.). (2011). *Poverty mosaics: Realities and prospects in small-scale fisheries*. Springer.

Jensen, C. F. (1999). A critical review of the Common Fisheries Policy. IME Working Paper 6/99, Department of Environmental and Business Economics, University of Southern Denmark.

Johannes, R. E., Freeman, M. M. R., & Hamilton, R. J. (2000). Ignore fishers' knowledge and miss the boat. *Fish and Fisheries, 1,* 257–271.

Kennedy, M. (2000). *Division and consensus: The politics of Irish cross-border relations, 1925–1969*. Institute of Public Administration.

Kooiman, J., & Bavinck, M. (2005). The governance persoective. In Kooiman et al. (Eds.), *Fish for life: Interactive governance for fisheries*. Amsterdam: Amsterdam University Press.

Kurien, J. (2011). Governance arrangement for small-scale fisheries: some reflections from negotiating small-scale fisheries governance arrangements in real life, Chapter 21: 323–331. In R. Chuenpagdee (Ed.), *World small-scale fisheries: Contemporary visions*. Deflt: Eburon Academic Publishers.

Lister, R. (2004). *Poverty*. Cambridge: Polity Press.

Loughs Agency (2009). *Annual Catch (No.s) of salmon of driftnetlicence holders 1999-2009*. Loughs Agency, L'derry.

Loughs Agency (2009b). *Figures concerning hardship fund*. L'derry: Loughs Agency.

Majone, G. (1989). *Evidence, argument, and persuasion in the policy process*. Yale University Press.

Marshall, N. A. (2007). Can policy perception influence social resilience to policy change? *Fisheries Research, 86,* 216–227.

McClanahan, T. R., Castilla, J. C., White, A. T., & Defeo, O. (2009). Healing small-scale fisheries by facilitating complex socio-ecological systems. *Rev Fish Biol Fisheries, 19,* 33–47.

McClanahan, T. R., & Cinner, J. E. (2012). *Adapting to a changing environment: Confronting the consequences of climate change*. NY: Oxford University Press.

McGoodwin, J. R. (2001). *Understanding the cultures of fishing communities: A key to fisheries management and food security*. FAO Fisheries Technical Paper 401. Food and Agriculture Organization of the United Nations, Rome: 287.

McGregor, J. A. (2007). Researching wellbeing: From concepts to methodology. In I. Gough & J. A. McGregor (Eds.), *Wellbeing in developing countries: From theory to research*. Cambridge: Cambridge University Press.

McGregor, J. A. (2008). *Wellbeing, poverty and conflict*. WeD Policy Briefing 01/08. 2008. http://www.bath.ac.uk/econ-dev/wellbeing/research/bp/bp1-08.pdf.

McGregor, J. A. (2009). *Human wellbeing in fishing communities*. Paper prepared for ESPA Workshop 1, April 2009, Institute for Ocean Management, Chennai, India. Project: Building Capacity for Sustainable Governance in South Asian Fisheries: Poverty, Wellbeing and Deliberative Policy Networks. Available online at www.wellcoast.org. pp. 1–8.

Morrissey, K., & O'Donoghue, C. (2012). The role of the marine sector in the Irish national economy: And input-output analysis. Marine Policy, In Press. http://dx.doi.org/10.10.16/j.marpol.2012.05.006.

NASCO. (2008). Atlantic salmon management strategy for Northern Ireland and the cross border Foyle and Carlingford catchments to meet the objectives of NASCO resolutions and agreements 2008–2012. *Ad hoc* review group Ip(07)11 final *Implementation plan European union—uk (northern ireland)*.

NI Assembly. (2012). *Committee for Culture, Arts and Leisure: North Atlantic Wild Sea Salmon Fishermen's Association*. Official Report (Hansard) 16 February 2012. Northern Ireland Assembly, Belfast.

Pita, C., Dickey, H., Pierce, G. J., Mente, E., & Theodossiou, I. (2010). Willingness for mobility amongst European fishermen. *Journal of Rural Studies, 26*(3), 308–319.

Pollnac, R. B., Pomeroy, R. S., & Harkes, I. H. T. (2001). Fishery policy and job satisfaction in three southeast Asian fisheries. *Journal of Ocean and Coastal Management, 44*(7–8), 531–544.
Raakjær, J., & Hegland, T. J. (2012). Introduction. Regionalising the Common Fisheries Policy. *Maritime Studies, 11*, 5.
Raakjær Nielsen, J., & Mathiesen, C. (2003). Important factors influencing rule compliance in fisheries—Lessons from Denmark. *Marine Policy, 27,* 409–416.
Smith, S., Jacob, S., & Jepson, M. (2003). After the Florida net ban: The impacts on commercial fishing families. *Society and Natural Resources, 16,* 39–59.
Sweeting, C. J., & Polunin, N. V. C. (2005). *Marine protected areas for management of temperate North Atlantic fisheries: Lessons learned in MPA use for sustainable fisheries exploitation and stock recovery*. London, DEFRA. www.defra.gov.uk/marine/pdf/science/mpareportnorthatlantic.pdf.
Symes, D., & Phillipson, J. (2009). Whatever became of social objectives in fisheries policy? *Fisheries Research, 95,* 1–5.
Tingley, D. (2006). Northern Ireland Fleet Futures Analysis (2004–2013)—Methodology and Analysis. *Final report*, DARD, April 2006. http://www.dardni.gov.uk/ni-fleet-futures-analysis.pdf. Accessed 26 Aug 2011.
Trimble, M., & Johnson, D. (2012). Artisanal fishing as an undesirable way of life? The implications for governance of fishers' wellbeing aspirations in coastal Uruguay and southeastern Brazil. *Marine Policy*. http://dx.doi.org/10.1016/j.marpol.2012.04.002.
Turner, N. J., Gregory, R., Brooks, C., Failing, L., & Satterfield, T. (2008). From invisibility to transparency: identifying the implications. *Ecology and Society*, *13*(2), 7.
van Ginkel, R. (2007). *Coastal cultures: An anthropology of fishing and whaling traditions*. Apeldoorn: Spinhuis Publishers.
WeD. (2008). Wellbeing indicators: Measuring what matters most. *WeD briefing paper* 2/08. http://www.welldev.org.uk.
White, P. (2005). *Decommissioning Requirements for Ireland's Demersal and Shellfish Fleets*. A report to Marine Minister, Pat the Cope Gallagher T.D. Department of Communications, Marine and Natural Resources, Dublin.
White S.C. 2009. Bringing wellbeing into development practice. *WeD Working Paper* 50. http://www.welldev.org.uk/wed-new/workingpapers/workingpapers/WeDWP_09_50.pdf
White S.C. 2010. Analysing wellbeing: a framework for development practice. *Development in Practice 20* (2):158-172.
White, S. C. (2011). *Reflections on our first year*. Wellbeing and Poverty Pathways ESRC/DfID Research Project, Interview September 2011. Available online: http://www.wellbeingpathways.org/news-a-events/project-blog/91-year-ine-reflection.
White, S., & Ellison, M. (2006). *Wellbeing, livelihoods and resources in social practice*. WeD Working Paper 23. ESRC Research Group on Wellbeing in Developing Countries.
Wilen, J. (2004). Spatial management of fisheries. *Marine Resource Economic, 19,* 7–19.
World Bank, & FAO. (2009). *The sunken Billions: Economic justification for fisheries reform. Agriculture and rural development*. Washington: The World Bank.

Chapter 9
Fisheries Diversification: A Case Study of French and English Fishers in the Channel

Richard Morgan, Marie Lesueur and Laura-Mars Henichart

Keywords Fisheries diversification • Pluriactivity • Analytic Hierarchy Process • Channel fishery • European Fisheries Fund

9.1 Introduction

The marine fisheries of Europe have witnessed considerable change over the last forty years—characterised by changing access agreements, the overfishing and depletion of key fish stocks, rising operating costs, and the introduction of management measures aimed at restricting fishing effort and output. The fishing industry has restructured accordingly through a process of modernisation and rationalisation of the catching sector (Symes 2000), leading to significant concentration and contraction of the European fishing fleet (Brookfield et al. 2005, p. 57). Despite the limited contribution that fishing makes to Gross Domestic Product (GDP) at the European and national level, the impact of these developments may be significant; particularly where fishing remains a significant component of regional/local economies, and among independent fishers who find themselves unable to compete with large-scale fishing operations (Brookfield et al. 2005, p. 58).

One response by fishers to these difficulties is to diversify their income source into other activities. The inherent uncertainty associated with harvesting a wild resource means that fishers have long practised horizontal diversification with respect to the species targeted and methods used—responding to variables such as season, species distribution and market price. This is particularly prevalent among the Channel inshore fleet where vessels engage in a range of fishing activities and

R. Morgan (✉)
Centre for the Economics and Management of Aquatic Resources,
Portsmouth Business School, University of Portsmouth, Portsmouth, UK
e-mail: richard.morgan@myport.ac.uk

M. Lesueur · L.-M. Henichart
Fisheries and Aquaculture Sciences Centre, Agrocampus Ouest, Rennes, France
e-mail: marie.lesueur@agrocampus-ouest.fr

L.-M. Henichart
e-mail: laura-mars@hotmail.fr

J. Urquhart et al. (eds.), *Social Issues in Sustainable Fisheries Management*,
MARE Publication Series 9, DOI 10.1007/978-94-007-7911-2_9,

gears throughout the year (Ulrich et al. 2002, p. 381). However, in recent years fishers have found such opportunities restricted by administrative and economic constraints. Thus, it has been argued that diversification of the employment base and the creation of alternative opportunities is necessary to tackle declining employment and low incomes within the industry (Whitmarsh 1998; Symes 2000).

While evidence may be found of European fishers diversifying into non-fishing related activities (e.g. Pettersen 2000; Salmi 2005), opportunities for such pluriactivity will be subject to a range of economic and social factors. It is important to note that while fishers are highly skilled in the work they undertake, few of these skills are directly transferable to onshore occupations. Additionally, the notion that many fishers view their profession as a 'way of life' presents a further deterrent to pursuing employment outside of this industry. In response to such constraints, fishers may adopt an alternative approach of diversifying into activities that maintain a link with fishing. This form of 'fisheries diversification' has a number of potential benefits. In many cases fishers can exploit their skills, knowledge and social networks gained through fishing, without the requirement of retraining. Furthermore, undertaking these activities in addition to fishing may enable fishers to increase or stabilise incomes, and reduce the risk associated with their primary occupation. Significantly, the contribution of diversification to the sustainable development of fisheries areas is recognised within Axis 4 of the European Fisheries Fund (EFF) (2007–2013) which provides structural assistance to the fisheries sector. In addition to the promotion of multiple employment and job creation outside the fisheries sector, financial support is available under Axis 4 for adding value to fisheries products; supporting fisheries and tourism related infrastructure; promoting eco-tourism; and enhancing the natural and architectural heritage of fisheries areas (Council Regulation (EC) No 1198/2006). A comprehensive discussion of Axis 4 is provided by Budzich-Szukala in Chap. 10.

Anecdotal evidence reveals that diversification of the fishing business typically occurs in one of two directions: horizontal diversification into new products and/or markets at the same stage of the production process; or vertical integration into preceding (upstream) or succeeding (downstream) stages of production. Examples include the supply of new or differentiated seafood products to the market (horizontal), and diversification into direct selling (downstream vertical). However, despite its policy relevance and potential contribution to the sustainability of European fisheries, the body of literature on this subject remains relatively limited.

A number of studies in the Channel and adjacent waters provide some indication of the different types of activities being practised and the constraints that fishers face in their adoption. The implementation of product quality schemes as a response to the recession in the French fishing industry during the early 1990s is examined by Charles et al. (2003), who found that the use of labels to specify line-caught sea bass in Brittany did generate price increases in favour of fishers. However, the level of support among fishers for such schemes was influenced by a number of factors including boat size; place of sale; and perceptions of cost effectiveness—including the potential threat posed by free-riders. Attitudes of fishers towards diversification into the provision of tourist boat trips was explored by Alban and Boncoeur (2004) in their study of the Iroise Sea in Western Brittany. Around one-quarter (27 %) of surveyed fishers expressed potential interest in chartering their vessels for recreational purposes; although interest was greater among respondents aged under 30, and those with vessels of less than 10 m in length. In addition to financial viability, the authors identify French administrative and fiscal rules as key constraints upon diversification, together with the existence of a "difference of culture between fishing and tourism" (2004, p. 197). A more recent study of diversification among Breton fishers identified a range of activities currently being practised, including tourism; ecological activities; and adding value to products (Merrien et al. 2008). Interestingly, nearly three-quarters (72 %) of respondents believed that fishing businesses would adopt a diversification strategy in the future, although consistent with previous findings, a number of obstacles to diversification are cited. These include technical and regulatory constraints associated with using vessels for non-fishing purposes; hesitancy among fishers to take risks; and lack of time (Henichart et al. 2010).

The practice of fishers diversifying into activities that complement their principal role of fishing is comparable with the concept of multifunctional agriculture that emerged from academic and policy discourses during the early 1990s (Wilson 2007). In recognition of the existence of multiple interpretation of multifunctionality, the Organisation for Economic Co-operation and Development (OECD 2001, p. 7) provide a working definition acknowledging "the existence of multiple commodity and non-commodity outputs that are jointly produced by agriculture"; some of which exhibit the characteristics of externalities or public goods "with the result that markets for these goods do not exist or function poorly". A notable shift towards the provision of non-commodity outputs was evident from the late 1980s following EC reforms to tackle overproduction through the Common Agricultural Policy. The subsequent introduction of grant funding schemes facilitated the widespread development of on-farm diversification activities, including visitor accommodation and retail. In comparison with the fisheries sector, the body of literature on this subject is comprehensive, and despite some inherent differences remains relevant to the study of fisheries diversification. Of particular interest is the notion that diversification is subject to a range of factors working 'externally' and 'internally' to the farm household (Ilbery et al. 1997). As such, the decision to diversify extends beyond financial viability to include the personality and commitment of the farmer (Ilbery 1991); household composition (Halliday 1989; Walford 2001); and the anticipated loss of identity and socio-cultural rewards (Burton 2004). Such findings support the work conducted within fisheries that indicates that diversification is subject to multiple constraints encompassing economic, social and administrative factors.

In light of the above observations, the aim of this chapter is to develop an initial understanding of fisheries diversification activities in the Channel using a combination of primary and secondary data sources. In the first instance, an inventory is compiled to examine the nature and extent of existing activities. Survey data is then analysed to identify the opinions of fisheries stakeholders with respect to the opportunities, motives, and likelihood of French and English fishers diversifying. Through the application of the Analytic Hierarchy Process, it is possible to determine the relative importance that stakeholders attribute to different types of constraints faced by fishers when diversifying.

First, though, we need to determine how we define 'diversification' for the purposes of this chapter. Economic theory dictates that diversification can take a range of forms and directions. A firm may diversify into new (related or unrelated) product areas or different geographical markets (Andreosso and Jacobson 2005); either at the same stage of production (i.e. horizontal diversification), or into preceding (upstream) or succeeding (downstream) stages of the production process (i.e. vertical integration). Given that the focus of this research is upon complementary activities that maintain a link with fishing, it was necessary to agree a specific definition of fisheries diversification prior to data collection. A decision was taken to adopt the definition provided by Merrien et al. (2008, p. 11) in their study of Breton fishers: "Complementary activities to production, in link with the product, the profession or the business that fishers practice to have an additional income but also to promote products, profession or territory."

This definition is based upon three key attributes: firstly, that fishers undertake such activities *in addition* to fishing; secondly, that these activities are directly or indirectly related to fishing; and thirdly that they may be practised for reasons other than immediate financial gain. On this basis, the practice of fishers diversifying their fishing activity with respect to species and/or gear type does not constitute fisheries diversification. Similarly, the adoption of pluriactive working strategies involving non-fishing related employment is not included in this definition.

9.2 Methodology

The study area of this research comprises the Channel fishery—defined by the International Council for the Exploration of the Seas (ICES) management areas VIId (eastern Channel) and VIIe (western Channel), but extended to include Channel ports on the periphery of these areas. Covering an area of approximately 75,000 km^2 the Channel contains approximately eighty commercially-caught species of fish, shellfish (crustaceans and molluscs), and seaweed (Boncoeur et al. 2000, p. 106), although the majority of landings are dominated by a smaller number of higher-value fish and shellfish species.

The Channel fishery is exploited predominantly by English and French fishers. Analysis of vessel data reveals that approximately 1,900 vessels are registered to Channel administrative ports in England, with around 1,600 vessels in France (MMO 2012; Phélippé et al. 2011). However, it should be noted that not all of these vessels are active in the fishery, and vessels from ports outside the Channel also fish within this area. For these reasons, the research population is defined as fishers operating vessels from Channel ports in France and England—including the Channel Islands.

Given the limited research that has been conducted on fisheries diversification in the Channel, a three-year study was funded through the EU INTERREG IVA programme. The study consisted of three phases, commencing with an inventory of activities to establish the nature and extent of diversification among English and French fishers. This was followed by a survey of stakeholders to explore opinions regarding opportunities and motives for diversification; the likelihood of fishers diversifying into different activities; and the relative influence of different constraints that fishers face when diversifying. The final phase consisted of a survey of fishers, incorporating key elements of the stakeholder survey in addition to more specific questions on diversification behaviour. The chapter details the key findings of the first two phases of the study.

9.2.1 *Inventory of Fisheries Diversification*

The aim of the inventory was to develop a general overview of existing fisheries diversification activities by collecting data on attributes such as type, participation, location and organisation. This preliminary work enabled a typology of activities to be defined and provided the general context in which the research was embedded.

The first stage of the inventory involved the examination of secondary data, consisting primarily of online sources, but including print media, non-academic publications and television/radio coverage. Given the assumed likelihood that some secondary data sources may be incomplete and/or inaccurate, it was necessary to complement these findings with primary data. Primary data was obtained by contacting stakeholders directly by telephone, or in person. For the purposes of this research a stakeholder was defined as an individual or organisation whose work involves the management/regulation, representation, or exploitation of the Channel fishery.

9.2.2 *Survey of Stakeholders*

Upon completion of the inventory, a survey was conducted to explore stakeholders' perceptions of existing diversification activities, and identify constraints upon diversification development in the study area. A questionnaire consisting of two sections was designed. The first section contained a series of closed-questions, using rating scales to ascertain strength of opinion with respect to a range of fisheries-related diversification activities. The second section sought to determine the relative importance of five key constraint types upon diversification, as identified from supporting literature and discussions with stakeholders: economic factors; social factors; lack of information; lack of opportunities; and administrative constraints. In order to understand the relative priorities that stakeholders attributed to each of these constraint types, the Analytic Hierarchy Process (AHP) was used.

Developed by Saaty (1977), the AHP technique is a form of multi-criteria decision analysis that works by presenting respondents with a series of paired attributes or objectives arranged at opposing ends of a numerical scale. In each case, the

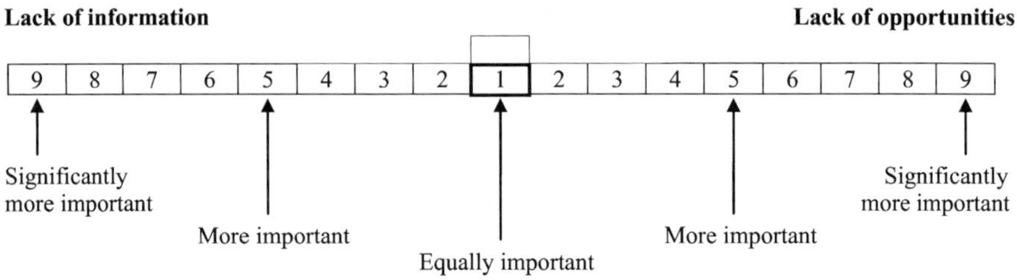

Fig. 9.1 Example of 9-point pairwise comparison scale

respondent is asked to make a pairwise comparison by selecting the position on the scale that best represents the importance/preference of one attribute relative to the other. Typically a 9-point scale is used where 9 represents the extreme importance of one attribute over the other, and 1 represents equal importance between the two attributes (Fig. 9.1). One of the advantages of AHP over more traditional techniques of respondent elicitation is that the use of simple pairwise comparisons reduces the cognitive burden of prioritising decision-making problems (Himes 2007). Furthermore, given that AHP is not a statistically-based procedure, it is well suited to interviewing small samples of key stakeholders (Herath 2004).

In both England and France, a sample of stakeholders was identified from those working within the Channel fishing industry and related sectors. In England, the zones of jurisdiction of the five regional Inshore Fisheries and Conservation Authorities (IFCAs) covering the Channel were used as broad sampling areas, within which respondents were drawn from fisheries management; industry representatives; fishers' representatives; and related organisations (e.g. harbour masters, environmental organisations). The sub-sample of English fishers' representatives was composed of 15 individuals; 11 of whom were active fishers. In France, sampling was undertaken within the seven maritime districts of Boulogne-sur-Mer, Dieppe, Fécamp, Le Havre, Cherbourg, Paimpol and Morlaix. Stakeholders were identified from professional organisations; administrative organisations; local authorities; the tourism industry; and other organisations with links to the sector. The sub-sample of professional organisations included 27 fishers' representatives. In contrast to the English sample none of these individuals were active fishers, although 8 had previously worked in this occupation. The questionnaire was administered by face-to-face interviewing to maximise the response rate and ensure that the AHP component was clearly understood by respondents. Fieldwork commenced in the summer of 2010; 83 stakeholders were interviewed in France and 38 interviewed in England.

For the survey questions relating to opportunities, motives, and likelihood of diversification, data was analysed using standard quantitative techniques. Analysis of AHP data involved two stages: the conversion of individual scores into a pairwise comparison matrix; followed by the calculation (and subsequent aggregation) of normalised priority weights. The output is an overall weighting for each of the five attributes, which collectively sum to 100%. Priority weights were calculated using the eigenvalue method, which also provides a measure of inconsistency of individual responses in the form of the consistency ratio (CR) (see Whitmarsh and Wattage 2006).

9.3 Key Findings

The following section provides an overview of preliminary findings from the two-phases of data collection outlined in the methodology. Where the inventory provides a largely descriptive account of diversification, the survey expands upon these findings by exploring the opportunities, likelihood and motives of diversification, and the constraints faced by fishers adopting this strategy.

9.3.1 Inventory of Fisheries Diversification

The inventory of fisheries diversification reveals that a range of activities are being practised on the English and French coastlines. For the purposes of analysis, activities are categorised by type under the following headings: marketing initiatives; vertical integration; leisure and tourism; contract work; and voluntary activities.

Marketing initiatives constitute a form of horizontal diversification whereby fishers have sought to increase awareness and add value to their products through the use of initiatives that promote provenance, traceability or sustainability. In England, the most widely known of these initiatives is the ecolabel introduced by the Marine Stewardship Council (MSC), of which seven fisheries were certified at the time of writing, with a further two undergoing assessment. In France, two saithe fisheries involving the Channel fleet were certified (Euronor and Scapêche and Compagnie de Pêche de St. Malo saithe) but the location of fishing activity is not within the Channel. Lobster fishers from Normandy, members of the 'Comité Régional des Pêches Maritimes de Basse Normandie' and the Jersey Fishers' Association were certified as sustainable by the MSC in June 2011. In addition to the MSC scheme, product labelling schemes have been introduced in England by regional seafood bodies including Seafood Cornwall and South East Seafood. In France, several labelling schemes exist along the coastline (Roussel et al. 2011). These include schemes that are specific to regions (Filière Opale, Normandie Fraîcheur Mer, Bretagne Qualité Mer), and associations that have created specific schemes (e.g. Association des ligneurs de la pointe Bretagne).

Fishing firms may undertake *vertical integration* by diversifying upstream into the building and maintenance of fishing vessels, supply of equipment and bait, and the provision of training/information services; or downstream into seafood processing, wholesale and retail. On the English coastline, examples of upstream integration were relatively limited and included examples of diversifying into vessel building and repairs, and the supply of fishing equipment. In contrast, downstream integration was more common with a number of fishers selling their catch through mobile and fixed retail units—including stalls and shop premises, and, to a lesser extent, restaurants. Typically this involves the selling of fresh fish, although examples were also found of processing/preparing seafood for consumption—particularly shellfish. In France, direct selling by fishers is considerably more prevalent and represents the most developed form of diversification activity. In some ports, stands have been built adjacent to the docks which fishers rent for direct selling.

The practice of fishers diversifying into the *leisure and tourism* sector is well established in parts of the Channel, with evidence found of both English and French fishers operating sightseeing trips and recreational angling charters. Sightseeing trips are typically 1–3 h in duration and provide participants with the opportunity to view local landmarks and wildlife. In the case of recreational angling, trips are generally 1–3 h in length targeting mackerel using rods or handlines, or longer trips of up to 8 h targeting a range of species. In both cases, fishing tackle and tuition are usually included in the price, thereby appealing to members of the public who would not consider themselves 'recreational anglers'. An additional benefit of operating recreational angling charters is the potential for fishers to sell customers' unwanted catches commercially.

While the majority of leisure and tourism activities are conducted by fishers using commercial fishing vessels, the practice of taking tourists onboard to observe fishing activity is notably limited. Only one example was identified in the Channel—in the case of an English inshore fisher offering tourists the opportunity to observe pot fishing for crab and lobster. This finding is not altogether unsurprising, given the safety implications of taking passengers onboard active fishing vessels.

The practice of using fishing vessels for *contract work* is not uncommon in the UK, particularly on the east coast of Scotland where fishers undertake guardship duties for the North Sea oil and gas industries. Fishers can be well suited to these activities, having the vessels and skills to operate in unfavourable sea conditions, together with experience of towing equipment. On the English coastline, a number of examples were found of diversification into contract work for the utilities sector—including telecommunications, water and renewable energy. Roles include fishers using their vessels for attendant/guardship duties and working in an advisory capacity, for example, liaising with other fishers to inform them of the work being undertaken and its potential impact upon their fishing activity. In France, examples of contract work for the utilities sector were less prevalent, although fishers have exploited such opportunities in recent years.

The other main source of contract work for fishers in the Channel is the environmental sector. On the English coastline, a number of examples were found of diversification into activities related to the marine environment—principally chartering their vessels to environmental organisations for research purposes. The source of much of this work is the UK Centre for Environment, Fisheries and Aquaculture Science (Cefas) which employ fishers for surveying through the Fisheries Science Partnership (FSP). Such contracts are open to tender and advertised in the trade press. As with other types of contract work, environmental activities can be lucrative, particularly during quiet fishing periods or where fishing activity is constrained by lack of quota. However, this work is generally sporadic and can be selective with respect to the types of vessel required and the duration of contracts. On the French coastline, fishers participate in local and national research programmes including initiatives with Ifremer (the French Research Institute for Exploration of the Sea) and national authority work. Activities identified through the inventory include participation in scientific work and surveying; allowing observers onboard vessels; and various forms of data collection. Other examples of environmental contract work, particularly the collection of waste at sea, are practised within the framework of 'Contrat bleu'. This is a contract between fishers and the French state, which develops the environmental involvement of fishers beyond regulation and previous practices. Participants receive indemnity for loss of turnover and incurred costs.

While the majority of diversification activities are undertaken for financial gain, examples were also found of Channel fishers undertaking activities on a *voluntary* basis. In both England and France, fishers participate in local maritime and fishing festivals to promote the fishing industry and/or individual species of fish or shellfish, and to raise money for charitable purposes. Typical activities include demonstrations of net making, answering questions from the public and allowing tourists onto moored fishing vessels.

In addition to festivals, English fishers also participate voluntarily in the 'Fishing for Litter South West' project; part of a wider international initiative which aims to reduce marine litter through the involvement of the fishing industry. At the time of writing, 86 vessels from 7 ports in the western Channel were participating in this project. In contrast to waste collection in France under the Contrat bleu framework, fishers do not receive financial compensation although administrative costs (including waste collection and disposal upon return to port) are covered by the project.

9.3.2 Opportunities and Likelihood of Diversification: Survey of Stakeholders[1]

English and French stakeholders were asked to identify the existence and extent of current opportunities for fishers to diversify into a range of activities, using a four-point scale (many; some; limited; none). A range of opportunities for diversification in the Channel fishery were identified, although predictably the extent of these opportunities was seen to vary by activity type. Combining responses for 'many opportunities' and 'some opportunities' reveals the greatest opportunities for English fishers in marketing initiatives (66 %); direct selling (55 %); allowing scientists onboard the fishing vessel (53 %); and participating in maritime/fishing festivals (53 %). Conversely, the majority of stakeholders identified limited or no opportunities to diversify at present into contract work for telecommunications companies (87 %); contract work for energy companies (79 %); or participating in exhibitions/trade shows (71 %).

[1] It is important to stress that the majority of stakeholders interviewed have an exterior view of the fishing industry, and their 'reality' may differ from that of individual fishers. Consequently, the expressed opinions of stakeholders may not necessarily reflect those of fishers in the Channel.

In contrast, French stakeholders identified the greatest opportunities in direct selling on docks (71 %); direct selling at markets (54 %); contract work for energy companies (46 %); and contract work for environmental organisations (22 %). Limited or no opportunities were identified for diversification into delivering products (65 %); contract work for environmental organisations (47 %); or collecting litter/waste at sea (45 %). The perceived opportunities in direct selling may be seen to reflect the fact that this activity is historically well-established in France, and is anticipated to develop further in response to consumer demand (Henichart and Lesueur 2011). Similarly, the identification of opportunities linked to the environment and renewable energy may be seen to reflect the higher profile of these sectors. However, additional discussion with fishers suggests that opportunities are greater in activities such as waste collection, where their participation is encouraged through existing policy initiatives.

After revealing their opinions on existing opportunities for different fisheries diversification activities, stakeholders were asked to consider the likelihood of fishers diversifying into different activity types in the future, using a five-point scale (very likely; likely; neither/nor; not very likely; not at all likely). These results, therefore, reflect both the existence of current opportunities, and opportunities that are envisaged as becoming more prevalent in the future. However, it is important to reiterate that the opinions of stakeholders may not necessary reflect those of fishers in the Channel. Combining responses for 'very likely' and 'likely' reveals the greatest likelihood of English fishers diversifying into taking scientists onboard the fishing vessel (71 %); contract work for environmental organisations (61 %); and marketing initiatives (55 %). Conversely, stakeholders identified diversification into contract work for telecommunications companies (61 %); participating in exhibitions/trade shows (53 %); and providing training (47 %) as unlikely in the future.

French stakeholders identified the greatest likelihood of fishers participating in maritime/fishing festivals (72 %); collecting litter/waste at sea (70 %); contract work for energy companies (63 %); and direct selling at markets (62 %). While the energy sector is an area where opportunities are clearly present, some respondents believed that conflicts between companies and fishers may prevent diversification into these activities. Stakeholders identified delivering products (60 %); leisure/tourism activities (40 %); and contract work for telecommunications companies (40 %) as the least likely options for diversification in the future.

When asked to identify the main motives for diversification (Table 9.1), the majority of English stakeholders selected 'business survival' (87 %) as the principal factor; followed by 'increased profit' (50 %); and 'maintaining the traditions of fishing' (40 %). The dominance of 'business survival' as a principal motive was evident among the responses of fishers' representatives, with fourteen of the fifteen respondents citing this factor. The main 'other' motives cited were 'maintaining fishing communities' (29 %) and 'improving working conditions' (11 %).

In France this question was adapted to the national context with different options, although the importance of business-oriented motives remains consistent. French stakeholders selected 'increased profit' (59 %); 'business survival' (41 %); and 'maintaining the traditions of fishing' (29 %) as the main motives, although for

Table 9.1 Key motives for fisheries diversification

Motive	England (%) (n = 38)	France (%) (n = 83)
Business survival	87	41
Increase profit	50	59
Maintain traditions of fishing	40	29
Spread/minimise risk	26	13
Promote the fishing industry	21	5
Reduce environmental pressure	11	13
Other	47	24

Table 9.2 Summary of normalised priority weights for key constraints

Constraint type	England (%) (n = 36)	France (%) (n = 70)
Economic	20	26
Social	14	15
Lack of information	11	12
Lack of opportunities	27	14
Administrative	27	34

French fishers the aim is not to increase profit but rather to maintain profit. The main 'other' motive cited was 'demand coming from society' (16 %).

Stakeholders were asked to assess the relative importance of five key constraints upon the decision of fishers to diversify into related activities: economic factors; social factors (defined as individual attitudes and preferences); lack of information; lack of opportunities; and administrative constraints. The number of valid responses was slightly less than that of the other survey questions due to participants declining participation for the AHP component.

As noted previously, one of the advantages of analysing data using the eigenvalue method is the generation of a consistency ratio (CR) score for each respondent. The standard practice is to accept responses with a CR score of 10 % or less (Himes 2007), but it was found that the majority of responses among both English and French stakeholders fell outside this threshold. A number of reasons may be proposed for inconsistency responses, including lack of understanding; difficulty in determining the relative importance of different constraints; or respondents simply being indifferent (Ishizaka et al. 2011). In determining the most appropriate strategy for dealing with these responses, a decision was taken to include them within the dataset on the basis that they represent a stated preference which is arguably valid on its own terms and thus should be retained (Whitmarsh and Wattage 2006, p. 120).

Comparison of normalised priority weights reveals both similarities and differences between English and French stakeholders (Table 9.2). Of the five constraint types, English stakeholders attributed the greatest importance to administrative constraints (27 %) and lack of opportunities (27 %), followed by economic factors (20 %). Social constraints (14 %) and lack of information (11 %) were considered relatively less important. French stakeholders also identified the importance of administrative constraints (34 %) and economic factors (26 %) as barriers to diversification. However, in contrast to English stakeholders, lack of opportunities (14 %) was not deemed to be particularly important. French stakeholders attributed relatively little importance to social constraints (15 %) and lack of information (12 %)—a finding that retains consistency with English stakeholders.

Analysing AHP results by stakeholder type reveals a number of notable observations, although caution is advised due to the small number of responses in some categories. The opinions of fishers' representatives in the English sample are largely supportive of the aggregated findings, although slightly greater importance is attributed to administrative constraints (36 %) than lack of opportunities (24 %). Fishers' representatives also attributed relatively little importance to social constraints (13 %) and lack of information (7 %). For the French dataset, stakeholders working within fisheries administration attributed less importance to administrative constraints (24 %) than other stakeholders—an observation that is not entirely unexpected. Conversely, the importance attributed to administrative constraints was greatest among stakeholders working within the tourism industry (37 %).

9.4 Discussion

The findings of this study have a number of implications for the development of fisheries diversification, and its potential contribution to sustainability of fishing as an occupation. Although financial data was not collected at an individual level during this phase of research, it is clear that fishers in the Channel are currently supplementing their fishing income with earnings from a range of complementary activities. However, these individuals constitute only a minority of those engaged in fishing. Further examination also reveals that diversification is practised mainly by inshore fishers, for which a number of explanations may be proposed. It is possible that inshore fishers are more inclined to consider a strategy of diversification, given that they typically diversify their fishing methods throughout the fishing season (Ulrich et al. 2002). As such, these individuals accept flexibility and adaptability as inherent aspects of their profession. Furthermore, the characteristics of inshore vessels may make them better suited to alternative activities; particularly vessels that return to port daily and have the available time ashore to develop activities such as direct selling. For these fishers, diversification is almost invariably practised in addition to fishing rather than as an alternative to it: fishers may diversify if they have available time, but they are unlikely to reduce their fishing time to diversify. In this context, diversification represents a complementary activity rather than an activity that diminishes fishing effort.

While some of the activities identified are well established, many have developed more recently in response to changing market conditions. Thus, the practice of diversification is seen to reflect localised supply and demand—which may contribute to explaining why fisheries diversification is currently a minority activity. For example, diversification into tourism/leisure activities is more prominent in regions that receive large numbers of tourists during the summer months, and more specifically in ports where demand is not being met by specialised providers. Similarly, while non-fishing contract work is often financially lucrative, opportunities to diversify into such activities are supply-led and invariably sporadic. For example, stakeholders reported that opportunities within the telecommunications industry have become less frequent following the installation of underwater fibre-optic cables for broadband internet services. In contrast it is possible that new opportunities will arise on both sides of the Channel with the development of offshore wind farm zones at Hastings, the Isle of Wight, Le Tréport, Fécamp, Courseulles-sur-Mer and Saint-Brieuc. However, stakeholders also suggested that the administrative process of tendering for such work may represent a potential constraint upon fishers applying.

The observation that opportunities for diversification may be highly variable both within, and between, regions is supported by the findings of the survey results. On the English coastline, the identification of opportunities in marketing and taking scientists onboard the fishing vessel reflects a growing public awareness of the marine environment. Consumer demand for traceable and sustainable seafood products has increased in recent years, and the fishing industry has responded with the introduction of labelling schemes. Similarly, the introduction of the first UK Marine and Coastal Access Act (2009) represents a shift towards increased protection and enhancement of the marine environment, which may create new opportunities for fisheries diversification. In France, the opportunities identified by stakeholders may similarly be seen to reflect growth areas—particularly with respect to tourism and leisure through the development of 'blue-tourism'. A notable distinction is also observed between the two survey populations with respect to direct selling which is prevalent on the French coastline but practised less in England. While opportunities for direct selling are influenced less by geographical constraints than other activity types, this distinction may be seen to reflect wider cultural differences in the buying and selling of fresh fish. For example, direct selling at the dockside is a common and traditional practice in some regions in France (e.g. Nord-Pas-de-Calais, Haute-Normandie), and supported in some cases through the provision of dedicated selling areas (e.g. Boulogne-sur-Mer, Le Havre). In contrast, English stakeholders identified a number of administrative obstacles that deterred fishers from developing this activity, including restricted public access and health and safety legislation. The existence of current opportunities and the potential development of growth areas will clearly influence the likelihood of future diversification in the Channel. Thus, based on the opinions of stakeholders, it is possible that English fishers will continue to diversify into science/environmental activities and marketing initiatives; and those in France will move into tourism, collecting litter/waste at sea, and to a lesser extent the offshore energy sector—while continuing with the practice of direct selling.

However, it is the influence of different constraints that will ultimately determine the likelihood of diversifying. The focus that AHP places upon broad constraint types represents a limitation with respect to understanding the reasoning behind individual responses. Nonetheless, the application of this technique provides some general insight into how stakeholders conceptualise key obstacles to diversification. The relatively low importance attributed to social constraints suggests that fishers are not perceived to be averse to diversifying into non-fishing activities. However, given that many of the activities identified maintain a strong link with the fishing industry, the potential loss of identity and socio-cultural rewards is arguably less relevant than diversifying into unrelated onshore employment. With the exception of lack of opportunities—which was clearly perceived to be a greater constraint among English fishers, the two survey populations share a number of characteristics. Significantly, both populations identified administrative and economic factors as major constraints upon diversifying, a finding that supports previous work in the Channel (e.g. Alban and Boncoeur 2004). Anecdotal evidence collated through the research process indicates that the main administrative constraints include legislation and bureaucracy. Examples include the certification required to take non-fishers onboard fishing vessels and the administrative process of tendering for contract work.

Importantly, the identification of economic constraints highlights the potential risk associated with diversification as a strategy. The nature and extent of existing diversification activities suggests that Channel fishers seek to minimise risk by diversifying into activities where financial outlay is minimised, and the application of existing knowledge and skills is maximised. The growth of market-based initiatives is one such example; many of the labelling initiatives have been developed by collective bodies that shoulder the financial and administrative costs, thereby reducing the risks of participation. In contrast, downstream integration into retail-based activities carries higher risks but may generate greater profits by eliminating the 'middleman', and ensuring reliability of supply. Furthermore, in addition to financial risks, diversification into downstream sectors requires knowledge, skills, and time that fishers may lack, or be unwilling to invest.

While this chapter has only outlined preliminary findings with respect to fisheries diversification in the Channel, it is possible to draw a number of parallels with previous work—within both fisheries and agriculture. Stakeholders attribute importance to multiple constraints upon diversification, which remains consistent with the findings of Alban and Boncoeur (2004) and Merrien et al. (2008). Of particular note is the recurrent theme of administrative or regulatory constraints that restrict fishers in developing new activities. These constraints may operate at a national, regional, or local level; and can be particularly problematic for new activities that lack defined 'regulatory status'. The finding that stakeholders attribute relatively less importance to social constraints is also noteworthy, by challenging the assertion that fishers are resistant to undertaking activities that fall outside their perceived role of catching fish. However, individual stakeholders who attributed greater importance to this constraint provided a number of justifications for their decision; including reluctance to adopt alternative working practices, and the perceived loss of identity that may result from diversifying. These opinions reflect those found in agriculture (e.g. Burton 2004) and support the notion that diversification is subject to both external and internal factors (Ilbery et al. 1997).

9.4.1 Implications for Sustainability

Applying the findings of this preliminary research study to understand the contribution of diversification to the sustainability of fisheries is inherently complex; due in part to the scale and scope of activities in the Channel, and the absence of data on their financial contribution to fishing businesses. While the intention is to address the latter with a succeeding phase of research with fishers, it is possible to draw a number of observations from research conducted to date. The finding that fisheries diversification is well-established and practised along both English and French coastlines indicates that its contribution to the sustainability of individual businesses is not insignificant; although this is clearly dependent upon the type of activity being practised. The findings of the inventory also suggest that the financial benefits of diversification are typically greater where fishers display individualistic entrepreneurial behaviour, e.g. contract working or direct selling. However, the observation that these types of activity are only practised by a minority of fishers is reflective of the constraints faced in their adoption.

The findings also suggest that the potential contribution of fisheries diversification to the sustainability of individual fishing businesses is also greater in the inshore sector. As noted previously, the flexibility and adaptability that characterises both fishers' behaviour and their vessels may explain the comparative prevalence of diversification within this sector. However, anecdotal evidence also suggests that the relative earnings of diversification compared with fishing activity is typically greater for smaller vessels, thus making it a more attractive prospect.

Despite these observations, the implications of diversification for the wider sustainability of the fishing industry are less clear. If we accept Symes' (2001) observation that the contribution of inshore fishing remains important to employment and wealth creation in the local economy, then strategies such as diversification that allow fishers to continue fishing can make a positive contribution to sustainability. It is important to also consider the impact of diversification beyond the level of the individual fisher. For example, activities such as boat trips, maritime festivals and product labelling can serve to raise the profile of fishing communities, encouraging tourism and investment that stimulates wider regional benefits (see Chap. 15 and 16). For such reasons, the wider socioeconomic significance of sustaining fishing activity at a localised level should not be overlooked.

Through further research it will be possible to improve our understanding of fisheries diversification, and support fishers adopting this strategy. A key objective of this forthcoming work will be to examine whether fishers hold alternative perceptions of diversification compared with non-fishing stakeholders. While the findings of this research indicate some differences in opinion between these two groups, these observations are limited by the small number of active fishers interviewed and an absence of data on respondent characteristics. However, it remains important to acknowledge that diversification represents just one of a range of strategies that fishers can adopt in response to difficulties. While fisheries diversification is anticipated to increase in response to the challenges facing the fishing industry, the established trend suggests that the majority of fishers will seek to develop solutions within the catching sector rather than diversifying. As such, the key to sustaining fishing and fisheries dependent regions may lie in the reform of fisheries policy, or ultimately in the development of alternative employment prospects as previously suggested (Whitmarsh 1998; Symes 2000).

9.5 Conclusion

The findings of this study demonstrate that both English and French fishers have diversified into activities that complement their principal role of fishing. A range of different activities is currently practised, although their prevalence is determined largely by geographical opportunities. Furthermore, the majority of these activities are practised by inshore fishers—which may result from the suitability of their vessels, the nature of fishing they undertake and their attitudes towards diversification. The scale and scope of diversification in the Channel suggests that the principal motive for fishers adopting this strategy is one of business survival, rather than increased profit and growth.

While opportunities to diversify into certain activities are anticipated to increase in the future, the ultimate feasibility of diversification will be determined by the existence and influence of a range of different constraints. Given this observation, it is unlikely that fisheries diversification will provide a wide-scale solution to the challenges faced by fishers—despite the provision of financial support through Axis 4 of the EFF (see Chap. 11). Instead, the sustainability of fishing and fisheries dependent regions may lie elsewhere; in the reform of fisheries policy and the development of alternative employment opportunities. Nonetheless, in some parts of the Channel diversification into fishing-related activities continues to provide fishers with an additional source of income that can offset the impacts of unprofitable periods of fishing. However, our present knowledge of the financial contribution of diversification to fishing businesses, communities and regions remains limited. It is envisaged that further research will seek to address this gap in knowledge, in addition to generating a greater understanding of fishers' attitudes towards diversification and the obstacles they face in its adoption.

Acknowledgements Research for this chapter was conducted as part of the project: CHARM 3 (CHannel integrated Approach to marine Resource Management), selected within the scope of the INTERREG IVA France (Channel)—England cross-border European cooperation programme, co-financed by the ERDF.

References

Alban, F., & Boncoeur, J. (2004). An assessment of the potential interest of fishermen to engage in boat-chartering in the context of a marine park: The case of the Iroise Sea, Western Brittany, France. In J. Boissevain & T. Selwyn (Eds.), *Contesting the foreshore: Tourism, society, and politics on the coast* (pp. 185–203). Amsterdam: Amsterdam University Press.

Andreosso, B., & Jacobson, D. (2005). *Industrial economics and organization: A European perspective* (2nd ed.). Maidenhead: McGraw-Hill.

Boncoeur, J., Coglan, L., Le Gallic, B., & Pascoe, S. (2000). On the (ir)relevance of rates of return measures of economic performance to small boats. *Fisheries Research, 49*, 105–115.

Brookfield, K., Gray, T., & Hatchard, J. (2005). The concept of fisheries-dependent communities: A comparative analysis of four UK case studies: Shetland, Peterhead, North Shields and Lowestoft. *Fisheries Research, 72*(1), 55–69.

Burton, R. (2004). Seeing through the 'good farmer's' eyes: Towards developing an understanding of the social symbolic value of 'productivist' behaviour. *Sociologica Ruralis, 44*(2), 195–215.

Charles, E., Boude, J., Murray, A., & Paquotte, P. (2003). Coastal fishing: Resource's enhancement and preservation. *Ocean & Coastal Management, 46*, 421–437.

Council Regulation (EC) No. 1198/2006 of 27 July 2006 on the European Fisheries Fund [2003] OJ L223/1.

Halliday, J. (1989). Attitudes towards farm diversification: Results from a survey of Devon farms. *Journal of Agricultural Economics, 40*(1), 93–100.

Henichart, L.-M., & Lesueur, M. (2011).*Synthesis: Direct-to-Consumer sales on the English channel (French side)-Strategies and development factors in the fishing industry. Interreg program (English Channel) CHARM 3*. Pôle halieutique: AGROCAMPUS OUEST.

Henichart, L.-M., Lesueur, M., Fontenelle, G., Boude, J.-P., & Ropars-Collet, C. (2010). Diversification of fisheries activities and construction of sustainability. IIFET Conference 13-16 July 2010, Montpellier. http://halieutique.agrocampus-ouest.fr/pdf/323.pdf. Accessed 8 Mar 2012.

Herath, G. (2004). Incorporating community objectives in improved wetland management: The use of the analytic hierarchy process. *Journal of Environmental Management, 70*, 263–273.

Himes, A. H. (2007). Performance indicator importance in MPA management using a multi-criteria approach. *Coastal Management, 35*, 601–618.

Ilbery, B. W. (1991). Farm diversification as an adjustment strategy on the urban fringe of the West Midlands. *Journal of Rural Studies, 7*(3), 207–218.

Ilbery, B., Healey, M., & Higginbottom, J. (1997). On and off-farm business diversification by farm households in England. In B. Ilbery, W. Chiotti, & T. Rickard (Eds), *Agricultural restructuring and sustainability: A geographical perspective* (pp. 135–151). Wallingford: CAB International.

Ishizaka, A., Balkenborg, D., & Kaplan, T. (2011). Does AHP help us to make a choice? An experimental evaluation. *Journal of the Operational Research Society, 62*, 1801–1812.

MMO (Marine Management Organisation). (2012). UK Vessel Lists 2012. http://www.marinemanagement.org.uk/fisheries/statistics/vessel.htm. Accessed 1 Feb 2012.

Merrien, V., Lesueur, M., Boude, J., & Folliard, G. (2008). *Diversification des activités de pêche en Bretagne: Acceptabilités et conditions de développement*. Rennes: AGROCAMPUS OUEST.

OECD. (2001). Multifunctionality: Towards an Analytical Framework. http://www.oecd.org/agriculture/agriculturalpoliciesandsupport/1894469.pdf. Accessed 8 Mar 2012.

Pettersen, L. T. (2000). Household adaptations and gender differences in inshore fishing communities in Northern Norway. In D. Symes (Ed.), *Fisheries dependent regions* (pp. 82–91). London: Fishing News Books.

Phélippé, D., Daurès, F., & Le Gallic, B. (2011). Panorama des caractéristiques d'exploitation des ressources halieutiques par la flotte de pêche française en manche. Projet Interreg IVa-CHARM III. http://www.umr-amure.fr/electro_rapports_amure/R_19_2011.pdf. Accessed 8 Mar 2012.

Roussel, F., Sérazin, T., Hénichart, L.-M., Ropars, C., & Lesueur, M. (2011). *Diversification des activités de pêche en Manche: Etat des lieux et conditions de développement*. Rennes: AGRO-CAMPUS OUEST.

Saaty, T. L. (1977). A scaling method for priorities in hierarchical structures. *Journal of Mathematical Psychology*, *15*(3), 234–281.

Salmi, P. (2005). Rural pluriactivity as a coping strategy in small-scale fisheries. *Sociologica Ruralis*, *45*(1), 22–36.

Symes, D. (2000). Fisheries Dependent Regions: Scoping the Problem. In D. Symes (Ed.), *Fisheries dependent regions* (pp. 3–14). London: Fishing News Books.

Symes, D. (2001). Inshore fisheries in Europe at the turn of the century. In D. Symes & J. Phillipson (Eds.), *Inshore fisheries management* (pp. 3–23). Dordrecht: Kluwer Academic Publishers.

Ulrich, C., Le Gallic, B., Dunn, M. R., & Gascuel, D. (2002). A multi-species, multi-fleet bioeconomic simulation model for the English Channel artisanal fisheries. *Fisheries Research*, *58*, 379–401.

Walford, N. (2001). Patterns of development in tourist accommodation enterprises on farms in England and Wales. *Applied Geography*, *21*, 331–345.

Whitmarsh, D. (1998). The fisheries treadmill. *Land Economics*, *74*(3), 422–427.

Whitmarsh, D., & Wattage, P. (2006). Public attitudes towards the environmental impact of salmon aquaculture in Scotland. *European Environment*, *16*, 108–121.

Wilson, G. A. (2007). *Multifunctional agriculture: A transition theory perspective*. Wallingford: CABI Publishing.

Chapter 10
Area-based Local Development—A New Opportunity for European Fisheries Areas

Urszula Budzich-Tabor

Keywords Local development in fisheries area • Axis 4 EFF • Diversifying fisheries activity • Bottom-up approach • Fisheries Local Action Group (FLAG)

10.1 Area-based Development in the European Fisheries Fund

The European Fisheries Fund (EFF) provides for a wide range of actions to help the European fisheries sector meet challenges such as depleted fish stocks, loss of markets, increasing dependency on imports, as well as rising costs of fuel and other inputs. Representatives of the fisheries sector have benefited from EFF support, for instance reducing fishing capacity, modernising boats, investing in aquaculture or undertaking common activities. However, the impact of the changes goes beyond the fisheries sector itself, and affects also families of fishermen and the wider community. This is particularly true in areas that have been traditionally dependent on fisheries activity as a main source of employment and income.

European policies are, therefore, also targeting those communities which can no longer rely primarily on the fishing sector. These communities are encouraged to find their own ways of development, i.e. to apply elements of the so-called 'area-based' approach. This approach, also known as the 'Leader approach'—the LEADER Community Initiative implemented since the early 1990s[1] -, has already proved successful in dealing with problems of many European rural areas, such as

[1] LEADER stands for (in French) Links between Actions for the Development of the Rural Economy, and involves a set of characteristics, including a bottom-up, 'area-based' (or territorial) perspective, integrated (as opposed to sectoral) approach, a strong role for the local partnership between public, private and voluntary sectors and support for mutual learning and cooperation. See for instance ec.europa.eu/agriculture/rur/leader2.

U. Budzich-Tabor (✉)
FARNET Support Unit, 1040, Brussels, 38 rue de la Loi, Belgium
e-mail: urszula.budzich-szukala@farnet.eu

J. Urquhart et al. (eds.), *Social Issues in Sustainable Fisheries Management*, MARE Publication Series 9, DOI 10.1007/978-94-007-7911-2_10,

the diminishing importance of agriculture, lack of alternative economic activities, ageing and depopulation of rural communities[2].

The application of the 'area-based' approach as a solution to structural problems focused in specific areas (such as rural or fisheries areas) is consistent with the findings of many recent studies, notably by the OECD within the framework of its 'new rural paradigm' (OECD 2006). Based on the analysis of the impact of traditional, sectoral policies on rural areas, OECD experts argue for the implementation of place-based rather than sectoral approaches, focused on such issues as transport and ICT infrastructure, public service provision, valorisation of rural amenities (natural and cultural) and rural enterprise promotion.

Such an area-based or local development approach requires a combination of three key elements: area, partnership and strategy (Soto 2004). Thus, support is provided not to a single sector or group of beneficiaries, but rather to an area as a whole—provided that it is coherent and that its key stakeholders have formed a partnership (involving the public, private and NGO sector), working together to design a long term development plan, i.e. a local development strategy. The most characteristic element, which distinguishes this approach from other activities carried out at the local level by local authorities, is the local partnership, which drives forward the implementation of the area's strategy. Such a local partnership can be seen as a form of self-organisation for the management of common assets in the sense of Ostrom's seminal work (Ostrom 1990). The positive impact of local partnerships in rural development has also been demonstrated in a wide-ranging study of Leader-type partnerships in Europe (see final report of the PRIDE project in Moseley 2003).

On the basis of this experience, starting with the programming period 2007–2013, a new instrument of support has been put in place in the European Fisheries Fund, the so-called Axis 4. This Axis focuses on those local communities which are affected by changes in the fisheries sector, and helps them achieve more sustainable growth. There is no EU-wide definition of "fisheries-dependent areas" and each Member State developed its own approach. In Southern Europe the focus was mainly on coastal areas (some countries, as France, declared the whole coast eligible, others—as Portugal—pre-selected certain areas[3]), while in many Northern countries inland areas with significant concentration of fisheries or aquaculture (usually measured by percentage of active population employed in the sector) could also apply for Axis 4[4].

[2] This is not to say that all the applications of Leader are equally successful in addressing rural problems, see for instance Macken-Walsh (2012) for a critical analysis of the Leader method applied in Western Ireland. As discussed in Sect. 10.4 below, applications of the Leader approach can vary greatly between countries (and sometimes regions), as Member States try to adapt it to the national administrative contexts. In the period 2014–2020 the European Commission is preparing a series of guidance documents to ensure that local communities can benefit from the full application of the Leader principles.

[3] For instance in the case of Portugal, the preselected areas had to have population density below 120 persons per square km, declining catches and employment in the fisheries sector above 3 %

[4] In all cases the interested areas had to express interest in participatin in the programme and submit an application in a competitive call.

Using the area-based methodology, Axis 4 seeks to enable communities to develop and implement local strategies that take into account the needs of the fisheries sector as well as the specific challenges and opportunities of their area. Once the strategy is approved by the programme authorities, local actors can submit project proposals which are evaluated and selected for funding by the local partnership. Through Axis 4 support, local partnerships, including strong representation from the fishing industry, are encouraged to find their own solutions to the economic, social and environmental problems faced by fisheries communities.

Local strategies developed by these partnerships seek to strengthen and diversify local economies and to provide alternative or additional sources of income to fishermen and their families. Such additional income can come from activities based around fishing aiming, for instance, to increase the value added of fisheries products, as well as from other sectors, as diverse as tourism, social services, arts and culture, renewable energies, information technologies or environmental stewardship. The key role in this approach is played by the Fisheries Local Action Group (FLAG), which ensures community involvement in developing the local strategy, provides local actors with skills and advice and ensures complementarity of the projects selected for funding.

Out of the 27 EU Member States, 21 have decided to use this approach in their fisheries communities. In February 2012 there were about 240 Fisheries Local Action Groups (FLAGs) in operation, with an additional 59 planned to be set up by the end of the year. The Member States allocated over € 800 million of total public funding to this approach. Of this, over € 580 million comes from EFF and the rest from national co-funding (see Table 10.1).

Several important differences can be noted between Member States. The first is the importance of Axis 4 in the EFF, as measured by the proportion of EFF funding allocated to Axis 4 (ranging from 2.6 % in France to 32.5 % in Romania), the number of FLAGs (from 1 FLAG in small countries such as Belgium, Cyprus and Slovenia, through 20–25 FLAGs in Latvia and Spain, to 36 FLAGs in Italy and 48 in Poland), as well as average budgets per FLAG, varying from under € 1 million in Lithuania to over € 7 million in Romania. There are examples where FLAGs have budgets of less than € 100,000 (Latvia, Germany), which raises the question of their viability. On the other hand, very large FLAGs with budgets of € 15 or even 18 million in Poland and Romania may encounter problems with absorption and ability to deal with the administrative burden. It is interesting to note that countries with a strong fisheries sector tend to be less interested in this approach than countries where fisheries are weak or in decline.

Secondly, there is a difference in the speed of implementation, with some Member States taking a long time to select their FLAGs and bring them into operation; this is particularly true of Bulgaria, Romania and Ireland, as well as regionalised countries such as Italy and UK. At the time of writing (March 2012), out of the 240 FLAGs already in place, only about 150 were fully operational in that they are selecting and financing local projects. The remaining groups, even if formally approved, are still at a very early stage of development. This raises the question whether these late starters will be capable of delivering their strategy objectives by the end of the programme period.

Table 10.1 Overview of Axis 4 in EU member states. (Source: Own compilation on the basis of data received from the European Commission and Member States, spring 2012)

Member State	Axis 4 budget—EFF contribution (€)	Axis 4 budget as percentage of total EFF	No. of FLAGs planned	Average budget per FLAG—total public contribution (€)	No. of projects approved by February 2012	national FLAG network
Belgium	2,900,000	7.2%	1	5,800,000	3	no
Bulgaria	12,001,456	15.0%	6	4,000,000	0	no
Cyprus	1,000,000	5.0%	1	2,000,000	0	no
Denmark	12,461,279	9.3%	18	2,300,000	302	yes
Estonia	19,281,513	22.8%	8	3,200,000	245	yes
Finland	3,606,000	9.1%	8	1,000,000	207	yes
France	5,699,644	2.6%	11	1,000,000	143	yes
Germany	18,553,519	11.9%	23	1,000,000	50	yes
Greece	33,300,000	16.0%	13	5,000,000	0	no
Ireland	1,500,603	3.5%	6	1,200,000	0	no
Italy	43,420,936	10.9%	36	1,800,000	0	no
Latvia	17,172,786	13.7%	24	1,000,000	228	yes
Lithuania	6,693,770	12.2%	10	900,000	1	yes
Netherlands	4,987,125	10.3%	6	1,700,000	24	no
Poland	234,909,624	32.0%	48	6,500,000	125	yes
Portugal	17,403,406	7.0%	7	3,400,000	45	yes
Romania	75,000,000	32.5%	14	7,100,000	0	no
Slovenia	2,164,029	10.0%	1	3,000,000	0	no
Spain	49,336,048	4.4%	29	2,600,000	188	yes
Sweden	8,199,720	15.0%	14	1,500,000	64	no
UK	11,598,450	8.4%	15	1,000,000	0	no
Total	581,189,508	13.5%	299	3,200,000	1625	

Thirdly, information gathered by the FARNET Support Unit from individual FLAGs (see FARNET 2012c) shows also that there are significant differences with respect to the balance between the focus on the fisheries sector and the wider territorial development. Usually FLAGs with lower budgets tend to focus more on fisheries-related activities, while those with more funding look for wider diversification opportunities, but there are exceptions: for example, in Latvia many FLAGs with relatively small budgets tend to carry out a wide range of non-fisheries activities. Some FLAGs use their small budgets for preparatory activities to lever additional funds from other sources, thus facilitating access to investment funding for the fisheries sector (France, Finland).

10.2 The Expectations of an Area-based Approach in Fisheries

The application of a 'territorial' instrument, or the area-based approach, in a 'sectoral' policy dealing with a single sector (i.e. fisheries) is a relatively innovative solution which is expected to have a strong impact on the fisheries areas in which it is applied. The overall objective of Axis 4, as formulated in the EFF Regulation, is "to encourage sustainable development and the improvement of the quality of life in areas with activities in the fisheries sector" (Council Regulation 1198/2006, Article 4), in particular through adding value to fisheries products, creating jobs to diversify fisheries areas, and improving the quality of coastal environment (*ibid.*, Article 43). However, due to the delays mentioned above, it is still too early to evaluate to what extent these objectives have been reached. It is thus more appropriate to consider the rationale for applying the Axis 4 approach in terms of the expectations of public authorities, FLAGs and other stakeholders, trying to see how local development is expected to contribute to the desired changes in fisheries areas, and in what way these changes differ from what might have been achieved using traditional approaches.

The following are among the more common expectations expressed by stakeholders in their contacts with the FARNET Support Unit:[5]

1. *New solutions to better address local needs.* Fisheries areas vary greatly, even within a single Member State, and so do the needs and opportunities facing the fisheries communities. To design and implement a meaningful set of actions from the national or even regional level would be extremely difficult. Local strategies that take into account the specific characteristics of a given area and its fisheries as well as local knowledge are likely to come up with innovative, long-term solutions for strengthening the economy and improving quality of life.
2. *Achieving synergies.* The participatory method of preparing the strategy and joint decision-making by the local partners means that there is greater chance to achieve linkages and synergies between projects. In traditional, centralised funding systems there is a risk of supporting projects without ensuring that they are coherent with the overall development of the area (e.g. too many hotels with not enough restaurants, bicycle routes not connected to each other), or can even be contradictory (e.g. industrial development in a touristically valuable part of the area). This risk is reduced in the local development approach, where the involvement of partners from different sectors in the preparation and evaluation of projects is required. The partnership approach can also help to maintain coherence between projects financed from other sources, and even to lever in additional funding for projects of strategic importance.
3. *Strengthening links between the fisheries sector and the local community.* In many areas, fisheries—especially the catching sector—have relatively few links even with other members of the sector or with the upstream and downstream sectors, not to mention other local actors. Axis 4 offers the possibility for joint activities of fishermen together with other local actors and encourages linkages, for example, between the catching sector, processing and distribution, including the small-scale players.

[5] Source: author's analysis on the basis of documentation of FARNET activities, including country fact-sheets describing priorities for Axis 4 of individual Member States (available at FARNET Tools b 2012b), as well as internal material such as FLAG focus groups and Managing Authorities meetings.

This integration does not happen of its own accord, simply as a result of forming the local partnership and applying for Axis 4 funding; considerable effort is required to ensure ownership of the strategy by fishermen as well as other actors. Even in countries where a specific proportion of fisheries representatives in the FLAG decision-making body is required, it is not always easy to ensure a genuine involvement of the sector and to build trust between fisheries and non-fisheries partners. With the introduction of Axis 4, however, this process has been started in many places.

4. *Local ownership*. The bottom-up approach of implementing Axis 4 with a locally developed strategy combined with local decisions on financing projects, as well as the strong focus on local identity, mean that a wide range of local actors start to identify with the programme, support it and contribute to it in various ways, for example, helping identify local assets, proposing new activities and helping to reach out to those not yet involved. The amount of unpaid, voluntary work by local leaders in the process of strategy development, identification and encouragement to beneficiaries, and even in monitoring and evaluation of results is often greater than the paid work carried out by the FLAG employees and can be a significant source of value added in the local development approach. Such local ownership can also help lever in additional resources.

10.3 Some Examples of Area-based Approach in Fisheries Areas

According to recent estimates, by February 2012 approximately 1625 projects have been selected for implementation within Axis 4 (Table 10.1). While the full list is not available, the FARNET Support Unit monitors examples of good practice (FARNET 2012d) from those countries where Axis 4 has been fully operational for some time. It is not easy to develop a common classification of these projects, since Member States tend to use different typologies of eligible activities. However, a possible classification involves the following three main thematic categories:

- *adding value to fisheries products* covering activities such as direct sales and short supply chains, new fisheries products, support to local fishery and processing businesses, promotion and awareness raising among of fish consumers;
- *diversification of fisheries areas* including by-products from fishing activities, tourism, pescatourism, angling, recreation fishing, gastronomy and culinary heritage, new activities, products and skills and promotion of the area; and
- *environment, society and culture* which covers a wide range of activities linked to the environment and natural resource conservation, as well as promoting the involvement of women and youth, culture and heritage. This classification is not clear-cut with many activities falling into more than one category—partly because of their integrated or multi-sectoral character.

Fig. 10.1 Edible seaweed products. Photo courtesy of Danish small islands FLAG and Bornholm FLAG

As an indication of the level of interest of fisheries areas in these themes, an analysis of proposals for cooperation projects submitted by FLAGs during transnational events organised by FARNET in 2010 and 2011 has shown that out of a total of 72 proposals, 17 can be classified as adding value to fisheries products, 25 as diversification of fisheries areas and 30 as environment, society and culture (FARNET 2012a).

10.3.1 New Solutions to Better Address Local Needs

Many projects financed from Axis 4 are based on assets and opportunities that were identified by the local actors themselves; solutions proposed to address these opportunities can be very specific and often innovative. It is also possible to implement complex projects involving several stages, or combining different types of funding, which could be difficult under traditional methods of support.

One example is the production of edible seaweed (FARNET 2012d), a cooperative project between stakeholders of two Danish FLAGs, one on the island of Bornholm and the other covering several smaller islands (Fig. 10.1). These remote areas are threatened by depopulation and loss of business opportunities. The growing popularity of sushi and the development of the New Nordic Cuisine opens up new possibilities for local entrepreneurs including fishermen. Seaweed production in Denmark can be of better quality than imported dried seaweeds, and it could help to broaden the range of products offered locally and lead to increased earnings for local food producers and restaurants.

The project combines seaweed cultivation with harvesting wild seaweed to develop a range of innovative products, and involves 16 producers coming from a variety of backgrounds, such as mussel growers, fish farmers, gardeners, as well as wine, cider, jam and ice-cream producers. It is expected that the project will result in new products, the creation of at least four new jobs, as well as raising the profile of the islands and improving their attractiveness to tourists. Thanks to the

Fig. 10.2 Fisherman involved in the 'Fish from the Boat' scheme. Photo courtesy of FLAG Ostseeküste

well-planned, strategic approach of the local actors, the project combines various sources of funding, including the Danish Ministry of Food, Agriculture and Fisheries support to pre-project market analysis, Axis 4 funding (€ 84,500) as well as other national and regional funders.

Another example of a project adapted to the specific local condition is Fish from the Boat (FARNET 2012d), a project supported by the Ostseeküste FLAG from Germany, aiming to address the issue of price fluctuation of fish species such as cod or flounder, generally sold through wholesalers (Fig. 10.2). The project enables fishermen to sell part of their catch directly to the final customer, successfully exploiting the increasing demand for fresh local fish and progress in information technology. Fishermen can send details of their catch from their boat by SMS to an interactive website through which customers can see where, when and what fish are available. Coupled with an information brochure and boards at landing sites, the project is increasing the amount and value of the direct sale of freshly caught fish in the region and fostering links between producers and consumers. The website attracts about 21,000 visitors a month, and it is estimated that the economic stability and survival capacity of the 11 fishing companies involved has been enhanced. The project cost € 20,000, of which 48 % was provided by the FLAG.

10.3.2 Achieving Synergies

The local development methodology of involving different partners enables synergies between activities that would be difficult to achieve otherwise. Many Axis 4 projects involve cooperation between partners that are not used to working with each other. This collaborative method also makes it possible to combine different sources of funding, where relatively small Axis 4 support enables the local actors to lever in higher amounts from other funders.

One example is the French FLAG Pays d'Auray in Brittany, which has supported a project aimed at tackling water pollution (FARNET 2012d)—a significant problem in the area, partly arising from lack of coordination between the activities of different local stakeholders and conflicts between farmers and shellfish producers. The FLAG has supported (with a small grant of € 7,200) the creation of local groups that bring together fisheries professionals, local authorities and other stakeholders in order to coordinate attempts to identify and reduce the sources of bacteriological pollution that impact on local shellfish producing areas. The ultimate goal is to help local authorities develop action plans to deal with the identified pollution, assigning specific responsibilities by sector. The project started in April 2011 and has already contributed to a better understanding among the local stakeholders and a growing consciousness as to the role they can play in identifying and minimising pollution.

In an even more striking example, activities funded by Axis 4 to develop a fish handling centre (FARNET 2012d) in Kuusamo through a grant of € 23,250 from the Kainuu-Koilismaa FLAG (Finland) played a crucial role in levering in a larger investment from Axis 2 of the EFF and local municipal funds, giving a total of € 2,700,000. The fish-handling centre financed in this way combines the needs of both fishermen and the processing industry under the same roof. Based on the positive results of the feasibility study funded by Axis 4, cooperation among local stakeholders (fishermen, fish farmers and fish processors) was established, leading eventually to a joint venture for the larger project. The logistics centre in Kuusamo provides new opportunities for local actors to develop their business operations; it has already contributed to the creation of six new jobs in fisheries companies and two new jobs in processing.

10.3.3 Strengthening Links Between the Fisheries Sector and the Local Community

Many activities financed by Axis 4 aim at a broader diversification of the fisheries area, that is at activities that go beyond the fisheries sector. However, the participatory approach, with a significant proportion of fisheries representatives taking part in decision-making, makes it possible to develop activities that bring benefit both to the fishermen or their families and to the wider community. In many instances they also contribute to strengthening the vertical links within the fish distribution chains.

For instance, in the project 'Tourism training for fishermen' (FARNET 2012d), the Finnish FLAG of Northern and Eastern Lapland has supported an integrated package of training courses tailormade for fishermen who want to diversify into tourism. The 20 professional fishermen remaining in the town of Sodankylä were looking for ways to complement their income by diversifying into tourism, but they lacked the skills and licences to make this a reality. At their initiative, the local

Fig. 10.3 Training of fishermen in tourist skills. Photo courtesy of Northern and Eastern Lapland FLAG

FLAG organised a group of specialists to develop and deliver a tailormade training package, including courses and exams in safety issues as well as study visits to tourism companies, allowing fishermen to establish contacts with tourism operators and learn from experts in the field. Training in product development, pricing and customer service (Fig. 10.3) was complemented by personalised study and guidance. A total of seven marketable products have been developed, and the nearest big tourist resort has started to market winter and summer fishing trips with the local fishermen. The project cost € 78,000.

The French association 'Navicule Bleue', with the support of the Marennes Oléron FLAG, has developed a number of tools, including training and advice, in order to enable fishermen injured at sea to be reintegrated into the labour market. The key elements of the project were ensuring the involvement of beneficiaries as agents of change and adapting support to the specific abilities of the workers. A social enterprise, in the form of an adapted workshop (Atelier des Gens de Mer, see FARNET 2012d), was created, with six full-time-equivalent jobs (Fig. 10.4). The project cost was € 89,000, used, *inter alia*, for the purchase of equipment adapted to the needs of the disabled workers. The workshop provides skilled labour services (including net mending, net making, catch sorting, vessel refurbishment etc.) to customers in the harbour, ensuring an appropriate quality of service while maintaining communication and awareness-raising among customers. A similar approach is being applied in another French FLAG (Arcachon) with the aim of creating 60 full-time jobs.

In the Netherlands, a renovated old building in the heart of the fishing village of Urk (FLAG Flevoland) will become a multifunctional space centred on a famous

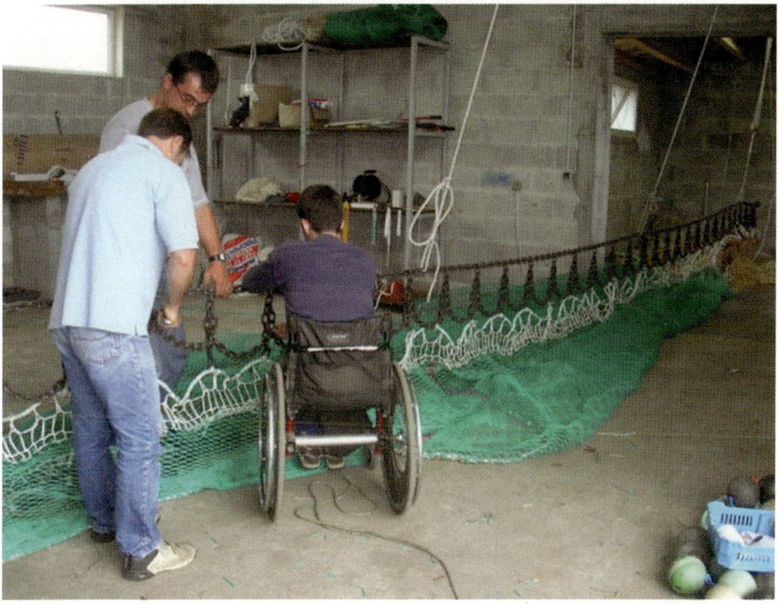

Fig. 10.4 Adapted workplace for fishermen injured at sea. Photo courtesy of FLAG Marennes Oléron

local fish restaurant, 'De Boet' (FARNET 2012d). The restaurant will be integrated into a wider tourism package, where customers will be allowed to bring back fish bought from the local fish auction, learn how to cook it and then enjoy it in the special atmosphere of the building, thereby reinforcing the links between the tourism and fisheries sectors. The project is expected to regenerate Urk's port area, bring new revenue and create 4–8 new jobs. The total investment is € 1,700,000, of which 10% was provided by Axis 4.

10.3.4 Local Ownership

While the mode of operation of the FLAGs and types of project financed vary greatly between countries, the commitment and enthusiasm of a variety of local actors in planning and implementing local development projects is probably the most striking characteristic that most FLAGs have in common.

For instance in the Stockholm Archipelago in Sweden, the activities of the FLAG are seen first of all as a forum of dialogue (FARNET 2011), where all stakeholders in fishing—professionals and amateurs, guides, landowners, environmental organisations and local authorities sit around the same table and talk to each other. They meet each month and discuss joint action, for instance, how best to restore fish stocks to the area. In this way even relatively small-scale support from the FLAG can be used to maximise the effect—for instance a contribution of € 9,000 to the

Fig. 10.4 Supporting local community initiatives. Photo courtesy of FLAG Obra-Warta

participation of 25 fishermen in a fish promotion event, or restoration of wetland areas at € 5,000 per hectare.

In Poland, the inland FLAG Obra-Warta in the West of the country has started innovative activities to ensure the involvement of local communities in Axis 4 (LGR Obra-Warta 2012): it has invited village groups to submit proposals for animation and training activities, for which it provides funding of up to approximately € 5,000 (Fig. 10.4). In spite of the small level of FLAG support, the projects have a strong multiplier effect, due to the voluntary work of local actors and to additional resources levered in from the special village fund. In this way the FLAG is hoping to get wider community involvement, raise the profile of the fisheries sector and encourage innovative thinking about Axis 4, hoping eventually for better quality projects in line with local needs and ideas.

10.4 Conditions for Success

The projects listed above are just a few examples of activities in over 240 fisheries areas across the EU, thanks to the application of the area-based approach. Although most of them are still in the early stages, they indicate the potential that this approach has to contribute to the sustainable development of EU fisheries areas. For this potential to be realised, a certain number of conditions must be met. As pointed out by Martinos (2012), the European model of local development requires a balance between local initiatives and external top-down support; therefore appropriate conditions must be in place both at the programme level (national or regional) and at the local (FLAG) level.

At the national level, the primary responsibility of the programme authorities is to create a delivery framework adapted to the specificity of the area-based approach. Attempts to apply the same rules and procedures to local development as for standard, top-down funding schemes may restrict the autonomy of the local partnership and thus reduce the benefits of the approach. A recent assessment of the LEADER approach carried out by the European Court of Auditors (ECA 2010) indicates that in many cases regulations imposed on the local partnerships by national or regional programme authorities prevent the value added of the area-based approach from being realised. Thus a distinction should be drawn between the LEADER (or area-based) concept—setting out the principles of the approach -and LEADER practice, which can in some cases be considerably different (Budzich-Szukala 2012).

At the local level, the most important role in ensuring that the right conditions exist to realise the full potential of the area-based approach is played by the FLAG. This organisation must ensure that the projects selected for funding contribute to the objectives of the strategy, through providing training and business support to the entities that carry out the projects, including local SMEs and NGOs. The FLAG's role towards the local community is even more important: activities aimed at involving local actors in developing the strategy, and later in decision-making on which projects are to be supported, can help stimulate their motivation and creativity. The FLAG is thus responsible for a wide range of activities, from animating local project promoters, enlisting support of key stakeholders, to mediating conflicts and mobilising the wider community.

Another important task of the FLAG is ensuring cooperation between the various local actors, for example, between local producers if a local brand is to be created; or between providers of accommodation and other tourist attractions to ensure a coherent package. The cooperation of the local authorities which can implement projects aiming to develop local tourist infrastructure or promotional activity is also crucial.

Thus, the application of the area-based approach in fisheries areas requires a considerable effort of capacity building at all levels to ensure that the potential benefits of Axis 4 are realised. In particular, capacity building is needed for fishermen, local authorities and other community actors, to understand the benefits of Axis 4 and to get involved in its implementation; for partners and employees of the FLAGs, to learn how they can drive the strategy forward and make the best use of Axis 4 resources; and for the programme authorities, to put in place rules and conditions that are adapted to the partnership approach of Axis 4 and will allow the local actors the necessary flexibility to innovate.

Recognising the key role of the local partnership in the area-based approach, much attention is paid to capacity building for the FLAGs. In some Member States this is assured by the national networks of FLAGs and by the programme managing authorities. Some form of national FLAG network exists in 10 countries, and they ensure mutual learning and encourage joint activities. This is complemented by activities of the FARNET Support Unit, which also provides various forms of support and advice to managing authorities and national networks themselves, thus helping to improve the delivery framework of Axis 4.

The FARNET Support Unit was set up in 2009 to facilitate the implementation of Axis 4 with the following priority objectives: to build capacity in integrated territorial development by providing guidance and support to FLAGs as well as managing authorities; to identify, test and transfer successful responses (promising and good practices) to the challenges facing fisheries areas; and to create a platform and a voice for fisheries areas by helping to connect local learning and innovation to the broader European and national policy debates. The activities of the FARNET Support Unit, such as organisation of transnational events, promoting exchange between FLAGs, managing authorities and national networks, publication of guides, magazines and newsletters, web-based discussion groups focus on the main themes related to sustainable development of fisheries areas, involving adding value to fisheries products, the diversification of fisheries areas, and the enhancement of environment, society and culture. The Unit also aims to stimulate exchange between the FLAGs and other stakeholders across Europe. Recent activities of the FARNET team include an analysis of the delivery systems of the area-based approach and their impact on the fisheries areas (FARNET 2012).

In preparation for the next period of EU funding (2014–2020) the European Commission has published its proposals for the new European Maritime and Fisheries Fund. One of the main pillars of this Fund is 'sustainable development of fisheries areas', which envisages a strengthening of the area-based approach following on from the positive early experience of Axis 4 EFF. This approach, called in recent Commission proposals 'community-led local development' (CLLD, see for instance (EC 2012)) will also be applied more widely in other EU funds, and—depending on Member State decisions—some local partnerships will be able to combine different sources of funding into a single strategy. In this way, the local development approach has a chance to be even more widely applied in areas dependent on fisheries. Further studies will be needed to analyse the potential benefits and risks associated with this new opportunity for the fisheries communities and to provide a further insight into the impact of area-based approaches on fisheries areas.

Acknowledgements This paper is based on material and information collected by the author in the course of her work for the FARNET Support Unit (FARNET is the European Fisheries Areas Network initiated by the European Commission to support the implementation of Axis 4 of the European Fisheries Fund). It does not in any way represent the views of the European Commission or the FARNET Support Unit. The author would like to thank Paul Soto and Gilles van de Walle from the FARNET Support Unit for sharing their comments and ideas in the process of preparing this chapter.

References

Budzich-Szukala, U. (2012). *Lessons from local development in practice*. (Martinos 2012). http://ldnet.eu/Lessons+from+LD+in+Practice & structure=The+Local+Development+E-book. Accessed 18 Dec 2012.

EFF Regulation. (2006). Council Regulation (EC) No. 1198/2006 of 27 July 2006 on the European Fisheries Fund. Official Journal of the European Union, L 232.

EC. (2012). *EU Cohesion Policy 2014–2020: legislative proposals*, European Commission—Regional Policy. http://ec.europa.eu/regional_policy/what/future/proposals_2014_2020_en.cfm#1. Accessed 18 Dec 2012.

ECA. (2010). *Implementation of the leader approach for rural development*. Special report of the European Court of Auditors no. 5/2010. http://eca.europa.eu/portal/pls/portal/docs/1/7912812.PDF. Accessed 18 Dec 2012.

FARNET. (2011). *Back to the islands. Upstream project to diversify fishing in the Stockholm archipelago*. In Farnet Magazine no. 4, Spring-Summer 2011 (pp. 17–21). https://webgate.ec.europa.eu/fpfis/cms/farnet/sites/default/files/documents/FARNET_Magazine_04_EN.pdf. Accessed 18 Dec 2012.

FARNET. (2012). *Making the system work for local communities—key issues for Axis 4 delivery*. In Farnet Magazine No. 7, Autumn-Winter 2012 (pp. 28–30). https://webgate.ec.europa.eu/fpfis/cms/farnet/files/documents/FARNET_Magazine_07_EN.pdf. Accessed 18 Dec 2012.

FARNET. (2012a). Tools: Cooperation. https://webgate.ec.europa.eu/fpfis/cms/farnet/tools/cooperation. Accessed 18 Dec 2012.

FARNET. (2012b). Tools: Country Information. https://webgate.ec.europa.eu/fpfis/cms/farnet/tools/country-information. Accessed 18 December 2012.

FARNET. (2012c). Tools: FLAGs. https://webgate.ec.europa.eu/fpfis/cms/farnet/tools/flags. Accessed 18 Dec 2012.

FARNET. (2012d). Tools: Good Practices. https://webgate.ec.europa.eu/fpfis/cms/farnet/tools/good-practices. Accessed 18 Dec 2012.

LGR. (2012). Obra-Warta. http://www.lgrow.pl. Accessed 18 Dec 2012.

Macken-Walsh, A. (2012). Operationalising contemporary rural development: Socio-cultural determinants arising from a strong local fishing culture. *Human Ecology, 40,* 199–211.

Martinos, H. (2012). The e-book overview, introduction to The e-book on Local Development, published by LDNet. http://ldnet.eu/The%20E-book%20Overview. Accessed 18 Dec 2012.

Moseley, M. J. (Ed.). (2003). *Local partnerships for rural development—The European Experience*. Wallingford: CABI Publishing.

OECD. (2006). *The new rural paradigm: Policies and governance*. OECD Publishing.

Ostrom, E. (1990). *Governing the commons: The evolution of institutions for collective Action*. Cambridge, UK: Cambridge University Press.

Soto, P. H. (2004). El programa LEADER: una revisión analítica. In J. G. Regidor (Ed.), *Análisis del Desarollo Rural in Extremadura: Bases para una Estrategia*. Junta de Extremadura.

Part IV
Issues in Gender Roles

Chapter 11
Flemish Fishermen's Wives: Their Lives and Roles in Fisheries

Katrien Vervaele

Keywords Flemish fishery • Fishing communities • Relationship • Education children • Brown shrimp • Fish market • Accordion women • Coastal fishery

11.1 Introduction

When you find yourself in the Flemish ports walking along the quays and looking at the boats, you will be surprised by the many women's names you read on the vessels' bows. How visible are the names? Yet how invisible are the women themselves? The fishery is a closed world where few outsiders are allowed, but the fishermen's wives themselves prefer to stay anonymous, in the shadow of their fishermen. However, once you gain their trust, once you get them to talk, a fascinating and unknown world opens up. This is what happened to the author when she interviewed 26 women about their lives with fishermen for the book *Vissersvrouwen* (Fisher Wives) in 2009. The youngest was a girl of 18, the girlfriend of a young fisher. The oldest was a woman of 92 years and widow of a fisher who had dedicated his entire life to the distant water fisheries. It was a very diverse group: wives of fishermen, vessel owners, company owners and women who sell fish on the quay. Yet, they had one thing in common: they were all independent, strong women who 'can and had to put up with a lot'. Although their stories were personal and unique, there was much similarity between their stories. Throughout the interviews the women and their narratives became entwined in a single story—their collective story of a difficult and changing life.

K. Vervaele (✉)
Groenestraat 22, 8377 Zuienkerke, Belgium
e-mail: katrien.vervaele@telenet.be

J. Urquhart et al. (eds.), *Social Issues in Sustainable Fisheries Management*, MARE Publication Series 9, DOI 10.1007/978-94-007-7911-2_11,

11.2 Intangible Maritime Heritage

Interviews were conducted with 26 fishermen's wives in the Belgian coastal towns and villages in 2009 for my book *Fisher Wives, 26 Lives One Story* (2009). From 2009 to 2012 I spoke with many other fishers' wives and with women working in the fishery sector, some of whom were interviewed again in the summer of 2012. It was not easy to find fishermen's wives who wanted to be interviewed. Many said that they had 'nothing to tell'. Others would rather not participate because their husbands would not agree, yet others no longer wanted to recall stories of the past because they were afraid to face the pain and sadness once again. But after the first contacts and the first interviews, there was a snowball effect with responses from women who wanted to participate.

At first, it was mainly the older women who wanted to talk about the past, about the time that their husbands were at sea. These interviews yielded interesting and valuable stories to be preserved as part of our maritime heritage. For a variety of reasons it proved more difficult to find young women whose husbands were active fishers. Firstly, most of the young women had jobs and did not want to take time out for the interview. Secondly, there are not that many young women who are fishermen's wives. In the three fishing ports, Zeebrugge, Oostende and Nieuwpoort together, there are only about 80 active vessels. In total, there are no more than 500 men that are still enrolled as fishers. Some of them have no partner, while some are still quite young. It is, therefore, not easy to find a fishermen's wife. Thirdly, some women were less than happy that their husbands were fishermen and, therefore, did not want to talk about it. Some were having difficulties in their relationships and would rather not talk, except for one who spoke openly about her divorce. Fourthly, younger women tended to live in different areas of the city and maintained little contact with each other, somewhat at odds with the popular image of the fishing community. Therefore the 'snowball effect' experienced while interviewing older women did not occur with the younger women. A fifth reason came as a surprise: a few fishermen's wives that had been interviewed indicated that they did not want their daughter-in-laws to participate in the study. These mother-in-laws were unhappy because the young women urged their husbands to stop fishing. The mother-in-laws were reluctant to make this public. As a result, the majority of the interviews were with older women, but the results are relevant in the context of oral maritime heritage.

11.3 Fishing Communities Then and Now

In former times fishing communities were well connected and based on strong family ties. Today the fishing communities tend to be much more diffuse and often non-existent in the sense of a close knit set of social relationships. Women that do not come from a fishing family do not understand what it means to be married to a fisherman. When they look around them, they see little or no examples to refer to. Previously this would not be the case: there were coastal fishing communities in Heist, Oostende, Blankenberge and Nieuwpoort where the fishers' families would live in the same streets and neighbourhoods. Every port had a typical fishermen's quarter. These were quite closed communities in which fishermen's sons would marry fishermen's daughters. In this way traditions and customs were continued. The daughters looked at their mothers as a role model, as outlined by Daisy:

> In the fishery everyone knows each other … Women who marry a fisherman are always very welcome in the community.

With the arrival of coastal tourism in the early twentieth century, quite a few girls came from rural areas to find a job in hotels and restaurants and met young fishermen. Gradually the closed communities began to accept these marriages between fishermen and girls from outside the community. It was not always easy for the girls to adapt though. Conversely, more girls from fishermen's families understood the difficulties of being married to a fisherman, and preferred a man who had a job ashore.

This is one of the reasons, perhaps the most important one, why these tight and closed communities no longer exist and the same is true for the fishers' districts in the larger coastal towns. Younger wives hardly keep contact with each other but they do tend to live fairly close to their own family. This way they can get some help from their mother, and possibly from their sisters, when their husband is at sea, particularly when they have a job and children to take care of. Support from the family is important because the fisherman's wife spends most of her time alone when her husband is at sea and this can, at times, result in feelings of loneliness, as expressed by Rita:

> Before my son was going to marry, I told my future daughter-in-law that the best thing to do would be to live near to her parents, because a fisherman is always away for weeks. I told my son that he had to accept that, because for a fisherman's wife it is very important that she can count on her family. I also suggested that my daughter-in-law kept her job. It was a wise decision, because she is doing very well, even though he is away sometimes for more than one month. She understands his love for the sea.

Another reason why the fishing districts no longer exist is that purchasing a property on the coast is becoming increasingly expensive, so many fishing families have moved inland where the price of property is somewhat lower. In fact most fishing families no longer live on the coast or in port cities but in small villages a few kilometres or further from the coast. This has led to a gradual but complete disintegration of the fishing communities, particularly for the women and families. However, the men still meet each other on the quay, in a café or around the fish market.

11.4 Relationships Between Fishermen and Their 'Girls'

Seventeen-year-old Kimberley found it difficult as the girlfriend of a fishermen and struggled when he was away at seas for weeks on end (Fig. 11.1):

Fig. 11.1 Young women find it difficult not seeing their men for weeks. (Photo: Katrien Vervaele)

> He was studying his last year at the fisherman's school when I met him. He was going to complete his studies during the summer holidays and he had the intention to go off fishing immediately after finishing school. I didn't really realise how it would be. At first, I saw him every day after school, but as soon as he went to sea he would be away for over a month, sometimes as long as three months. I then realised was it was like to date a fisherman!

Although she could not really relate to any life experience as a fishermen's wife, she explained how her friends somehow looked down on her because she was having a relationship with a fisherman and how she missed him when he was away, especially at weekends. She found it hard to go to parties by herself or having to go to school when he was back home. She found it also very difficult to keep saying goodbye and she was not sure that she would be able to live this kind of life. The relationship, therefore, only lasted 18 months and then she broke up with him. The young man was totally heartbroken and ended up with a severe depression and even suicidal tendencies. Fortunately, he recovered and joined the crew of his brother sailing on their father's boat.

Stories like this are not uncommon and they do not always have a good ending. Last year, a young man aged 20, jumped overboard and drowned. He ignored the lifebuoy that the crew threw overboard to save him. The reason was the same: his sweetheart had left him.

It is a common story in the fishery: young fishers have difficulty in finding wives. The young women are interested for a while, but when it comes to sharing their life with a fisherman, things look rather different. Young women are attracted to the macho image of the young fisherman and the fact that he makes a good living is a bonus. But living with a fisherman is not so easy. Very few women find this appealing, especially if they do not come from a fishing family themselves. They have no knowledge or experience of how to deal with this way of life. But even girls who come from a fishing family choose not to have this life and it becomes increasingly difficult for a young fisherman to find a partner. "My father and grandfather were fisherman", tells Mary-Lou. "When I was young, I said to myself never to take a fisherman as husband, but I fell in love. We married and we are already together for 40 years! I didn't regret, but it wasn't so easy when he was at sea."

Fig. 11.2 Selling shrimps at the fish market in Oostende. (Photo: Katrien Vervaele)

11.5 Involvement of Women in Fisheries

At a first glance, the involvement of the fisherman's wife in the offshore fishing sector seems of little importance, except when her husband is the owner of the vessel, as running the ship has a large financial impact on the family budget. However, the involvement of the vessel owner's wife seems greater nowadays than in earlier years. Previously it was the vessel owner who arranged everything, joined the crew on the fishing trips and had a lot of work to do. The woman was seldom or not involved at all. She would even be, for superstitious reasons, banned from the ship and, in most cases, from the quay, as Dorine pointed out:

> I've never been on his boat. Sometimes, when we went out for a walk, I saw his vessel. But I stayed on the quay and I never got in his boat, because that could bring bad luck! I never even waved him out when his boat took off. No never. Even that could bring bad luck…

'No women on board' is a typical fishermen's saying that was still respected up until the 1960s. Gradually, however, women got more involved and also took on more responsibilities. Nowadays, many women do the accounting and arrange, for example, the transport of the fish when it is landed in a foreign port or the transport of the crew when it needs to be relieved. When the ship calls in at a Belgian port, women are there to inspect the catch and, in some cases, even to help carry the fish boxes. When the ship is in port, they clean the galley and sleeping areas onboard the vessel, and wash the kitchen towels and sheets. Some even help out when the ship is in dry dock for repair or painting works, as indicated by Kelly (Fig. 11.2):

> My sister-in-law was more involved with the paperwork of the boat. Myself, I prefer to do manual work, such as repairing and painting the boat. I painted the bunks with little hearts and flowers, just a bit of flowerpower and every one had it's own meaning. There was one bunk with *love* for my husband, *miss you* for my brother …. At the very beginning I also helped unloading, but one time I tore a muscle. To lift up 40 kilos and throw it on the sorting-machine is not really a woman's job!

A skippers' wife will also usually be involved with the vessel and the events on board, as the skipper is responsible for the ship and the crew. If there is a problem, she is likely to be informed earlier than the wife of a crew member. The fact that the skipper takes great responsibility for the catch makes her feel more involved. Because the pay is a percentage of the catch, it is in everyone's interest that the catch is as large as possible, but it is the skipper's main responsibility, as suggested by Irene:

> It is not just the vessel owners who work under a lot of pressure, but the skippers also do, because they are responsible for what happens to the vessel. The vessel owners put pressure on them to catch as much as possible in the shortest possible time, because there is much at stake, but one can only do his very best in order to catch as much as possible. In fact the fisher is a hunter and he gives his job 100 %, but the vessel owner keeps him under a lot of pressure. If anything irregular happens on board, it is often the responsibility of the skipper.

For inshore fisheries, however, things appear to be quite the opposite. Where in previous times the coastal fishermen's wives had a much stronger presence, this is less the case today. This is noticeable both in Nieuwpoort and on the fish market in Oostende. Formerly, almost all the women selling fish at the market were married to a skipper who, in most cases, would also be the owner of the ship.

This brings us to the interview with Yvette, the wife of a retired coastal fisherman who used to land his catch directly at the fish market in the centre of Oostende. Every day she would be ready before dawn when the boat arrived to help them unload the fish and prepare it for the market. She would be involved with selling the fish and shrimp all morning and on a few occasions she even joined the crew when a crew member was ill. Yvette did her job with great enthusiasm and enjoyed giving fish recipes to her customers:

> When my husband came in with his boat, I was already standing on the quay in order to help him unload the fish. We brought the fish up in a big basket. I sorted out the same fish, size by size in metal sieves. Then we put the sieves on the ground. I wrote the name of the fish on beer mats, and later on, on white painted boards. I put the price on it with a pencil. Above the fish I had a sign on which our boat was painted and also my name *chez Yvette*. I distribute papers with recipes and how to prepare fish and I also give brochures with the most important events of Oostende. In fact, my stall was also a kind of tourist information office.

Another interviewee, on the other hand, set up a small shop to sell the fresh catch of fish and shrimp from her husband's boat. The shop was located in her garage at first, then later in the fish market of Nieuwpoort, but this practice was later banned.

The daily fresh fish market in Oostende still flourishes, albeit to a lesser extent than in earlier days. Most of the shop owners are from the fishing community, although they are usually not related to the boat owner. There are a few exceptions: Dini sells fish from her father's boat, together with her mother. Although her parents were divorced a few years ago, they have a good working relationship. This is something that cannot be said about the relationships between the saleswomen themselves, which reflects the competition and day-to-day struggle for survival—women against women, shop against shop. Each is very proud of their good quality, fresh produce and they understand and respect what the fishers have to go through every night to earn it. Viviane has been here on the quay since her fourteenth birthday and now, 30 years later, she is still there:

> This is my life. I lost my soul here on the quay, by the boats… I wouldn't do anything else.

Fig. 11.3 Accordion women: fishers' wives welcome their men home and a few days later say goodbye again. (Photo: Katrien Vervaele)

11.6 Accordion Women

Whether or not the woman is involved in the fishery itself, she has a big role in the home. She runs most of the household, does the paperwork and many odd jobs and household chores herself. She is usually a very independent, strong woman who can deal with the loneliness. It is also very important for her to create a happy, warm and comfortable home, both for herself and her children, but it is equally important for the fisherman to know he has a safe home to return to. This can even impact on his performance and wellbeing on board. A fisherman who has problems at home cannot function well on the ship and this can affect the whole crew.

Being together as a couple is limited to a week or a few days, but it is very intense. Even after the fisherman departs, the sense of belonging continues while the fisherman is at sea and the woman ashore. Unfortunately, not all relationships run so smoothly. Many women find the life of a fisherman's wife too hard to endure and ask their husbands to look for a job ashore. Sometimes he will accept this in order to maintain the relationship, but often with a persistent urge to return to the sea. However, there are many instances where the relationship breaks down and the woman leaves her husband. Divorce in the fishery community is commonplace. Even sadder are the situations where the woman stays with her fisherman because of financial reasons. Some of these women do lead their own lives and while their husband is at sea, they may have another partner. But infidelity can occur both ways and when visiting a foreign port, the husband can have a 'one-night stand'. It can be difficult for women to fully trust their husbands. They understand after the toil, the cold, and working in all weathers, fishermen are looking for some warmth when having a pint in a pub. But they hope that it remains with drinking pints, as indicated by Nadine (Fig. 11.3):

> A fisherman who is far away from home and goes into a pub and gets a little bit of attention… much more isn't needed!

She is not jealous and she is very understanding:

> When they do two journeys and they dock in England and have two days to wait, it is impossible to wash properly on board. You can't expect them to sit there two days in the cold. They go ashore to have a wash and, of course, they go to the pub for a drink or to a strip-club, sometimes to a brothel. The temptation is often very big there. I understand this very well. It is normal after being on the sea in frightful conditions, it is absolutely normal that they long for a bit of tenderness. But that hasn't anything to do with love, does it? Somehow you have to accept that a fisherman is a hunter, an adventurer….

The women have to compromise and tolerate a lot. And in this rather passive role of 'understanding', they have the important task of making sure that the man knows he can return to a warm nest and safe haven, as illustrated by Jeanne:

> Our men are at sea and our lives exist in waiting. Even when they are back, they are not at home. They don't sleep or eat at regular times. First unloading the fish then the maintenance of the boat… It is always waiting until he comes back. It is not unusual having to fry chips and prepare a steak in the middle of the night!

Most women do appreciate the combination: the freedom and independence on the one hand, the joy and happiness of togetherness on the other. Yet they admit that it is not easy and they need to constantly adapt. There is the pain of saying goodbye and to adjust to being alone again. The first day is always very difficult, they all say. For there is not only the fact that they are alone again, there is also the knowledge that their husband will have to work very hard again, sometimes in very harsh conditions. There is, of course, always danger around the corner, because fishing is and will forever remain one of the world's most dangerous jobs.

When the husband comes home, it is the joy of reunion, but also the difficulty to adjust and adapt to each other. It is a challenge to insert a father figure in the family structure that has been functioning without him for weeks. And that is why they call themselves 'accordion women' (Fig. 11.3). As Kelly says:

> Of course there is a party when he arrives at home. But I notice that it's a bit too busy for him the first days. The children want much of the attention. The first couple of days I give him a little space. The day when he is leaving is also difficult. He is a bit nervous. On departure day it can also be stressfull. I notice it especially at night while he is sleeping. The first two days he is very nervous and wakes up several times during the night. After a few days he is less restless, but after four or five days he has to leave again. We try to make it as cosy as possible for each other, but three days sometimes can be very short.

When their husband is due home, the women ensure that the house is cleaned, laundry done, and there are enough supplies in the house, so the time can be fully dedicated to him. After 3, 4, or 5 weeks of separation, the week together is spent as intensely as possible. While the woman adjusts her daily rhythm and the family gets used to the father figure again, it is also the man who must adapt. The free life at sea among men, and the hard work and little hours of sleep, must be swapped for the more regular life on land in a structured family setting. His wife will notice that all too well, if only because of his tossing and turning the first night ashore, and sometimes his extreme fatigue. She also senses that the day before leaving again he will become restless and nervous. For, although he is drawn to the sea, his departure will always be fraught with sadness and some reluctance. This may be the case especially if he is a skipper with a lot of responsibility or if he is the vessel owner, because of financial difficulties such as high fuel oil prices and uncertainty surrounding the next catch.

However, perhaps even more than her husband, the woman finds it difficult to deal with the approaching farewell, though she must take care not to show it. She must stay strong as he leaves, because tears only make it harder. And she knows that questions like "You'll surely be back by then" or "Do you know how long you'll be staying away?" are better left unsaid. When driving her husband to the quay and waiting for the departure of the ship, it is certainly not appropriate to show weakness. A quick kiss and a wave, but no signs of emotion as this would embarrass her husband in front of his fellow fishers.

11.7 Education of the Children

For the women it is a lot easier if they have children, since they take away the sense of loneliness. In earlier days, this loneliness was easier to bear when the wives were surrounded by members of the fishing community and living within a tight social network. The women would support each other, because they were all in this same situation and they understood each other:

> In earlier days it was very normal that you met other women on the streets, at the school-gate, in the shops. It wasn't necessary to be close friends, you just met each other and you chatted. The school gate was the most important meeting point. We were always chatting about the children, but also often about the men on sea, about the weather, about the fish. Of course there was also a lot of gossip and we heard the latest news. But most of the time we spoke about our children.

Fishermen's women play a very important role in the education of the children. They are basically on their own with this, and usually do it with good insight and assign a great deal of importance to this task. This also means that they can be very strict at times and will administer punishment where necessary. The women will seldom leave the punishing to her husband as, if she did this, the children would see their father as the bogeyman. So you will seldom hear a fisherman's wife say that 'they will tell dad all about it', as Viviane explained:

> When the children have been naughty, I never told him. No need to be punished when it already happened. It was too late. When the children did something wrong, you have to punish them immediately. You have to do it yourself, because if you wait until their father comes home, then punishment has no more effect!

Where the women in earlier days used to do the parenting alone, currently they understand how important it is that the father also takes on his role when he comes home so as to strengthen ties with the children. They will also make sure that they are not giving them too many treats, which is something that fishermen tend to do when they come home to see the children. This can often lead to tension between the parents. Evy has a son of 7 years old:

> When I am alone with him, it's much easier. When my husband arrives home, he spoils him. He buys a lot of toys, because he wants to make up for the lost time.

11.8 A Job of Your Own

Unlike previously, a lot of the younger fishermen's wives choose to have their own job. This is usually part-time, because they like to take a few days off when their husbands are at home. Women feel this is important for two reasons. First, there is the need for financial security. For, although the husbands generally make a good living from fishing, this is not always the case. There is also the need to build up some savings, because fishermen often retire early and have little pension compensation. Also, there is always the high risk of an accident to be considered. If the husband can no longer go to sea it is helpful if the woman is earning and the couple have some savings. Buying your own house is important for many fishermen's wives: they want some security as a counterpoint to the greater insecurity attached to their husbands' profession. This was illustrated by Cecile:

> Before we bought our boat, we first bought a house. It is a big house with lots of space, very cosy and full of light. Sometimes I am afraid that things may go wrong and that we will have to sell our house. Even my husband is afraid. He is often nervous about the financial situation. He is afraid that he might go bankrupt. If that happens, he will never go on the quay again, he says.

On the other hand, they often look for a job simply in order to create some extra social contact and puncture their often lonely lives. However fishermen's wives seldom look for a job in the fishing industry itself. They prefer to find work in the social sector, for example, or as a nurse, but not in fish processing companies. In fish shops you will rarely find the wife of a fisherman, mainly because it is very difficult to combine the working hours with childcare:

"Now that the children are in high-school and don't come home at lunchtime for a meal, I have gone out to work. It is not that I don't have work at home, but I needed to see more people. Nowadays I work part-time for a boat that returns to port every day from the offshore wind farm to bring staff or supplies. I organise a lot of things for this boat and that's my way of living with the sea. My husband supports me, because he understands that it isn't easy for me to sit at home waiting." (Louise)

11.9 Women Working in the Fisheries Sector

As mentioned earlier, where the husband is the vessel owner, his partner will usually share some of the responsibilities of the fishing experience, especially where the husband goes to sea on a regular basis. Here the fisher's wife will tend to have a lot of input. She may do the bookkeeping and help with the maintenance of equipment; she may arrange the transport of the fish when it is landed in a foreign port.

Fig. 11.4 Woman selling shrimps and fish from her husband's boat in Oostende. (Photo: Katrien Vervaele)

She may also arrange for the purchase of new equipment for the vessel, and the purchase of provisions for the crew. In the case of inshore fisheries, there are some women who sell fish from their husband's vessel, although this is mostly limited to the fish market in Oostende (Fig. 11.4).

Generally speaking, however, looking for women who have jobs in the fisheries sector is like looking for a needle in a haystack. In previous years there were many women who peeled shrimp for restaurants and shops, but since this was forbidden by European legislation in the 1990s, this practice is now rather dispersed and small-scale. There are a few women who still peel shrimp at home to earn some extra money, receiving around €5 per kilo for the peeled shrimp. They use the remaining unpeeled shrimps that are not bought by the Dutch companies for transporting to Morocco for peeling. But these women are rarely the wives of shrimp fishers. However, the wife of the owner of the vessel *Versluys* is involved. Their shrimp ('Purus' shrimp') is peeled mechanically while she inspects the quality of the peeled products.

Most of the women are employed in sales. There are fish shops that are run by couples where one is usually from a fishing family and maintains close links with the industry. There are also the distribution services to inland towns which is mostly done with mobile shops which service the village markets. In most cases, the wife of the shop owner will join her husband, who generally comes from a fishing family.

But at sea you will not find Flemish women. In the course of history, there have been two cases of women that actually went to sea as crew on their husbands' vessels. They were small coastal vessels crewed by two people and they would sail out every evening and back early the following morning. One of these two women helped her husband in the shrimp fisheries in the 1960s. The other, Carine Ulin, practised ecological fishing with entangling nets on the 0.369 *Attila*, a decade ago: "I worked together with my husband. We were fishing six nights out of seven and we slept by day. It was a pretty tough job for a woman, but it was what I liked to do. The fact that I had no children made this possible." Since then their boat has been sold because the fishery was no longer yielding enough income.

In deep-sea fishing, women are simply not accepted. Firstly, because it really is a man's world and the wives of fishermen may not appreciate the news that a woman would be joining the crew. Secondly, the fishing vessels do not meet the necessary requirements, such as a separate toilet, shower and sleeping area for women. Yet, during the interviews, there were women who admitted that they would have liked to become a fisher and almost certainly would have been a 'fisherman', 'if I were a man.' Each and every one of these were women from real fishing families. Women who, just like their brothers, fathers and grandfathers, also have the salty sea flowing in their veins.

11.10 Conclusion

Fisheries are commonly thought of as a man's world. The hard work at sea requires physical strength and the working conditions on a fishing vessel are seldom women friendly. However, women and, in particular, fishermen's wives play a much greater role in fisheries than is generally known or recognised. This may include anything from the purchase of equipment for the vessel, the careful keeping of the accounts, the arrangements for transport of the fish products or the crew that needs to be relieved, to the selling of the fish, but also the important and multiple roles the women play in the household and the education of children. The support provided by fishermen's wives mostly concerns activities on land, and women's jobs are mostly associated with the selling and processing of fish. Yet, they also take on roles that are crucial and complementary to the work at sea, on the fishing vessels. Throughout the 26 interviews and conversations with Flemish fishermen's wives, this wide diversity of roles and functions that women represent in fisheries has been described and discussed. The interviews bring out differences between generations and situations and illustrate very clearly how fishermen's wives are involved in the fisheries of this area. Most strikingly, however, is that different women and their different lives also tell a common story of the independence and strength of women in fisheries.

Such women may have their counterparts in other parts of Europe and, indeed, throughout the world. Their roles are vital to maintaining and sustaining the fishing industry and the fishing communities, though their contributions may often be unrecognised, undervalued and possibly under threat from the values and culture of modern youth.

Acknowledgments I would like to thank Ann-Katrien Lescrauwaet for her initial translation of this manuscript, along Christiane Knudde who kindly helped with the translation of the quotations from the interviews.

References

Vervaele, K. (2009). *Fishermen's wives, 26 lifes, one story*. Lannoo: Tielt.
Vervaele, K., & Fockedey, N. (2012). *Garnalen (gernoars) [Brown Shrimp]*. Lannoo: Tielt.
Vervaele, K. (2011). *To Island, Fishermen tell about the fishery in the waters of Iceland*. Lannoo: Tielt.

Chapter 12
Women's Organisations in Fisheries and Aquaculture in Europe: History and Future Propects

Katia Frangoudes, José J. Pascual-Fernández and Begoña Marugán-Pintos

Keywords Fisherwomen organisations in Europe • Current situation and prospects • Communities • Sustainable development

12.1 Introduction

In Europe, fisheries are considered a male activity simply because men make up the main labour force on fishing vessels. This stereotype has dominated the European public space despite social scientists' work highlighting women's tasks within fisheries enterprises and the family. Women's contributions in fisheries were, and are, often invisible because their efforts are rarely paid and because women themselves consider their work as part of their traditional home tasks (Frangoudes 2011). The desire to modify this common perception started in the mid-1990s when French fisherwomen[1] participated in fishers' movements against the economic crisis faced by the French fishing industry. This crisis resulted from the combination of two phenomena: catch reductions of several fish stocks (for example, hake, cod, monkfish,) and the reduction in fish prices (Piriou et al. 1995). Throughout these events women participated in demonstrations and established survival committees. Their objective

[1] The definition of the term 'fisherwomen' used in this chapter includes all women contributing to the family fishing enterprise (spouse, mother, sisters, etc.), women involved in fish harvesting and wives of crew members.

K. Frangoudes (✉)
UMR AMURE, IUEM, Université de Brest,12 rue Kergoat,
CS 93837, Cedex 3, 29238, Brest, France
e-mail: Katia.Frangoudes@univ-brest.fr

J. J. Pascual-Fernández
Instituto Universitario de Ciencias Políticas y Sociales, Facultad de Ciencias Políticas y Sociales, Campus de Guajara, Universidad de La Laguna, 38205, La Laguna, Tenerife, Spain
e-mail: jpascual@ull.es

B. Marugán-Pintos
Department of Social Analysis, Universidad Carlos III de Madrid,
Calle Madrid, 126, Getafe 28903, Madrid, Spain
e-mail: bmarugan@polsoc.uc3m.es

J. Urquhart et al. (eds.), *Social Issues in Sustainable Fisheries Management*,
MARE Publication Series 9, DOI 10.1007/978-94-007-7911-2_12,

was to financially assist fishing families in meeting their material needs during the crisis. During this period, women challenged decision makers on the fisheries situation and called for the survival of artisanal fisheries.

At the end of this chaotic period, women decided to create their own organisations to fight for the survival of their fishing communities. Based on the French model, many associations were established at a local level in other Member States with scientists, NGOs and even the European Commission supporting projects concerning women's contribution to fisheries. At the end of the 1990s and the beginning of 2000 more and more women's groups were established. Their action moved from the defence of the fishing industry, communities and families to claims related to women's rights and sustainable fisheries. Women claimed a legal status that recognised their invisible contribution and sought training and education to satisfy their need to participate in, for example, resource management. Women's organisations brought women into the public arena as they became one of the players in the decision making, not only in matters of women's rights but also in social issues concerning the fishing industry. In general women's history is based on transformation and changes (Fraise 1998) and the history of fisherwomen's organisation is no different. This summary of the activities of fisherwomen's organisations aims to highlight the main changes and transformations which have occurred during their existence. These women are concerned not just with change in their own situation, but also with that of the communities in which they live. It is hoped that the history and background of these organisations presented here will assist women in creating their own history.

An analysis of the current European fisherwomen's organisations shows that they are divided into two main groups: organisations consisting of women contributing to the fisheries enterprise; and organisations consisting of women directly involved in fish harvesting or in ancillary activities. In some countries women's groups are independent and in others they are part of a mainly male organisation. The differences between these groups and the benefits and constraints of the differing structures are issues discussed in this chapter. The data used to illustrate the chapter has arisen from workshops held by the FEMMES programme 'Women in fisheries and aquaculture in Europe'[2] and the European Network AKTEA[3].

12.2 First Generation of Fisherwomen's Associations: Supporting Fishers, Women and Families

Before the mid-1990s, women's organisations in fisheries were rare, with only two cases being documented. The reasons for their establishment were very similar to those of recent women's organisations and so it is pertinent to consider them here. One of the first organisations was established in 1946 in Kristiansund in Norway

[2] FEMMES is a thematic network funded by the Quality of Life programme of the European Union Commission-DG Fish, contract n°Q5TN-2002-5th.

[3] The European Network AKTEA regrouped women's organisations in fisheries and aquaculture was established in 2006. It is a result of the FEMMES programme.

and was related to *Damegruppen Havlimt* (the Shining Sea Women's Group). This group campaigned for improvement in the living conditions of fishermen's families. One of their main activities was to raise funds to help families of fishermen lost at sea, those who suffered from illnesses or had died (Nilsen 2004; Nielsen 2004). Following this example, women from other places in Norway established organisations with similar objectives in order to help fisher families in times of need. They were so successful that the *Norges Fiskerlag* (Norwegian Fishermen's Association) decided to regroup all women's organisations into one organisation. In 1953, the *Norges Fiskerlag Kvinnelag* (NFK), Norwegian Fishermen's Association Women's Group, was created at a national level with the aim of supporting the Norwegian Fishermen's Association (Nielsen 2004; Nilsen 2004). The active involvement of the NFK in fundraising resulted in the establishment of the *Fiskernes Hjelpefond* (Fishermen's Aid Fund), a form of health insurance and fund for widows of fishermen. This Fund operated until 1970 when the State took over this responsibility.

During the same period that women's movements in southern countries were becoming more important, feminist ideas were introduced into the NFK. Women began to discuss their own situation and demanded to have influence and to be seen to have influence. In 1983, NFK became an independent organisation, although it continues today to collaborate with fishermen's organisations and has the benefit of observer status within the *Norges Fiskerlag*. Besides social and welfare activities, the local women's groups of the NFK were involved in a large range of activities and issues, varying from safety at sea campaigns and enterprise development for women, to fisheries policy issues and defending coastal communities (Nielsen 2004).

In County Cork, in the south of Ireland, Castletownbere is the largest white fish port in the country, and there has always been a very strong sense of community there. In 1964, a group of women formed themselves into an association to support each other when the men were at sea and communications were difficult. When radio communications improved in the 1980s the women's group activities decreased. Their contact was mutually supportive and often social but rarely focused on the political, managerial or administrative aspect of the fisheries. In the late 1990s, the group in Castletownbere re-energised themselves and began to realise that there were indeed political, managerial and administrative agendas within fisheries to be tackled. They made contact with other fishing ports and an inaugural meeting of a national body was held in Enniscorthy in January 2003. The group assumed the name *Mna na Mara*, which was the name of a group of women who had first gathered together in Donegal in 1991 to campaign against the arrest of their salmon fishing spouses.

12.3 Second Generation of Fisherwomen's Associations: Diversification of Goals and Strategies

This second generation of fisherwomen's organisations is characterized by their diversity, with a wider range of goals, geographical scope and representativeness. These organizations arose as a response to a range of factors that are analyzed in the following sections. These factors range from responses to crises, working conditions, resource management, formal recognition, etc. This exemplifies the diverse situations in different European countries, and the variety of roles and tasks women's organisations have developed.

12.3.1 Women's Organisations and the Defence of Fishing Communities

The second generation of women's organisations started in the 1990s as a result of the crisis faced by the fishing industry in Europe (France, Netherlands, United Kingdom and Ireland). During this period fish prices in France were low and family income was considerably reduced. The women rallied together for the survival of their families and their communities, meeting with the authorities to advocate their cause and organising numerous publicity events. They distributed free fish to the public to attract attention, collected donations, organised the distribution of food packages to help fishers' families who had been without any income for months. They also negotiated the non-payment of electricity, water and other utility bills with the respective companies (Esteban 2004). These initiatives resulted in the establishment of various women's organisations at a local level and eventually these formed two national federations: Fédération des Femmes et Familles de Marins du Milieu Maritime (3FM, Federation of Seafarer's Wives and Families) in 1998, based in the south-west part of Atlantic coast; and the Fédération Interrégionale des Femmes du Littoral (FIFEL, Interregional Federation of Women of Littoral) in 1998, based in Brittany, Normandy and the Mediterranean Sea. These two federations merged into one national federation in 2003 and established the 2FM (Femmes et Familles de Marins du Milieu Maritime) (Morin 2004; Esteban 2004).

In the Netherlands, Dutch fisherwomen reacted to two particular fisheries crises through their national network *VinVis* (Women in Fisheries Network 2000). The first was the cod crisis of 2001, instigated by the EU decision to ban fishing in certain areas of the North Sea as a protective measure to slow stock shrinkage, and the second, the shrimp crisis of 2003 following the end of the production-regulating agreement by the Dutch anti-cartel authorities. These crises promoted a real sense of purpose for the women. Members of *VinVis* came forward in defence of their family enterprises and championed the survival of the Dutch fishing community (Quist 2004).

In the UK, women also formed groups at a local level in response to fisheries crises, which affected the stability of the fishing community's income. These organisations were formed initially in the Shetland Islands, in Scotland, and in Northern Ireland. One such group, Moray Makes Waves (MMW), is a group of local wives of fishermen formed in January 2003 to support the Scottish fishing industry during what many regarded as the worst ever crisis the industry had ever endured. Its aim was to raise public awareness of the abundance of high quality seafood landed by Scottish vessels in Scotland and to encourage consumers to purchase Scottish-caught seafood. The group wanted to protect the fishing heritage of local communities that had been badly affected by decommissioning. Members of MMW attended demonstrations, travelled to the Scottish Parliament to talk with the Fisheries Minister, gave evidence at the Scottish Parliament's Rural Development Committee and organised a meeting 'Fishing In Our Community' attended by members of the public, fishermen, fishing leaders, Members of Parliament (MPs) and Members of Scottish Parliament (MSPs). MMW generally sought to raise public awareness of the effect the decline of the industry was having on their communities and aimed to act in a constructive and responsible way to keep the plight of fishing communities high on the political agenda.

In April 2003, the group launched a campaign to promote Scottish fish and produced a leaflet outlining the difference between the prices fishermen receive for their catch and those paid by the consumer. They canvassed local supermarkets to determine whether Scottish vessels caught the fish being sold in their stores. For those that sold Scottish-caught produce, the MMW created special certificates of authenticity confirming their support for the Scottish fisheries industry. The certificate of authenticity campaign caught on and awards were also presented to fishmongers and restaurants.

12.3.2 Women's Role in Managing Natural Resources and Their Work

Shellfish gathering on foot is an activity carried out on European South Atlantic coasts. This activity takes place in Spain (Galicia), Portugal (Ria de Aveiro, Algarve) and France, but it is only in Galicia that shellfish gatherers have created their own organisations. In the Algarve, women are involved in shellfish gathering but they have not established their own organisations since they are members of the national network of Portuguese women in fisheries '*Estrela do Mar*'.

Groups of professional shellfish gatherers were common in Galicia in the late 1990s. These groups were formed in response to a paucity of resources, the invisibility of their work and as a result of education and training courses offered by the regional authorities. Shellfish gathering has traditionally been a women's activity because it is easily combined with domestic chores and childcare. More than 90% of the shellfish gatherers were women, but they were not considered to be professional workers, received no social security benefits and had no access to professional fishers' organisations such as the *cofradías* (Marugán Pintos 2004).

Historically, access to shellfish beds was free and shellfish gathering was practised in an informal and unregulated manner. The need to regulate this traditional fishing activity arose, in the late 1970s, when shellfish stocks were at risk of collapse (Meltzoff 1995, 1996). With the support of the Galician government, women gatherers of shellfish on foot succeeded in regulating their activity through a licensing system that provided each permit-holder with a quota. Professional training programmes were organised by the regional authorities to improve *mariscadoras* knowledge and skills. The women's organisation, with the help of biologists, established exploitation plans and succeeded in regulating the market and improving their incomes. Moreover, the women succeeded in introducing the word *mariscadora,* which is the female form of the Spanish word for shellfisher; thereby gaining dignity, confidence and public status (Marugán Pintos 2004, 2005; Frangoudes et al. 2008, 2013).

By joining the official male-dominated fishers' organisations, the *cofradías,* the women succeeded in obtaining formal professional recognition and gained access to decision-making bodies. Members of the *mariscadoras* now receive benefits under the marine social security system. In 2002, a federation of shellfish women's associations was formed with the name *Asociación de Profesionales de Marisqueo a Pie de Galicia* (AREAL). Through this organisation *mariscadoras* were aiming to improve working conditions, to add value to their products through labelling, to promote their products as well as their professional status and to strengthen their representation on official government bodies. A few years earlier, in 1998, the *Asociación Galega de Mariscadoras (*AGAMAR) was also established. The objectives of these two organisations are different[4]. AGAMAR targets individuals, rather than organisations, and represents people associated with shell fishing throughout the Galician region of Spain. It seeks to improve the situation of all shellfish gatherers, both women and men.

The success of the *mariscadoras* encouraged other women to form their own organisations. An example of this are the women involved in net mending in Galicia (*rederas)*. The work of the *rederas* was very poorly paid with no access to opportunities that would help them to improve their professional skills and, as a result, many of them gave up the work. With the support of the Galician authorities, the *rederas* organised meetings at the regional level where they discussed their work and their ambitions for the future. Out of these discussions arose a regional federation of Galician *Rederas,* created in 2004, with the aim of gaining professional recognition and improving working conditions. Through government-sponsored training programmes[5] the *rederas* aimed to improve their skills and to explore new business opportunities.

12.3.3 Improving Safety at Sea and Working Conditions

Issues related to the working conditions of their husbands at sea have also motivated wives of fishermen to form organisations. Wives of fishermen live in constant fear of accidents at sea, and in many European countries (Finland, France, Spain and Ireland) they have undertaken initiatives to improve safety on board vessels and working conditions at sea.

Finnish fisherwomen, together with Swedish and Norwegian women, were involved in safety at sea campaigns. These associations conducted a study on safety

[4] http://www.agamar.es.

[5] Projects NEXUS-REDE, PIC- EQUAL, C.E.E.

regulations and social security benefits (such as accident compensation, pensions and wage regulations) and presented recommendations for the harmonisation of fishing policy and regulations to the Nordic Council of Ministers (Talvitie 2004).

In Spain, wives of crew grouped together to combat the bad working conditions on board industrial fishing vessels and they have negotiated with ship-owners for collective labour agreements. The Galician crew wives' association *Rosa dos Ventos* (Compass Rose; Galicia 1991) campaigns for the implementation of protection of labour and social welfare reforms for deep-sea crew members and their families. This association has been actively involved in lobbying the public authorities for improved working conditions and safety at sea. As a result of their campaigns they have succeeded in attracting the attention of policy-makers at a national level and also at European level. One specific result of their efforts to date is that rescue methods on board deep-sea fishing vessels have been strengthened (Paredes Soto 2004).

In France, organisations of fisherwomen have participated in safety training programmes organised by the regional and local fisheries committees and have been actively involved in safety awareness-raising campaigns for fishermen. In 1999, members of the association *Femmes Entre Terre et Mer* (Women Between Land and Sea) in Brittany undertook a campaign to ensure that all fishermen wear personal flotation devices (PFDs) while on-board. Unfortunately, a lack of sufficient funds made it impossible for the association to complete this project with the result that French fishers continue to work on-board without using the safety equipment that is available to them (Carriou 2004; Bhouris 2004).

For the Irish fisherwomen, safety at sea and improvement of rescue services were two important issues of action for members of the Irish association *Mna Na Mara*.

12.3.4 Access to Fishing Rights

In some cases, female discrimination and female access to fishing rights have prompted women to unite to campaign for equal rights between men and women in fisheries. In Spain, the women of El Palmar, a fishing community near Valencia, have been fighting, through their association *Tyrius*, against the local community's customary law, which deprives women of hereditary rights. This law precludes fishing rights being inherited by female descendants and denies women any entitlement to fish in the local lagoon. After years of struggle, the *Tyrius* association won a court case against the local fishers' organisation. The legal decision given in 2003 grants fishing rights to women, although the decision has not been respected by the male fishing community of El Palmar (Serrano Soler 2004). In 2007, another court judgement[6] ruled the practice as being in contravention of the Spanish Constitution[7]

[6] Sentencia No. 262/07 juicio ordinario No. 610/07, Juzgado de Primera Instancia No. 23 de Valencia.

[7] The Spanish Constitution is the main law in Spain.

and contrary to the principle of equal opportunities in the workplace. The case of the *Tyrius* association is an example of gender discrimination culturally entrenched in the economic fibre of society.

12.3.5 Formal Recognition

In the process of mobilisation for issues that concerned the survival of their community, working conditions of their husbands and equal access to fishing rights, spouses of fishermen have become aware of the need to request formal recognition of their contribution to the family fishing enterprise.

In France, in 1995, following the fisheries crisis of 1993–94, a new Fishery Law, concerned with the status of collaborative spouse[8], was debated. Spouses of fishermen initiated the issue of formal recognition of the woman's contribution to the family fishing enterprise. In their view, the legalisation of the Collaborative Spouse Status (CSS) would not only give women an opportunity to represent their family enterprise and community in representative bodies but also give them access to social security. In the agricultural industry, spouses were already entitled to such a status. Women's associations' lobbied politicians and administrators at the national level and they obtained the CSS with the Fishery Law of 1997. Although this 1997 Fishery Law was a major break-through for the formal recognition of women's contribution to the family fishing enterprise, many regard the definition of the status as too narrow and the conditions too difficult to implement (Ancel 2003). Therefore, only a small percentage of fishermen's spouses actually opt for this status (AKTEA newsletter 2003). Improvement in the legal standing of the CSS was top of the agenda for the two federations, along with the issues of safety onboard fishing vessels and training for wives of fishermen.

The wives of shellfish farmers worked with the wives of fishermen to obtain the CSS and they joined forces to create associations. Shellfish farming in France is predominantly a small-scale family enterprise where the husband and wife work together. Only a few women are registered as shellfish farmers. The division of labour is very strict, with men working in the sea and women on the land. Women generally perform the following tasks: cleaning oysters, boxing shellfish, customer service, administrative work and various management tasks (Simon-Goulletquer 2000).

Other women's organisations around Europe have followed the French example and are requesting formal recognition of their contribution to the family fishing enterprise. Dutch wives of fishermen (*VinVis*) also claimed legal status without success. Later the Italian and Greek women's organisations also requested access to CSS but until now their national authorities have not replied positively to their demands.

In Ireland, in 1994, following two seminars on health and safety issues in the fishing industry, the wives and partners of fishermen in Greencastle decided to

[8] Council Directive 86/613, Article 6, on the application of the principle of equal treatment between men and women engaged in an activity, including agriculture, in a self-employed capacity and on the protection of self-employed women during pregnancy and motherhood.

organise themselves. They formed an association called Inishowen Women of the Sea. Their primary concern was to bring about change to the social welfare legislation of the day. Much lobbying took place and in January 1997 the Minister of Social Welfare announced a new scheme which entitled fishermen paid by share system to be recognised as employees with the resulting social welfare benefits (however four years later a high court ruling reversed these changes).

12.3.6 Training and Education Needs

The need for greater education and training are other factors that have motivated spouses of fishermen to become organised. In general, women consider that training and education would contribute to the overall improvement of their knowledge and skills, resulting in an enhanced family enterprise.

In most countries, women's organisations have encouraged their members and other women in their community to participate in training and educational courses. These courses have been related not only to their tasks in the family enterprise (management, communication and marketing), but also to the establishment of complementary activities that add value to the fishery enterprise and increase their opportunities in other sectors of the labour market.

French fisherwomen, after the experience of different types of training related to enterprise management, decided to seek training that provided them with a specific state-recognised diploma. Training programmes that provide certification without a specific nationally-recognised diploma do not give them access to the wider labour market. Following a successful experience, conducted in Sables d'Olonnes from 2004–2005, the 2FM asked the French authorities to add a maritime course to the existing Certificate of Collaborative Spouse (*Brevet de Conjoint Collaborateur d'Entreprise Artisanale,* BCCEA). The introduction of modules, focusing on fisheries and aquaculture in the curriculum, serves to provide wives of fishermen with access to an official state diploma that enables them, if need be, to seek employment in other sectors. Once it is established that this training is associated with an official national accreditation, it becomes much easier to obtain funding for the training programme.

The case made by French fisherwomen was successful on the 31 December 2007 with the issue of a Ministerial decree and the creation of a diploma called the 'Certificate of Collaborative Spouse for Maritime Family Enterprises' which is equivalent to the high school diploma. Women who opt for Collaborative Spouse Status and wish to obtain this certificate should have at least two years of previous professional experience and have the achieved an education level of secondary school. The programme involves intensive training and preparation, geared specifically for wives of fishermen, and the diploma is issued following the successful completion of a written exam[9].

[9] Arrêté du 31 décembre 2007 relatif à la délivrance du brevet de collaborateur de chef d'entreprise artisanale m maritime, J.O. du 8 février 2008.

Members of the Greek Union of Women in Fisheries, PEGA, have also benefited from training programmes. The first training course in 2006 focused on fishing enterprise management and provided some information concerning Europe's Common Fisheries Policy (CFP). The second training course in 2007 introduced the use of computers and the Internet. According to the participants "the training was a good experience" and they would like to have the opportunity to progress. Both training courses occurred with the cooperation of the European fisheries structural funds, PEGA and the district authorities. In Italy, members of the Penelope association, based in Ancona, participated in training on enterprise management. The Portuguese network also trained its members.

12.4 Sustainable Fisheries and Women's Organisations

The success of the initiatives put forward by the various fisherwomen's organisations has made women aware of their power to influence decision-makers and they are now becoming involved in policy-making. By studying policy documents and formulating responses and proposals, organisations of spouses of fishermen are contributing actively to consultancy meetings and hearings of policy-making bodies from local levels up to the European Union level. Several women's organisations have responded to the Green Paper on fisheries launched by the European Union to review its Common Fishery Policy (2001–2002). The Dutch *VinVis* network studied and discussed the Green Paper and formulated a response suggesting that the "fishery management problem is not to be solved by technical measures as such. This can only happen with the genuine support and participation of the fisher community, including women" (Quist 2002, 2004).

In some countries (ES, FR, FI) women were able to enter the traditionally male-dominated organisations due to their legal status. The *mariscadoras,* by obtaining professional status, not only established their own organisations, but also began to take part in the management of *cofradías* and actively participated in sustainable management of shellfish stocks. The Collaborative Spouse Status offered French fisherwomen the opportunity to become members of fishermen and shellfish farmers' organisations. They have also been elected to bank boards and cooperatives. In France and in Finland, some women have chaired male organisations such as the Local Fisheries Committee or a trawlers fishers' organisation.

Following the recommendations of AKTEA network[10], national or transnational organisations have joined the Regional Advisory Councils (RACs)[11] and participated in the decision-making concerning management of the resources, although their reasons for joining differ. Members of women's organisations face a key difficulty regarding resource management. They must struggle to avoid the paternalistic attitude of men towards them concerning this issue. Often men may feel women

[10] 2nd AKTEA Conference, Annalong April 2007.

[11] CE 2004/585/EC, J.O. 3.8.2004.

do not have the capacity to understand problems related to resource management because it concerns activity taking place at sea. It is probably for this reason that male organisations were against women organisations joining the RACs in the seats dedicated to the fisheries industries (2/3 of the seats). They considered that women 'stole' their seats. Only when DG MARE modified the regulation and provided the option for women's organisations to accede to the seats of the NGOs and consumer organisations (1/3) were women allowed to become part of the RAC board. One organisation member of AKTEA became a member of the board with the agreement of men because she could be an ally in cases of difficult decisions (Aktea newsletter 2009). By becoming members of the RAC, women strengthened their position within the decision-making process of resource management. For women involved in harvesting the situation is different. They need to prove their expertise at harvesting before they can qualify for a position in the decision making process.

The North Sea Women's Network (NSWN), which joined the North Sea RAC in 2004, is a member of both the Executive Committee and the General Assembly. The NSWN has spearheaded a campaign to focus the RAC's attention on socio-economic aspects in fisheries. As a result, a socio-economic working group was established, which the NSWN chairs. The main objective of this working group is to lobby for the integration of social and economic indicators in the fisheries management decision-making process. For the Network, management measures have an economic and social impact at the level of coastal communities, which should be taken into account before the implementation of the measure. The NSWN views itself as an advocate of coastal communities on the NSRAC.

Four other women's organisations became members of RACs: *Mna Na Mara* (IE) is a member of the North Western Waters RAC, *Estrela do Mar* (PT) is a member of the South Western Waters, the Scandinavian network is a member of the RAC of the Baltic Sea and Penelope Association has been a member of Mediterranean RAC for a year. These organisations do not play the same role within the RAC as the NSWN, which chairs a working group of its RAC, but they are members of one.

Estrela do Mar decided to participate in their RAC because Portuguese women considered that they could influence decisions about the management of fisheries resources, without which fisheries communities cannot survive. The *Estrela do Mar* is a member of the RAC working group 'traditional fisheries' where the interests of families and communities linked to these fleets are defended. Women play a more important role in artisanal fisheries than in industrial fisheries. By participating in the RAC, members of the Portuguese network contribute to the promotion of small-scale fisheries' interests and those of the fishing communities[12]. The Penelope Association participates in the MED RAC targeting the promotion of women's rights within the fishing industry and the defence of fishing communities.

The AKTEA network is currently lobbying, at European level, for the legal recognition of women's contribution to the fisheries industry and for the integration of women's organisations into the fisheries decision-making process. It is hoped that the actions of national organisations, transnational and trans-European networks

[12] Interview with the facilitator of Estrela do Mar, February 2008.

will increase awareness of women's contribution to fisheries and lead to legal recognition of this contribution. In the past, women's networks and organisations have contributed to the improvement of safety at sea and to the improvement of working conditions at sea. Now, they lobby for issues linked to women's empowerment and a place at the decision-making table. For these reasons the AKTEA network and other organisations have participated in the public consultation on the Green Paper for the reform of CFP. The AKTEA response covered different issues but it gave greater importance to the issue of resource-based individual rights (individual quotas or transferable fishing concessions) and on their impact on small-scale fisheries, women and communities. For AKTEA, the allocation of individual quotas (IQs) will considerably impact small-scale fisheries, in which women's participation is high, because small businesses are vulnerable to speculation from large businesses. Therefore, the AKTEA network asked that, in the case of implementation of an IQ system, the Commission should ensure women are not excluded from the benefits of the rights attached to the IQ, by including in the legislation the principle of co-ownership for husband and wife of the quota allocated to each enterprise. The network also considered that the implementation of an IQ system will impact fishing communities and it recommended a social study at community level to be undertaken before its implementation, and that this constitutes the social baseline.[13]

12.5 Functioning of Women's Organisations

This section describes how European fisherwomen organisations are organised into independent bodies, with women's groups associated with official male-dominated organisations in only two countries (the shellfish gatherers in Galicia -Spain- and the fishers' wives organisation of Ostrobothnia -Finland-). All the other organisations are independent of male organisations and, in some cases, have little contact with them. Women decided to establish independent organisations either because they had their own projects that were different from men's organisations or because the legal framework regulating the male organisations did not allow spouses participation in these organisations as they were open only to professional fishers.

12.5.1 Independent Organisations

The French fisheries committees provide an example of the constraints of women's organisations becoming associated with existing men's organisations; they did not open their doors to spouses of fishermen's groups because it was forbidden by law (1993). The French fisheries committees are now regrouping fishers and represen-

[13] European Network AKTEA: Response to the Green Paper "Reform of the Common Fisheries Policy, www//fishwomen.org.

tatives from other activities related to the fisheries industry. Women entitled to the collaborative spouse status are now allowed to become members of the committees, but only where the husband gives his seat to his spouse. In this case, it is an individual integration and the spouse is representing the family enterprise. In other European countries similar legal frameworks prevent spouses' groups from becoming part of male organisations. This is the case in Italy, where fishers' organisations are based on a cooperative system linked to unions and spouses cannot join in because they are not active fishers. Other reasons for not being part of male organisations are the lack of organisation or their weakness, as is the case in Greece and Portugal. Women's organisations from these two countries decided to advocate for the defence of sustainable fisheries activity and fishing communities as independent organisations. But both of them maintain strong links with fishers' (men) organisations because they consider that the future of the fisheries sector is based on the collaboration between women and men.

Independent organisations were and still are run on a voluntary basis by fisherwomen. Their scant financial resources are provided by territorial or national authorities and do not permit them to conduct their business as they would like. Many organisations devote more energy to raising funds to run their organisation than to implementing their projects and objectives (Laville et al. 1977). Women actively involved in the organisation contribute financially to the different expenditures related to their activism. This can only be done by women of financial means and excludes others. Thus, the democratic functioning of the organisation is challenged because women having this financial capacity represent the larger boats and they hold the power within the organisation, whereas women of small-scale fisheries and crew have less financial capacity and can become marginalised.

Aside from the financial problems, women's organisations have difficulty in keeping their membership alive. Only a few members are ready to give time to voluntary work. In some countries the participation at meetings demands too much time and not many women can provide it because they need a lot of flexibility and freedom to juggle the demands of their domestic responsibilities and the fisheries enterprise. In some cases husbands do not accept the involvement of their wives in organisations and in the public arena and, in some instances, ask them to stop their participation in women's groups. The personal situation of the different leaders indicates that most of them are greatly supported by their husbands. It is clear, though, that lack of finance and husband's support are the main obstacles to the ability of organisations to find active members. Another obstacle is women's lack of capacity to run organisations in a collective manner, with women more often concerned with the interests of their own enterprise. To overcome this problem the European network AKTEA asked for the establishment of training for leaders of organisations at the European level.

Another point to emerge is the need for women's organisations to have external assistance. The most successful organisations are those that receive help from people who are not members of fishing families, such as those involved in the fisheries industry but without a direct interest, such as scientists, social workers, NGO members. These members may be willing to give their time to build and run fisherwomen's organisation because they believe that it is the only way to obtain gender equality within this industry. They may become facilitators of organisations and assist women in defining their objectives and projects and help them conduct their activities. The organisations that do not benefit from such external support have greater difficulties and conflicts to overcome.

12.5.2 Women Organisations that are Part of Fishers' Organisations

Women's organisations that are members of fishermen's organisations have access to the financial support of male organisations. For example, a Finnish women's group integrated with a fishers' organisation in order to have access to the funds of this organisation. An employee was hired to help women to reinforce their organisation and to train them. The example of a Norwegian organisation demonstrates that women did not have any difficulty in accessing the funds of men's organisation when they were part of it. In this case, the group left Norges Fiskerlag only when it had secured funds through the national administration.

In Spain, the *mariscadoras* became members of the main fishers' organisation as soon they got the status of professional fishers. Their professionalisation gave them access to the services of biologists and training, for example. The *mariscadoras* groups gained more power within the fishers' organisation as soon they sold their products through auction and started to generate revenue for the whole organisation. At the beginning, the *mariscadoras* groups worked mainly for their own interests, but then they increased their power by taking over the leadership of some *cofradías* (Frangoudes et al. 2008, 2013). The modification of their situation within the male organisation progressed because women showed their capacity to lead and manage their own organisation and profession. The groups of shellfish gatherers became important representatives for public authorities and became members of the regional council of fisheries.

12.6 Conclusion

This overview shows that, in the main, the emergence of fisherwomen's organisations resulted from a fisheries crisis with their main aim being the defence of their fishing communities and their way of life. Through these organisations, the concerns of women have collectively evolved from social issues to more feminist issues, as they sought recognition of their own rights. Through their actions women's organisations have become players in the fishing industry and acted to influence policies linked to the sector. Fisherwomen have become advocates for the social aspect of fisheries and can play an important role in resource management. It can be stated that women's organisations are now in a secure position at all levels and take part in the decision making concerned with social aspects of fisheries and more recently with resources management. Women, by creating their own organisations, have claimed their own rights. But their organisations are still weak due to lack of money, capacity to lead, etc. Their 20 years of experience demonstrates that these organisations have a specific role to play within the fisheries sector but they need help to be maintained.

The authorities often consider women as more open than men in discussions of critical issues concerning fisheries. But even after 20 years of activity these important institutions are still facing difficulties. It has been outlined in this chapter how women's organisations function on a voluntary basis which does not permit them to build long-term projects. They require the support of authorities to enable them to more fully contribute to balanced sustainable fisheries policies, but are now recognised as important players in developing sustainable fisheries policies and, as such, require financial and managerial support. Without this support these organisations will cease to exist.

The implementation of European policies concerning gender equality opened the doors to structural funds for women. The current European Fisheries Fund (EFF) and the future European Maritime and Fisheries Fund give more attention to women's initiatives on the enterprise level and on the collective level. On the enterprise level, women can apply for funds to support their projects in the diversification domain (axis 4) and on the collective level for networking (axis 3). In practice only a few countries give attention to gender equality introduced by the EFF and women's initiatives financed by the structural funds are few. This lack of attention forces fisherwomen's organisations to question their role, existence and their acceptance in the public space. Without adequate support a great number of them will soon disappear because they cannot act for issues that are not recognised.

The actions to implement the European strategy for equality between women and men for the period 2010–2015 (COM 2010 491, SEC 2010-1080) also refer to women in fisheries. One of the proposed actions is the promotion of gender equality to "support MS [Members States] in promoting gender equality in the EFF programmes by drawing lessons from the mid-term evaluation" and the creation "of a pan-European network of women active in the fisheries sector and in coastal regions to improve the visibility of women in this sector and establish a platform for the exchange of best practices". Both of these proposed actions were claims made by the AKTEA network during the consultation process of the European roadmap of equality. Despite these proposed actions, promised by the European Commissionaire responsible for fisheries during a public hearing at the fisheries commission of the European Parliament (2010) to establish a European network of fisheries women at European level, fisherwomen's organisations are still waiting. This long wait will have a negative impact not only on the European network AKTEA but also for national organisations. This chapter aims to contribute to the history of women's organisations in fisheries in Europe and to highlight the need for the maintenance of these organisations. Their presence has contributed to the improvement of women's rights within this industry and to an improvement in socially informed European fisheries policies.

References

AKTEA Newsletter. (2003). No 2 September 2003, Brest. www//fishwomen.org.

AKTEA Newsletter. (2009). Issue n°12, Brest. www//fishwomen.org.

Ancel, J. (2003). Le statut du conjoint collaborateur, Le conseiller Juridique pour tous, 153, Editions du Puits Fleuri, Hérecy.

Bourhis, S. (2004). Collective actions, psychological support: comments from Sonia Bourhis. In K. Frangoudes, S. Mongruel (Eds.), *Women in Public Space* Proceedings of Vasa Workshop, Finland 25–27 Sept. 2003 (pp.104–106). Brest: CEDEM, University of Bretagne Occidentale. http://www.univ-brest.fr/gdr-amure/site-cedem/proceedings_vasa_en.pdf. Accessed 17 June 2013.

Carriou, L. (2004). Collective actions, safety at sea: comments from Liliane Carriou. In K. Frangoudes, S. Mongruel (Eds.), *Women in Public Space* Proceedings of Vasa Workshop, Finland 25–27 Sept. 2003 (pp. 90–92). Brest: CEDEM, University of Bretagne Occidentale. http://www.univ-brest.fr/gdr-amure/site-cedem/proceedings_vasa_en.pdf. Accessed 17 June 2013.

Esteban, A. M. (2004). Why joining together? Defending women's place, comments from Anne-Marie Esteban. In K. Frangoudes, S. Mongruel (Eds.), *Women in Public Space* Proceedings of Vasa Workshop, Finland 25–27 Sept. 2003, (pp. 24–28). Brest: CEDEM, University of Bretagne Occidentale, http://www.univ-brest.fr/gdr-amure/site-cedem/proceedings_vasa_en.pdf. Accessed 17 June 2013.

Fraise, G. (1998). Les femmes et leur histoire, editions Folio histoire, Paris.

Frangoudes, K., & Mongruel, S. (Eds.). (2004). *Women in Public Space*. Proceedings of Vasa Workshop, Finland 25–27 Sept. 2003. Brest: CEDEM, University of Bretagne Occidentale. In http://www.univ-brest.fr/gdr-amure/site-cedem/proceedings_vasa_en.pdf. Accessed 17 June 2013.

Frangoudes, K., Marugán-Pintos, B., & Pascual-Fernández, J. J. (2008). From open access o co-governance and conservation: The case of women shellfish collectors in Galician (Spain). *Marine Policy, 32,* 223–232.

Frangoudes, K. (2011). Women's contribution in small scale fisheries in the European Union. In R. Chuenpagdee (Ed.), *World Small-Scale Fisheries Contemporary visions* (pp. 101–116). Delft Eburon.

Frangoudes, K., Marugán-Pintos, B., & Pascual-Fernández, J. J. (2013). Gender in Galicia Fisheries: transforming for governability. In R. Chuenpagdee, M. Bavinck, S Jentoft, & J. Kooiman (Eds.), *Governability: Theory and Applications in Fisheries* (241–261) Dordrecht:. Springer.

Laville, J-L, Sainsaulieu, R. (1997). Sociologie de l'association. Editions Sociologie économique DDB. Paris.

Marugán Pintos, B. (2004). *Y cogieron ese tren... Profesionalización de las mariscadoras gallegas*. Santiago de Compostela: Xunta de Galicia, Consellería de Pesca e Asuntos Marítimos.

Marugán Pintos, B. (2005). Y cogieron ese tren… El proceso de profesionalización del marisqueo gallego. In J. J. Pascual-Fernández (Ed.), *Actas del Simposio Aktea: La mujer en la pesca, la acuicultura y el marisqueo en el contexto comunitario* (pp. 26–41). La Laguna: Red Europea FEMMES-Asociación Canaria de Antropología.

Meltzoff, S. K. (1995). "Marisquadoras of the shellfish revolution: The rise of women in co-management on Illa de Arousa, Galicia". *Journal of Political Ecology, 2,* 20–38.

Meltzoff, S. K., & Broad, K. (1996). "The rise of women in Fisheries Management: The marisquadoras of Illa da Arousa, Galicia". In R.M. Meyer, C. Zhang, M. L. Windsor, B. J. McCay, L. J. Hushak & R. M. Muth (Eds.), *Fisheries Utilization and Policy*. Proceedings of the World Fisheries Congress, Theme 2 (pp. 349–360). New Delhi: Oxford Lib. Pub. Co. Pvt. Ltd.

Morin, F-E. (2004). Why joining together? Political weight: Comments from Françoise-Edmonde Morin. In K. Frangoudes, S. Mongruel (Eds.), *Women in Public Space* Proceedings of Vasa Workshop, Finland 25–27 Sept. 2003 (pp. 44–46). Brest: CEDEM, University of Bretagne Occidentale, http://www.univ-brest.fr/gdr-amure/site-cedem/proceedings_vasa_en.pdf. Accessed 17 June 2013.

Nilsen, G. (2004). Collective actions, defending artisanal fisheries: comments from Gunda Nilsen. In K. Frangoudes, S. Mongruel (Eds.), *Women in Public Space* Proceedings of Vasa Workshop, Finland 25–27 Sept. 2003 (pp. 109–110). Brest: CEDEM, University of Bretagne Occidentale. http://www.univ-brest.fr/gdr-amure/site-cedem/proceedings_vasa_en.pdf. Accessed 17 June 2013.

Nielsen, R. M. (2004). Collective actions, defending artisanal fisheries: comments from Randi Marie Nielsen. In K. Frangoudes, S. Mongrue! (Eds.), *Women in Public Space* Proceedings of Vasa Workshop, Finland 25–27 Sept. 2003 (pp. 118–121). Brest: CEDEM, University of Bretagne Occidentale, http://www.univ-brest.fr/gdr-amure/site-cedem/proceedings_vasa_en.pdf. Accessed 17 June 2013.

Paredes Soto, P. (2004). Collective actions, safety at sea: comments from Patricia Paredes Soto. In K. Frangoudes, S. Mongruel (Eds.), *Women in Public Space* Proceedings of Vasa Workshop, Finland 25–27 Sept. 2003 (pp. 96–97). Brest: CEDEM, University of Bretagne Occidentale. http://www.univ-brest.fr/gdr-amure/site-cedem/proceedings_vasa_en.pdf. Accessed 17 June 2013.

Piriou, N., Coliou, J-R. (1995). De crise en crise, les pêche de Bretagne méridionale ne sont elles pas sur la voie de la déstructuration. Norois 165: 185–204.

Quist, C. (2002). Challenging stereotypes, Yemaya, 11: 7–9.

Quist, C. (2004). Why joining together? Political weight: comments from Cornelie Quist. In K. Frangoudes, S. Mongruel (Eds.), *Women in Public Space* Proceedings of Vasa Workshop, Finland 25–27 Sept. 2003. (pp. 50–58). Brest: CEDEM, University of Bretagne Occidentale. http://www.univ-brest.fr/gdr-amure/site-cedem/proceedings_vasa_en.pdf. Accessed 17 June 2013.

Serrano Soler, C. (2004). Why joining together? Defending women's place: comments from Carmen Serrano Soler. In K. Frangoudes, S. Mongruel (Eds.) *Women in Public Space* Proceedings of Vasa Workshop, Finland 25–27 Sept. 2003 (pp 22–23) Brest: CEDEM, University of Bretagne Occidentale. http://www.univ-brest.fr/gdr-amure/site-cedem/proceedings_vasa_en.pdf. Accessed 17 June 2013.

Simon-Goulletquer. C. (2000). Femme de la côte. Geste Editions/témoignages, La Crèche.

Talvitie, L. (2004). Why joining together? Management constraints: comments from Lena Talvitie. In K. Frangoudes, S. Mongruel (Eds.), *Women in Public Space* Proceedings of Vasa Workshop, Finland 25–27 Sept. 2003 (pp. 72–76). Brest: CEDEM, University of Bretagne Occidentale, http://www.univ-brest.fr/gdr-amure/site-cedem/proceedings_vasa_en.pdf. Accessed 17 June 2013.

Chapter 13
Women in English Fisheries: Roles, Contributions, Barriers and Prospects

Minghua Zhao, Marilyn Tyzack, Rodney Anderson and Estera Onoakpovike

Keywords Gender issues in fisheries • Women and social cohesion • Women's contribution to fishing communities in Europe • Social change in EU fisheries • Women and networking in fisheries

13.1 Introduction

Women throughout the world play a vital role in maintaining and sustaining the fishing industry and the fishing communities (Nadel-Klein and Davis 1998). They are directly involved in capture, processing, trading, financing or indirectly involved with various supporting roles in fisheries but many of these roles have been overlooked and unrecognised or under-recognised in fisheries management and policy development (Chapman 1987; Weeratunge et al. 2010). Although they are present at all levels and in all areas of the sector, the role of women in European fisheries has until recently remained largely unnoticed. They make a significant contribution to sustaining the fisheries and fishing communities from a variety of roles and positions, and yet this contribution has not been adequately acknowledged (Fishing in Europe 2003). This chapter has drawn on a research report from an externally funded study in 2010 in England. It aims to identify women's roles in and contribu-

M. Zhao (✉) · E. Onoakpovike
Greenwich Maritime Institute, University of Greenwich, Old Royal Naval College, Park Row, Greenwich, London SE10 9LS, UK
e-mail: M.Zhao@greenwich.ac.uk

E. Onoakpovike
e-mail: ejiromaf@yahoo.com

M. Tyzack · R. Anderson
IRIS Consulting, 5 Golden Square, Soho, London W1F 9BS, UK
e-mail: marilyn.tyzack@gmail.com

R. Anderson
e-mail: rb.anderson@btinternet.com

J. Urquhart et al. (eds.), *Social Issues in Sustainable Fisheries Management*,
MARE Publication Series 9, DOI 10.1007/978-94-007-7911-2_13,

tions to fisheries, the barriers they have encountered in their daily work and life, and their prospects for the future.[1]

13.2 Women in Contemporary Fisheries

There is a fair amount of information about the history of the roles played by women in fisheries and this makes for interesting reading for those interested with social change within the industry (Davies 2010). However, social changes have been rapid and far-reaching in recent years and, therefore, the focus of our literature survey places on more contemporary material. The discussion in this section is around two aspects: (1) the key roles identified by scholars and commentators with special research interest in women in fisheries; (2) the extent of women's participation in fisheries at national, regional and global levels.

Fishing, like mining, is considered as a "gendered, quintessentially masculine and very dangerous industry." This sector has historically been male dominant, but women have been found working as fishers on vessels today and in history in many countries, although always as a small minority (Hall 2004, p. 521).

Most women are seen to primarily work ashore, as traders, processors, and daughters and wives in fishing families and communities. They contribute significantly to fisheries through engagement in supporting, trading, processing and other fisheries-related activities (Frangoudes 2011; Frangoudes and O'Dohertty 2004; Nadel-Klein 2000; Munk-Madsen 1998). Thompson (1985, p. 3), as cited by Yodanis (2000, p. 286) notes that although on the whole, women do not fish, they "…gain power and status through the roles that they do within the fishing families and communities…The masculine image of the industry conceals the reality of an occupation which, by removing men to sea, makes them peculiarly dependent on the work of women ashore. And this dependence gives women not only more responsibility, but also the possibility of more power, both in the home and in the community."

The role of women is complementary and vital to the endeavours of fishermen and the fishing industry. Nadel-Klein (2000, p. 366) suggests that this complementarity of men and women's roles is responsible for the "…degree of economic autonomy and considerable domestic power" attributed to women. Hall (2004, p. 526) supports this concept of "women's power", as she mentions that women were responsible for managing the budgets and family savings, as well as being consulted during decision-making activities related to fishing equipment, such as buying boats, making repairs, etc. She also notes, 'despite the masculine character of fishing itself, wives played an essential part in their husbands' craft, so much so, that they defined themselves by their work in the fishing industry and were in every way their husbands' "business partners". Today, as in the past, women's income continues to supplement the family's earning (Davies 2010).

[1] The project was supported by grants from the Department for Environment Food and Rural Affairs (DEFRA) UK and the European Commission.

Table 13.1 UK employment in fisheries by gender and sub-sectors (2003), compiled from data in EC (2010)

UK	Total Fisheries	Fishing	Processing	Aquaculture
Men	24,637	11,774	9,999	2,864
Women	8,897	0	8,181	716
Total	33,534	11,774	18,180	3580

In her study on the role of women in the sustainable development of European fisheries, Frangoudes (2008, p. 32) notes "(F)ish processing is also an activity undertaken by wives with the objective of adding value to the catch." Many women are found in fish processing factories. Here women undertake various tasks, such as sorting, cutting and scaling. These tasks require a significant amount of skill, dexterity, complexity and difficulty. However, women are usually not recognised as skilled workers and their work not identified as skilled work. Davies (2010, p. 189) explains, "the concept of 'skill' as applied to work has been shown to be heavily gender-biased, with only long time served male labourers usually qualifying for the label of 'skilled workers'".

What then is the scale of women's participation in Fisheries? Searching for a gender based statistical picture of employment in fisheries and the related activities is difficult. Data does not exist in most cases. When they exist, they often differ from one source to another. This has been stressed by all the studies conducted at the European level (Frangoudes 2008, p. 5). The data presented here are largely estimates and should be used with caution.

Worldwide, women are found actively engaged in this sector of the economy. A report published by the UN's Food and Agricultural Organisation (FAO) in 2009 notes, "(W)omen play an important role both as workers in the fisheries sector and in ensuring household food security" (p. 26). According to this report, 44 % of the post-harvest activities in the world are as a result of contributions from women. Of these women, about 50 % are employed in small scale fisheries and the other 50 % in inland fisheries (FAO 2009).

In Europe, the World Fish Centre estimates that the participation rate of women in EU fisheries to be 6 % in marine fishing and 66 % in processing, with a total of 65,646 women employed in these two sectors combined (Rana and Choo 2001). In 2008, an EU-funded study covering 14 countries in the region found, "97,000 women are active in the fisheries sector, representing 25 % of the total 'visible' employment (no = 390,000) in the sector in 2003" (Frangoudes 2008, p. iv). Overall, "very few women are employed in marine fishing; women tend to be involved in other activities that are not onboard the vessel" (EC 2010, p. 10).

In the UK, during the same period, about 9,000 women were reported as employed in fisheries, representing 27 % of the total workforce. A breakdown of this figure found that all of these women were engaged in fish processing and fish farming, with none in capture fishing, as shown in Table 13.1.

It is worthwhile noting that women are always missing in official statistical calculations. One latest example can be found in the figure noting the trend of employment in the UK in the last 70 years provided by the Marine Management Organisation (MMO). According to these figures, the face of the workforce has been purely male. This is simply because women were not taken into consideration in data collection and computation.

In England, women have long been involved in fishing, carrying out various roles. Some of these roles, such as net braiding are discussed by Jeremy Tunstall, though quite briefly, in his sociological study of the fishermen and their families and communities in Hull, England (1962). Martin Wilcox, in his historical research of fishing and fishermen in Britain describes the lives of the English fishermen and their families during 1500–1815: "(W)hilst the men went to sea, women and children repaired nets or baited long-lines, gathered shellfish for bait and sometimes even sold the catch, hawking it around the neighbouring towns on foot." These activities are found continuing well into the twentieth century in the UK (2009, p. 46). Overall, however, studies on English fisheries tend to be from an historical approach focusing on fishermen usually with women only in the background or just mentioned in passing (Robinson 1987; Lummis 1985; Butcher 1982, 1979). There is inadequate systematic studies on women in contemporary English fisheries and we, therefore, hope that our research findings presented in this chapter would help to fill this gap in knowledge.

13.3 Research Methods

Extensive fieldwork was conducted for this research utilising primarily qualitative methods such as semi-structured interviews, focus groups and first-hand observation.

We focused the fieldwork in two regions in England: the South and South West and Northern England (Fig. 13.1). The southwest of England is the largest region in the UK, with a population of 5 million covering 23,828 km^2. It has the longest coastline of England's regions, totalling 1,130 km. There are very large differences in prosperity between the eastern parts of the region and the west (ODPM 2004), with Cornwall, one of our major research sites, as one of only four UK areas that qualifies for poverty related grants from the European Union (Cornwall Council 2011). Northern England has a population of around 14.5 million covering an area of 37,331 km^2 (Russell 2004, p. 15). This area is often referred to as 'the backbone of England'. This part of England was formerly dominated by heavy industry and mineral extraction and processing, also known as the 'grim up North' (*ibid*).

Our researchers were divided into two teams. One conducted fieldwork in the South and the other the North. Combined, a total of 13 fishing communities were covered by the study as shown in Table 13.2.

A total of 83 people were interviewed in early 2010. Fifty-four (65 %) were interviewed face to face and 29 (35 %) over the telephone. These were made up of 70 (84 %) women and 13 (16 %) men. They were people from a broad range of interests in the fisheries, including fisher women, fishermen's wives and partners, owners of vessels, skippers and crew, fish processors, shop owners, traders, se-

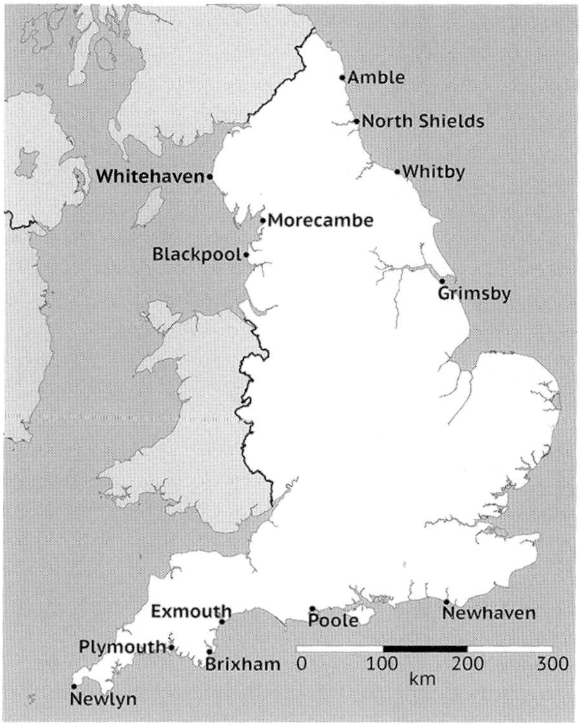

Fig. 13.1 Research sites covered by the project. (Map: Chris Bellamy)

Table 13.2 Places covered by the fieldwork in England

Area	Port cities/towns covered for fieldwork
S and SW England (6)	Newhaven (East Sussex), Poole (Dorset), Plymouth (Devon), Exmouth (Devon), Brixham (Devon), Newlyn (Cornwall)
Northern England (7)	Whitehaven (Cumbria), Blackpool (Lancashire), Morecambe Bay (Lancashire), Amble (Northumberland), North Shields (*North Tyneside*), Whitby (Yorkshire), Grimsby (Lincolnshire)

nior managers (public and private sectors) and public sector equality professionals (Fig. 13.2).

The International Collective in Support of Fish-workers (ICSF 2012) states that women play significant roles in all aspects of fisheries. Their roles span reproduction and production. Their reproductive roles are essential for nurturing and sustaining families and communities; the productive roles of women in fisheries contribute significantly to household incomes and local economies. Our findings

Fig. 13.2 Women's focus group, Newlyn, Cornwall, 4 March 2010. (Photo: Marilyn Tyzack)

confirm the importance of the contribution made by women and the way in which this contribution is often underestimated in the industry. The rest of the discussion presents the main findings from the primary data in four sectors: capture, families/communities, processing and trading. The United Nations, in one of its important documents on women and sustainable development points out: "(T)he expertise, knowledge and perspective of women have been overlooked for years and women are now demanding that their voices be heard" (UN 1997, p. 1). In order that the voices of those women (and men) who so willingly contributed to our understanding can be heard, we have included extensive quotes from the interviews in our field research in the rest of the discussion.

13.4 Women in the Catching Sector

As already noted, the number of women in capture fisheries is small. We found six women who have experiences of working on board fishing vessels or as cockle pickers, scallopers or crabbers. Although this indicates that there *are* women operating in the capture sector, they are certainly a small minority. Why don't women fish then? Yodanis (2000) asks this question in her well-cited study of rural fishing communities in Switzerland. Her research enabled her to respond to this, by focusing on the gender construction of women. Women who fish are usually classified as 'exceptional'. Even when helping on the boat, women did not expect or request remuneration as again they perceived this as an extension of their role – to help the man reduce his

Fig. 13.3 Fisherman and fisherwoman, Whitehaven, 2010. (Photo: Estera Onoakpovike)

expenses by using her services, rather than hiring another man to do the job. Interestingly, even though these women helped their husbands by fishing, they classify themselves neither as 'fishermen', nor their work as 'fishing'. That title was for the men. As a result, it is difficult to get rid of occupational segregation within the industry.

We find that women in this sector tend to have close family ties with fishing, especially inshore fishing. They are usually passionate and proud of their participation in this traditionally male-dominant sector of the economy. However, they also have a critical opinion of the harsh work and living conditions, in particular the strong masculine ship culture on board vessels, and that they have mixed feelings and attitudes towards their own identity as fisherwomen (Fig. 13.3).

13.4.1 Family Ties and Passion for Fishing

Women in this cohort usually had family members in inshore fishing. They were active in both fishing and running the family business. One woman fisher and vessel owner in Cornwall reported:

> I got into this industry because I met someone who was a fisherman. I used to go out with the boat because he had a boat. I went out a few times with them, the crew were all men, and eventually got the chance of investing some money to buy a boat.

The family link was also the introduction for a cockle picker and cockle trading business owner: "I have become part of the family business since I was 17 when I married into the business. I have been involved in all aspects of our cockle business, from picking to selling."

These women's passion for the job was particularly noticeable. One of the two female commercial fishers in the southwest, who both held a skipper's licence, reported: "I knew from a very young age that I never wanted a traditional job in an office…. I was so lucky that my partner introduced me to fishing. We now fish as far as France for cod, pollock and bass." The other woman described how she worked on a scalloper for seven years: "(I)t was my step-father's trawler and I helped build it. I passed my skipper's ticket. It was such an achievement!" One interviewee in Newhaven told us about her daughter's passion for fishing: "(M)y 23 year old daughter loves fishing and has her own boat. She catches mackerel. She goes out with her dad and her brother and has taken a job as a post-woman in order to have the time to do what she loves."

There were also a number of examples of women crewing for their fathers and husbands/partners. One young woman had been crewing for five and a half years since she was 13. Although she now runs her own beach-side tea business she continues to crew and said that she would be getting up at 4 am the next morning to fish with her father. The number of women who had developed a love for less traditional roles, either by working full-time as catchers or by continuing to fish with their fathers or husbands/partners on a part-time basis, points to the importance of family in the fishing industry.

As members of the fishing crew, women supported the family business in different ways. One, who had been married to a fisherman for 37 years in the south of England, described how, in the early days, she had collected shellfish and driven the seven-ton lorry with tons of mussels. Their business has now expanded and all the shellfish is now exported. Some women talked about actively working alongside their husbands/partners. One of the participants at the focus group in Poole described how she had lived with a fisherman for 10 years. As well as supporting her partner with the bookwork and licensing arrangements, she also fished locally and helped fix nets and pots. One woman, who was relatively new to the fishing industry, having recently moved in with her partner, described her work as: "We have two boats—one scalloper and one for crab meat. He says the crab meat one is mine! I pick the crabs three mornings a week and do the invoicing—all the paperwork. Lots of paperwork!" In these examples, women had dual roles of wives/partners and fisherwomen.

13.4.2 Work/living Conditions Aboard

As noted at the beginning of this section, Yodanis (2000) believes that women do not fish because they do not consider fishing as their sphere and, even when helping on the boat, women did not expect to be treated as fishers but as their husband/partners' helping "hands". While this may be true, we found that the poor working and living conditions including the hostility and harassment from a minority of the male crew, were a primary cause preventing women from participating in fishing production.

The female fishers interviewed reported no experiences of formal training in fishing. They learned by "doing it" alongside their male relatives on board. But, we heard about young women in fishing schools and working as trainees on board vessels. One such school in northern England has recruited both male and female trainees from across the UK in recent years. The school had been successful in finding work placement opportunities for their female trainees. Unfortunately though, the shipboard training was not successful. One of the young women left the school after six months' shipboard training; the other left abruptly after only two months. The wives of the skippers, on whose boats the women received their training, were of the view that "the girls left, because it was too tough at sea." We did not have the opportunity to explore these particular cases further, but the experiences of other female fishers as discussed in the next paragraphs, suggests that the trainees' departure could well have been closely linked with the harsh work and living conditions and an unfriendly male-dominated culture aboard.

Women who worked or wanted to work on fishing vessels experienced difficulties in various ways. First of all, it was always difficult for them to find their first onboard job. One senior woman fisheries officer shared her experience: "(A)fter doing my marine biology, for two years, I was desperately trying to get on fishing vessels as a volunteer, but with no luck, as they didn't want girls." One retired female skipper and crew member recalled her experience of going to sea: "I put my hair up in a woolen hat and persuaded an older man who used to take his granddaughter out, to take me."

For those women who succeed in working on the fishing vessel, the work and living conditions can be daunting. The fishing vessel is a confined space, constantly mobile at sea. It is both the workplace and the living space for fishers. The working conditions and risks were frequently cited as barriers to women seeking employment in the industry (Yodanis 2000; Hall 2004). It often means being exposed to poor and dangerous weather conditions, long and unsocial working hours and the lack of financial security ('no fish no pay'). Working at sea was generally perceived as 'physically unsuitable for women' because it demanded a great deal of physical energy and strength. "It is a hard and soul destroying job. Crabbing is particularly strenuous" (Fisherman's wife, Exmouth).

Lack of facilities to accommodate women's needs was clearly a serious barrier, for example the use of shared bunks and the toilet arrangements (or lack of them). One experienced woman fisher in Brixham said:

> You had to sleep in the same accommodation provided. Even as you are sleeping you are listening all the time, I usually had only a couple of hours of sleep…The toilet was a bucket, if you wanted privacy you would go to the engine room, the men would just pee on the side.

This problem was acknowledged to be an issue among some of the men interviewed. The director of a seafood and fishing training centre in Northern England, who had many years of fishing experience at sea commented: "The only barriers [for women] are working conditions: no toilets, sleeping arrangements…they [women] don't really want to take it [a job at sea] because of the lack of facilities catering to their needs."

The behaviour of some male shipmates can also have a negative impact on women fishers. In Plymouth, one woman described her experience as a crew member: "(T)he pranks they [the male crew members] played on women are not nice. They'd shit in your boots and then you need to go and wash it out and people are screaming at you because they are hauling and they need you to do things." Another women observed: "(T)hey gave me a lot of abuse. There was back stabbing. I wouldn't advise it for any girl."

Superstition against women remains an issue. One senior female director in charge of a fishing training school said: "Oh, yes, the idea is still very much alive. I know the local skippers very well, but I still have to be very careful, especially I have to avoid touching their boats, because I know they still believe women would bring bad luck to their boats if allowed aboard." One female fisheries administrator in the south west described how many boats had refused to go to sea with female members of a film crew.

13.4.3 Mixed Views About Women Fishers

There were mixed views as to whether women should be involved in the catching sector. The difficulties experienced by women who wanted to go to sea were highlighted by some participants at a focus group in Newlyn, Cornwall. However, those women who had worked on boats said it was possible to gain respect. It was generally agreed that women needed to be 'twice as good to be taken seriously'.

Many of those interviewed (men and women) commented on the problem of the ageing workforce and of the difficulty in recruiting young people. When asked whether attracting more women into the capture sector might be a solution, their response tended to be couched in pessimistic terms about the future of the industry, as indicated by one senior fisheries officer: "(A)s the industry is not looking to expand… there is really no particular interest in getting women more involved [in the catching sector]."

Some of the men interviewed were firmly of the view that working on board fishing vessels was not for women: "Honestly, I don't think fishing is women's job. I certainly don't want my women to do it. And, who will look after my kids and house when I am away?" (Fisheries officer in Northern England). Such remarks indicate the traditional view of women primarily as carers for the home, children and men.

How do women view themselves? Most of the women interviewed expressed reluctance when asked if they would like to join the capture sector. This reluctance was typically associated with their perception of the tough working and living conditions as well as the dangers involved. We could see a linkage between women not wanting to work on board vessels and their concerns about the decline of the industry and the lack of future prospects in the capture sector.

On the other hand, we also found progressive enthusiasm for more female participation. One male fisheries officer said passionately: "I don't think people should use [lack of] physical strength as an excuse to stop women from going to sea any more. Fishing vessels nowadays have all the modern gadgets. They don't need much strength to operate. Both men and women can do it." He even challenged the existing technology and practice in order to accommodate women's participation: "Of course the boxes may be too heavy for women. But, they can be redesigned, for example, from 40 to 20 k, so that women can carry them more easily."

Despite the many challenges, all the women interviewed in this study who had been involved on board vessels said that they did not regret their choice. They had enjoyed working at sea and they felt proud about having done so. They viewed themselves as 'strong women' who had 'made it' and they emphasised the importance of women being sufficiently tough and tenacious to get into and stay in the industry. At the same time, they also wanted change "to make things easier for other women who want to go fishing at sea" (Women's focus group, Newlyn, Cornwall).

13.5 Women in Fishing Families and Communities

Women are important in supporting fishing families and communities. Munk-Madsen (1998), in her study on the Norwegian fishing quota system, states that women are "fundamental in sustaining daily life and directly and indirectly supporting fishing activities" (p. 231). Hall (2004, p. 526), in her historical comparative investigation of women's roles in mining and fishing communities in Northumberland, England, reinforces this argument, asserting that women's contribution is so important that "a man could not practise the craft of fishing without a supportive wife/woman backing him", as "women's work is absolutely necessary" for the survival of the fishing industry and community (Fig. 13.4).

In England, women in fishing towns, as those throughout history and in other countries, are actively involved in various activities in fisheries and play multiple roles in families and communities.

13.5.1 Multiple Roles and Contributions

In fishing families, women, in addition to their caring roles as fishermen's wives/partners and mothers, fulfill multiple roles as accountants, book keepers, administrators, crew, cooks, drivers, community organisers, running the home, attending meetings and so on. Their contribution to households and to the industry is significant although many women themselves do not recognise their contribution, as shown in the following quote from a fisherman's wife in Grimsby:

Fig. 13.4 The fishing harbour at Newlyn, Corwall, 2010. (Photo: Marilyn Tyzack)

> I don't work at all. I run the shore side of things such as picking the crew up and driving them around, doing the shopping, the books and the paperwork, and I also attend meetings with Producer Organisations.

It is interesting to note that this fisherman's wife says at the beginning 'I don't work at all', but then she goes on to outline a whole host of 'work' activities that she is engaged in.

Some women had full time jobs separate from fishing but are also intensively involved in helping run the business of their male family members. For example, one was a full-time social worker but supported the business through handling the licensing arrangements and book keeping. Another was an accountant who also helped with the book keeping.

The restructuring and other changes in the fishing industry in recent years has had a significant impact on fishing towns and villages. Many fishing families have seen deep cuts in their household income. As money from catching and landing fish has dwindled, many families have sought second incomes. In most cases, these were from shore-based sectors of the industry, such as processing and trading. It was evident that many wives and partners have become significant second, or even first, bread winners in the households. One fisherman's wife from Grimsby said: "Money became hard. So, I got a job in the processing factory in order to help my husband."

> My husband was a fisherman but sold his boat two years ago, because the cost got too high and we just could not afford it any more. I began to help a few years ago. I made fish cakes and sold them on the market. At first, I made about £ 200 a month, and now I make about £ 500 a month, about half of what we need every month. (Wife of an ex-fisherman, Whitby)

Many of the wives and partners spoke about the difficulties they had in finding jobs. It was, therefore, necessary for them to start their own businesses. Where the fishing communities are in rural areas, these findings will need to be considered in the context of the limited nature of opportunities for women in the countryside and the fact that there are fewer local work choices than those in urban areas (Countryside Agency 2003). It is interesting to note that while the traditional gender division of labour remains fundamentally the same, the respective roles of men and women are changing in fishing families and communities in England. It appears that increasingly more women are becoming the primary bread winners, though still remaining primary carers in the home. Some of the men are now either unemployed or have become the secondary bread winner. This change is likely to have implications to other dimensions of the gender relationship in fishing families and communities in the long run.

13.5.2 Lack of Recognition

Women make direct and indirect contributions to family incomes and household and community decision-making, but these contributions are seriously under-recognized. Many women complained about a lack of recognition and appreciation of their roles by the government. At a focus group discussion in the south, a woman said:

> Women are not appreciated. It is hard to get yourself appreciated. There is no financial allowance and we could still be doing the VAT at 5am but still getting up to go to Billingsgate.[2] It would be good to have some recognition by the industry.

As already shown, sometimes women themselves were not aware of their own contribution. When asked about her involvement in fishing production, the wife of a trawler skipper in North Shields said: “I wasn’t involved at all. I only looked after the house and the kids, cooked for the boat, cleaned the crew’s clothes and drove them around sometimes”. Interestingly, women’s roles were usually better recognised by their husbands/partners. Nearly all the men interviewed said that they valued the contribution made by their female family members, as highlighted by one fisherman from Whitby:

> Women have a big part in running the business. In the 1950s-60s, my mother and aunts played a big part in picking mussels to put on the lines as bait. Currently my wife is also much involved, including decision making, even in aspects such as buying gear and equipment.

The National Federation of Fishermen’s Organisations (NFFO 2012) confirmed in the interview with us that the women who are shore-based are often influential through family involvement.

---

[2] Fish market in east London, the largest inland fish market in the UK.

13.5.3 Involvement in Decision Making

It was generally felt by most respondents that contributing to government consultation exercises was mainly the domain of the men. There were examples of strong women attempting to influence decisions through written challenges, calling for judicial reviews and joining campaigning groups to challenge Government policy. However, because of the individualised nature of the industry, these responses were not widely known. When meetings were held it was mainly the men who attended. It would appear that there are still only a few instances of women attending in their own right, unless they are vessel or business owners.

Participants in this study indicated that, among the barriers to greater involvement in decision making, was the strong masculine culture within the fishing industry. When asked to describe the culture of the industry descriptions such as 'macho', 'out of date' and 'complex' were given, along with 'competitive' and 'insular'. This was reflected in the attitudes of many of the women as well as the men. During a focus group the comment was made that government consultations were little more than paper exercises and there was no clear vision for fishing, so why bother?

Women gave a variety of reasons for not getting involved in aspects of the decision-making processes, including hopelessness, communications not understandable, communications not targeted, lack of confidence, feeling unwelcome at meetings, difficulty with time commitments because of child-care responsibilities, and so on. However, a lack of involvement in the traditional consultation processes does not mean that women had no interest in the wider concerns around the fishing industry. In fact, one of the strongest messages from the study was that they all cared passionately about the future of their industry and their community and they recognised that decisions regarding its future were being made elsewhere. All had very strong views on the way the industry had been treated by government and Europe and had feelings of frustration and disempowerment as a result. An interest in forming local community groups that had a specific purpose was expressed by a number of wives/partners, expressed by a fisherman's wife from Newhaven:

> Ideally I would like to see a body of women made up of fishermen's wives so that when decisions are made they are consulted as a separate group. We have the NFFO but it is not well known and yet it is supposed to be our voice. Everyone knows about the CPI, for example. The NFFO is mainly all men. What about us? Let women be the PR voice![3]

Strong evidence indicates that the barriers to participation in the decision-making process could be overcome to some extent by engaging women in existing

[3] NFFO refers to National Federation of Fishermen's Organisations. CPI stands for the Centre for the Promotion of Imports from developing countries, an agency of the Ministry of Foreign Affairs of the Netherlands, which provides technical support to exporters and BSOs (Business Support Organisations) in developing countries. One of CI's support activities is the provision of market intelligence reports, which contain all the information to keep exporters and staff of BSOs in step with the very latest developments on the EU market for their specific sector, including fisheries (www.cbi.eu/marketinfo).

institutions and organisations in fisheries. Also, the creation of community-based networking structures would help to overcome many of the difficulties that stem from isolation and lack of confidence.

Yodanis (2000) identifies the changing role of women in fisheries as they become "political representatives and lobbyists" (p. 281–282). In line with their political agenda, these women not only keep their men informed of changes in policies and regulations, but they also influence the policy making process by communicating with politicians, attending meeting and hearings in the and "giving testimony on the strength of the fishing stocks and the potential consequences of proposed regulations."

Similar evidence emerged in our study with the English women in fisheries. We found that some fishermen's wives are becoming more involved and, driven by concerns for family and communities, are overcoming some of the barriers identified above. One boat owner and a fisherman's wife observed: "I think the skippers' wives are getting more actively involved in the more political side of it. All the rules are getting so complex so women are trying to help their husbands."

In Whitby, a number of wives have recently set up their own organisation called 'Fishermen's Families and Friends'. In doing so, their primary objective was to "help enhance the image of our industry, and to save our community from disappearing." As a result of this the women have found new roles as networkers or activists, which could eventually lead to their more direct involvement in decision making.

13.6 Women in Processing

In her study of women's role in 14 European countries, Frangoudes (2008) also included women involved in fish processing. She found that fish processing is an activity undertaken by wives with the objective of adding value to the catch and that most of these processing businesses are "small scale family processing units that begin as small-scale endeavours, often in the family kitchen." In England, two processing factories participated in our study but both were much larger enterprises with one factory having a workforce of about 40 and the other over 400; in both cases more than half of the workforce were women (Fig. 13.5).

To reduce costs, factories in this sector take two main strategies. Firstly, they move all or part of the processing activities overseas to countries where the labour cost is low and the regulation relaxed; secondly, they use increasingly more foreign workers, women and men, from Eastern Europe and Asia. Foreign workers receive lower wages and have fewer rights compared with local workers. The participation of different nationalities in production in these factories has complicated the relationships regarding women's rights, interests and their part in decision making.

Fig. 13.5 Woman processor at work, Grimsby, 2010. (Photo: Minghua Zhao)

13.6.1 Women of Local and Foreign Origins

In English fish processing factories, most women are employed as processors, packers, quality controllers and other operators in the production line, working on three shifts to support the 24-hour operation of the facility, earning a weekly wage of around £ 200–£ 250. This cohort consists of two major groups: women locally employed and women recruited overseas through agencies.

We found that the local women workers tend to be school leavers, who married early with some as single mothers. Lack of confidence and ambition seemed a shared feature among these women. When asked about their views on their prospects for the future, most women in this group said that they would rather continue to do what they do 'on the line' and would expect little change or promotion, as indicated by one production controller: "Apply for promotion? Oh, no. I would rather stay as I am. It's too much responsibility for me." Their limited education and heavy family responsibilities are two major blocks to further personal and occupational development. Some women, however, expressed a strong wish for education and training.

Foreign workers, women and men, are increasingly employed in processing factories. According to a senior human resources officer in a processing factory in Northern England: "(S)even, eight years ago, it was the refugees from Iraq, then it was the Thais, the Filipinos, and now it is mostly workers from Eastern European countries: the Poles, the Bulgarians, the Latvians and so on." Compared with the local recruits, these women are generally better educated, most holding university degrees and many with professional employment histories in their home countries. Financial incentive seems to be the major motivation for them to leave their home countries and take jobs in the processing sector. One 43-year old woman with 18 years employment history in banking in Poland shared her experiences and views: "I take this job in the factory not because I love packing fish, but because of money. I have three children to support back in Poland. Here, the labour is hard, but the money is better. I work for my children."

13.6.2 Involvement in Decision Making

Shop-floor women workers, at the bottom of the factory hierarchy, have little influence but also demonstrated little interest in policy making. Poor education, lack of skills, language difficulties (for foreign women) and family responsibilities are the major obstacles behind their overall lack of confidence, interest and influence.

However, it is worth noting that many of these women, especially those from foreign countries, expressed a strong interest in joining a women's organisation. The foreign women's stronger interest in organisation can be explained by two reasons: first, their perception that the trade union was open only to local workers and not to foreign workers; secondly and more importantly, they felt isolated in a foreign land, and hence were eager to seek solidarity and networking

There are women who are supervisors and managers in processing factories, but they are still in the minority. Despite this, progress is being made and the number of women supervisors has increased considerably in recent years and more women are being promoted to managerial positions. The women in these positions have a better chance of contributing to decision making in the factory. They were noticeably more articulate and assertive in expressing their opinions. One quality manager in a large seafood processing factory said: "I always made sure that my voice is heard, even though I could see resistance in top managers. I had to use lots of meetings and paper work to prove my point." The human resources manager in this factory, also a women, had played a key role in introducing some important and popular personnel management policies, including English language training for foreign workers. In the small factory, there are women operators, administrators and supervisors, but decision making seems firmly in the hand of the owner, a man in his 40s with fishing and trading experience for many years.

13.7 Women in Trading

In her study of women's careers in Ghanaian canoe fisheries, Over argues that women play a crucial role as intermediaries in the distribution and exchange of fish after the processing process (Over 1999). In England, women are largely missing in the fish market, but they are found taking an active role in fisheries business, as entrepreneurs and small traders and their contribution is important in this sector (Fig. 13.6).

13.7.1 Entrepreneurs and Small Traders

There are successful business women engaged in the fish trade. In our interviews, one woman described how she had been in the industry for 37 years and managed a large number of beam trawlers. She described how she had been born into the

Fig. 13.6 Plymouth Trawler Agents Lorry, 2010. (Photo, Marilyn Tyzack)

business started by her great-great grandfather. She was clearly influential both in her community and through her membership of a wide range of bodies. Another had returned from a very different career in the United States to start a business in the industry. She had set up a scallop business three and a half years ago. It now employs 40 staff in the factory and 60 at sea. A third described how she had helped set up the Plymouth Trawler Agents (Fig. 13.5), which had been formed by fishermen for fishermen out of necessity. The old fish market had not met the required standards and there was no one available to run it, so this woman, as the wife of a local fisherman, had taken over that role. Unusually, the interviewee also acted as an auctioneer in a large fish market in southwest England.

Women traders also run small shops as retailers of fish products. These women seem to face more challenges. The competition from supermarkets was seen as a major issue by a number of women traders and shop owners. Supermarkets were well placed to enjoy the economies of scale and were able to source their fish from all over the country, or even the world. Small local shops, on the other hand, tend to be dependent on local landings, which means that the fish they sell are seasonal and even scarce when landings are low.

Of interest is the range of enterprising activities/businesses that some fishermen's wives/partners have established. This was particularly noticeable in the Penzance area. Although these were mainly linked to the fishing industry they have been chiefly set up as separate businesses. A focus group participant described how her husband had been a fisherman for 40 years: "I always did the book keeping but then started a business selling our own fish." Similarly one of the participants said that she had been married to a fisherman for 20 years. She 'picked crab' but got a grant to establish a business supplying crab meat to local restaurants. Another bought fish locally and sold it to restaurants throughout the UK. Many of these women continued to support their husbands/partners through managing their books and other business and support related functions.

13.7.2 Missing in Fish Market

In the fish market, where fish is sold and bought, supporting staff are often women, but women are rarely seen on the market which is a predominately male domain. In fact, the trading activities are often part of multiple responsibilities among the women, although largely invisible. One cockle picker and trader said: "I peel and pot the shrimps, prepare them. I butter them and spice them. I sell them. I do the book work. I also buy and sell cockles, arrange the transport and deal with customers in France. I sell cockles and mussels into France."

One of the barriers to women in this sector is the physical conditions, with the working hours inconvenient and sometimes irregular. This factor is often more of an issue with women than men because of women's child care and other family responsibilities. One woman with many years experience in the industry said: "I have never seen any women buyers or auctioneers; not even women standing at the auctions with the clipboards or even women secretaries at these auctions, because it's so early in the morning. 4am-7am latest [preparing the market] and the market usually starts at 6am."

Women's lack of training and opportunity to develop skills was also considered as a factor responsible for the lack of women's presence in this sector. One (male) owner of a small processing workshop observed: "the reason women are little involved in the direct sales of fish is due to their lack of experience.... Many of the men started off as fish porters or barrow boys—very physically demanding jobs, working their way up to auctioneer."

13.8 Discussion and Conclusion

The findings of this study would seem to confirm that in England, women are actively involved in various sectors of the fishing industry and the fishing community. They work as crew members; they support fishermen as ship-cleaners, cooks, accountants and book keepers; they sort, cut and package fish in processing factories as shop floor workers; they buy and sell fish at market as wholesalers and retailers, and much more. Women fulfill a range of roles from invisible, unrecognised and unpaid support functions to high profile jobs charged with considerable responsibilities such as skippers and successful entrepreneurs. The key issue, though, is that women are often under-represented, unrecognised, underpaid and seem to have limited say in decision making at various levels. This may well echo the experience elsewhere in Europe and in other parts of the world (Marthews 1993; Nadel-Klein and Davis 2000). This study has also documented the trend for increasing employment of women and men from Eastern Europe and other developing countries in processing factories and the socio-economic implications of this. It also sheds lights on the changed and still changing relationship between man and woman in fishing families resulting from wives/partners becoming primary bread winners as a response to the crisis of the declining industry. Further research will be needed to have a thorough investigation of these issues and their implications to the fishing industry and fishing communities.

Globally and across economic sectors, in spite of the significant progress already made in empowering women socially, economically and politically, achieving gender equality is recognised as a 'grindingly slow process' and barriers still exist, blocking women from furthering their participation in decision-making processes at family, community and state levels (Claros and Zahidi 2005). This implies that women have limited opportunity to influence the decisions that affect their lives and the decisions made are less likely to represent women's interests in some general issues that are of concern to women such as equal pay, employment conditions, child-care, and so on (Hoare and Gell 2009).

In this study, we found that the issue is complex in respect of women engaged in fisheries. This is because these women in England, as those in other parts of the world, are not only engaged as visible and paid workers, but also as 'invisible' and unpaid workers with a wide range of roles and responsibilities. They make significant social-economic contributions to sustain the fishing industry and the fishing community, but they are not fairly represented, recognised and they are always unpaid or underpaid with little say in decision making. This is unfair with negative social and economic implications for the sustainability of the sector.

What has been particularly noticeable is the passion and concern that women at all levels have about the future of their industry. This overrides any anxieties or interests in improving their position or status. The masculinity of the sector was mentioned by nearly all those interviewed, but this was only perceived in a negative way by a small minority. Although this may imply that women have merely absorbed the dominant cultural attitudes, their interest in exploring innovative ways to maintain the industry and their pride in their contribution to their business suggest otherwise. What was apparent was an unspoken understanding that the business depends on them for survival and sustainability. There was also enthusiasm in supporting the industry through finding new ways of working cooperatively and identifying new and imaginative ways of using their produce. One of the key features of this investigation is the passion and enthusiasm shown among many of the women contacted to establish groups and networks. This reflects the process of organising women as well as the success and the challenges involved in fisheries in Europe discussed extensively in Frangoudes's chapter in this book (see Chap. 13). Although finding women to attend focus groups was not always easy, primarily because of the short time period available for the study, we found that as the study became known there was growing interest. There were examples of women travelling many miles to attend meetings and most said they would welcome being part of a network so long as it had a specific purpose. Capitalising on the genuine interest in the sustainability of the industry and hence their families and their livelihoods would seem to be the most effective way of increasing women's awareness, involvement and participation.

Acknowledgments The authors would like to express their most sincere thanks to the Department for Environment Food and Rural Affairs in the UK and the European Commission for their financial support to this study, to Professor Chris Bellamy for helping with the map, and most importantly to many women and men whose participation in this research was most vital for our successful completion of the project.

References

Butcher, D. (1982). *Living from the sea: Memories of shore-side life in the days when fishermen's nets were often full, but their pockets usually empty.* Sulhampstead: Tops'l Books.

Butcher, D. (1979). *The driftermen: Life in the tough days of Britain's vanished herring fleets, recalled by the men who manned them*. Reading: Tops'l Books.

Chapman, M. D. (1987). Women fishing in Oceania. *Human Ecology, 15*, 267–288.

Claros, L, & Zahidi, S. (2005), 'Women's empowerment: measuring the global gender gap', World Economic Forum, 2005. https://members.weforum.org/pdf/Global_Competitiveness_Reports/Reports/gender_gap.pdf (p. 1). Accessed 7 Mar 2012.

Cornwall Council. (2011). Facts and figures. http://www.cornwall.gov.uk/default.aspx?page=18151. Accessed 22 Nov 2011.

Countryside Agency. (2003). Capturing richness. Countryside Agency Research Reports and Research Notes. http://publications.naturalengland.org.uk/publication/292346?category=60007. Accessed 22 Nov 2011.

Davies, S. (2010). A whirling vortex of women: The strikes of scots herring women in East Anglia in the 1930s and 1940s', *Labour History Review, 75*(2), 181–207 (Maney Publishing).

EC. (2010). *Employment in the fisheries sector: Current situation, 2006.* The European Commission. (Salz, P., Buisman, E., Smit, J., de Vos, B).

European Commission. (2003). Women in fisheries: An Unnoticed Role. *Fishing in Europe,* (17), 1–17.

Food and Agricultural Organisation (FAO). (2009). The State of World Fisheries and Aquaculture, 2008. http://www.greenfacts.org/en/fisheries/index.htm. Accessed 6 Feb 2010.

Frangoudes, K. (2008). The role of women in the sustainable development of European fisheries. European Parliament, Directorate-General for Internal Policies—Fisheries, pp. 1–62.

Frangoudes, K. (2011). Women can bring a lot to fisheries management if they are allowed to take their rightful place. *FARNET Magazine*, no.4, Spring-Summer, pp. 15–16.

Frangoudes, K., & O'Doherty, J. (2004). Legal recognition of women's contribution in fisheries and aquaculture in the European Union. Paper presented to World Fish Centre, Gender Workshop, the 7th Asian Fisheries Forum, Penang, 30 November–4 December.

Hall, V. G. (2004). Differing gender roles: women in mining and fishing communities in Northumberland, England, 1880–1914. *Women's Studies International Forum, 27*(5–6), 521–530.

Hoare, J, & Gell, F. (2009). *Women's leadership and participation: Case studies of learning for action* (Ed., p. 34), UK: Oxfam GB.

ICSF. (2012). Women in fisheries. http://wif.icsf.net/en/bibliography.html?themes=28&start=70. Accessed 15 Dec 2012.

Lummis, T. (1985). *Occupation and society: The East Anglian fishermen 1880–1914*. Cambridge: University Press.

Marine Management organisation (MMO). (2011). The UK Fishing Industry in 2011: Structure and Activity—Tables. http://www.marinemanagement.org.uk/fisheries/statistics/annual_structure2011.htm. Accessed 26 July 2012.

MacAlister, E. (2002), 'The Role of Women in the Fisheries Sector', Brussels: EC/DC Fisheries Final Report 1443/R.03/C.Makoloweka, S. and K. Shurcliff Ltd.

Mathews, E. (1993). Women and Fishing in Traditional Pacific island Cultures', Workshop on People, Society and Pacific Island Fisheries Development and Management: Selected Papers, August, Noumea, pp. 29–33.

Munk-Madsen, E. (1998). The Norwegian fishing quota system: Another patriarchal construction? *Society and Natural Resources, 11*(3), 229–240.

Nadel-Klein, J. (2000). Granny baited the lines: Perpetual crisis and the changing role of women in scottish fishing communities.*Women's Studies International Forum, 23*(3), 363–372.

Nadel-Klein, J., & Davis, D. L. (1998). *To work and to weep: Women in fishing economies*. Institute of Social and Economic Research, Memorial University of Newfoundland, Canada.

NFFO (The National Federation of Fishermen's Organisations). (2012). Sustainability Initiatives in Fishing. http://www.nffo.org.uk/sustainability.html. Accessed 15 Dec 2012.

ODPM (Office of the Deputy Prime Minister. (2004). Competitive European Cities: Where do the Core Cities Stand?. *Urban Research Summary, 13,* 12 (PDF).

Over, R. (1993). Wives and traders: Women's careers in Ghanaian Canoe Fisheries. *Maritime Anthropological Studies (MAST), 6*, 110–135

Rana, K., & Cho, P. S. (2001). Women in Fisheries in the European Union. Global Symposium on Women in Fisheries at the Sixth Asian fisheries Forum, 29 November, Kaohsiung, pp. 191–193.

Robinson, R. (1987). *A history of the Yorkshire Coast Fishing Industry, 1780–1914*. Hull: University Press.

Russell, D. (2004). *Looking north: Northern England and the national imagination*. Manchester: University Press.

Thompson, P. (1985, January). Women in the fishing: The roots of power between the sexes. *Comparative Studies in Society and History, 27*(01), 3–32.

Tunstall, J. (1962). *The Fishermen*. London: MacGibbon & Kee.

UN (United Nations). (1997). Women and Sustainable Development. Department of Public Information—DPI/1906/SD-June.

Weeratunge, N., Snyder, K. A, & Choo, P. S. (2010). Gleaner, fisher, trader, processor: Understanding Gendered Employment in Fisheries and Aquaculture. *Fish and Fisheries, 11*, 405–420.

Wilcox, M. (2009), *Fishing & fishermen: A guide for family historians*. Great Britain: Pen & Sword Family History.

Yodanis, C. (2000). Constructing gender and occupational segregation: A study of women and work in fishing communities. *Qualitative Sociology, 23*(3), 267–290.

Part V
Issues in Social Identity and Cultural Heritage

Chapter 14
Sense of Place and Socio-cultural Values in Fishing Communities Along the English Channel

Tim G. Acott and Julie Urquhart

Keywords Sense of place • Cultural ecosystem services • Social and cultural value of fisheries • English Channel • CHARM III • Sustainable communities • Co-construction • Place identity • Cultural heritage

14.1 Introduction

Coastal communities exist on the littoral boundary between land and sea, with the influence of the marine environment and its living resources spilling out onto the land in a range of heterogeneous associations. Marine fishing is often thought of as a primary industry providing a valuable source of food and jobs for those involved in the sector. However, marine fisheries also interface a largely hidden undersea world and a series of ecologic-socio-cultural-economic translations that help drive the creation of a particular sense of place in coastal towns that is often linked to fishing. Fishing activity is bound up in relational processes, with the act of catching fish taking place at sea, but resulting in the creation of networks of influence and activity on land, including fish landing, selling and processing through to the creation of particular individual and community identities linked to a fishing way of life. The process of fishing, therefore, creates a range of values that tie people, places and ecosystems into a network of relational encounters.

This chapter explores the importance of fisheries as a generator of social and cultural values for coastal communities by reference to a sense of place study in fishing towns and villages along the English Channel in England and France carried out as part of the Channel Integrated Approach for Marine Resource Management (CHARM) III Interreg IVa project. In fisheries management, the idea of community

T. G. Acott (✉) · J. Urquhart
School of Science, University of Greenwich, Central Avenue, Chatham Maritime, Chatham, Kent ME4 4TB, UK
e-mail: T.G.Acott@greenwich.ac.uk

J. Urquhart
e-mail: j.urquhart@greenwich.ac.uk

J. Urquhart et al. (eds.), *Social Issues in Sustainable Fisheries Management*, MARE Publication Series 9, DOI 10.1007/978-94-007-7911-2_14,

has often not been given a high priority in policy development (Jentoft 2000). However, if a sustainable future for marine fishing is to be realised, people and communities should form a central element of policy making. New perspectives and methods are needed that make visible the wide range of cultural and social values that are generated by marine fishing to stakeholders involved in policy making. In other areas of resource use there are attempts to raise the profile of the community concept as central to building more sustainable places (Marsden and Hines 2008). We suggest that sense of place is a useful approach to begin to understand the range of social and cultural values emerging from the relational process that connects marine fishing and terrestrial communities and is important in planning for a sustainable future.

Sustainability, however, is a notoriously difficult term to pin down, with many definitions and typologies that describe its meaning (Williams and Millington 2004). Agyeman and Evans (2004) define it as follows: "the need to ensure a better quality of life for all, now and into the future, in a just and equitable manner, whilst living within the limits of supporting ecosystems" (p. 157). This definition begins with the idea of quality of life and then casts that in terms of ecosystem limits (Blay-Palmer 2011). This has particular relevance for fisheries where approaches to management have tended to exclude explicit reference to well-being and focus on understanding ecosystems through the application of rational/science-centric paradigms to manage fish stocks and their resulting economic value.

This chapter explores how marine ecology, through the process of marine fishing, drives a series of terrestrial relationships that are woven into material and phenomenological worlds. Using sense of place as an approach, we explore how fishing practice is enrolled into community relationships and discuss how emergent cultural values contribute to developing a narrative about sustainable communities. Drawing on ideas of cultural ecosystem services (MEA 2005a; Mace and Bateman 2011) we show how sense of place can be used as an approach to identify a range of cultural values that are generated by marine fishing. This has applied implications for fisheries management by making visible relations that might otherwise be hidden in studies that focus on ecological and economic dimensions of fisheries management. The chapter begins by reviewing literature describing the importance of social and cultural research to marine fisheries management and examining the idea of sense of place and the importance of this concept for understanding fishing communities. A case example is then presented that details research into sense of place and fishing and demonstrates the utility in unpacking some of the natural, social and cultural values of marine fishing. The conclusion draws together the findings of the case example and describes the future research potential of bringing together ideas of sense of place and cultural ecosystem services for sustainable marine fisheries management. The use of non-dualistic conceptual theories to help understand relational processes resulting from marine fishing is commented on.

14.2 Social and Cultural Aspects of Fisheries: The Role of Sense of Place

It is well documented that policy makers have been slow to include social and cultural dimensions as part of the development of fisheries policy (Symes and Phillipson 2009). This is despite increasing evidence of the importance of developing holistic approaches to fisheries management (Firn Crichton Roberts Ltd 2000; Forst 2009) alongside understanding the importance of people's ethical values and perceptions (Jacquet 2009). In policy there is increasing recognition of the need to better understand the social, cultural and economic impacts of fishing (as, for example, in the recent proposals for the reform of the European Common Fisheries Policy (CFP), the European Integrated Maritime Strategy and the European approach to Integrated Coastal Zone Management). In the UK increasing interest in the social impacts of fisheries is evidenced by the recent Defra (Department for Environment, Food and Rural Affairs) project entitled Sustainable Access to Inshore Fisheries (SAIF) that examined the social impacts of England's inshore fisheries alongside economic and environmental impacts. A gap analysis on behalf of Defra identified the need for enhanced research effort to understand community and social cohesion in inshore fisheries (eftec 2010).

In Europe, the CFP is a major initiative that has influenced marine fishing since its inception; however, there are serious concerns as to its efficacy. Despite reform in 2002, the measures adopted yielded poor results and many fish stocks remain outside or almost outside safe biological limits with issues concerning poor enforcement of the CFP regulations, discarding, illegal landings, mesh sizes too small to protect juvenile fish and misallocation of catches (see, for example, EC 2009). A further major review of the CFP was announced in 2008 to be finalised by the end of 2012. Although there is recognition that such policy reviews need to be aware of the social relevance of the regulations they implement, there is little evidence-based research to inform the decision-making process. As Symes and Phillipson (2009) argue, policy makers seem to be largely unaware of the social context and relationships that make up fishing communities and the authors go as far as asserting that EU fisheries policy is largely to blame for many of the problems faced by fishing communities, as it tends to favour the more economically efficient operation of the large-scale sector.

The literature on social and cultural impacts of fishing is dispersed in a range of academic publications (Urquhart et al. 2011). Studies consider the lack of incorporation of social objectives into fisheries policy (Steelman and Wallace 2001; Symes 2005; Symes and Phillipson 2009) or focus on social impact assessments (Bradshaw et al. 2001; Pollnac et al. 2006; McClanahan et al. 2009). Processes of social representation and organisation of inshore fishing have been captured in more anthropological studies (Nadel-Klein and Davis 1988; Nadel-Klein 2000; Symes and Frangoudes 2001; Williams 2008) where an important element is the emphasis on fishing as not just an occupation or a means of earning a living (Nuttall 2000; Jacob et al. 2001; Brookfield et al. 2005). One challenge for fisheries management is how to capture the range of tangible and intangible cultural values that are associated with marine fishing. The idea of sense of place provides some methodological and conceptual possibilities in this area.

Understanding cultural values such as identity, heritage, attachment and social cohesion is not an easy task. It can be complex to try and unravel the relationships that people form with places; making them relevant in a policy-making context is perhaps even harder. However, sense of place is an approach that can help provide perspectives on community relationships and how people identify with a place. One of the potential benefits of sense of place is that it draws upon a range of academic disciplines including psychology, human geography, sociology and anthropology to understand complex human-environment relationships and, within the field of social science and human geography, there is an extensive literature on how places are socially constructed, the role of place in identity and how people become attached to place (Relph 1976; Tuan 1977; Proshansky et al. 1983; Altman and Low 1992; Massey and Jess 1995; Holloway and Hubbard 2001; Creswell 2004).

Drawing on phenomenological perspectives, humanistic geographers, such as Tuan (1974) and Relph (1976), suggest that place attachment or sense of place relies on the individual and the emotional meanings people associate with a place and is based on the social relationships and processes that occur in a particular setting. Thus, sense of place entails complex human-environment relationships and, as Kaltenborn (1998) contends, it is bound up with the meanings attributed to places interpreted and constructed by people. In his seminal work Relph (1976, p. 6–7) is particularly concerned with "exploring the various ways in which places manifest themselves in our experience or consciousness of the lived world." However, Malpas (2008) states that there is a common tendency to view culture as something that is additional to and separate from its materiality. As well as human experience and perception, places are defined by the physical environment (Relph 1976; Tuan 1977). Indeed, place attachment is related to the bonds that people form with their environment (Hummon 1992; Low and Altman 1992; Stedman 2003), with a number of scholars suggesting that physical attributes contribute to place satisfaction (Shumaker and Taylor 1983; Stedman 2003). Eisenhauer et al. (2000) assert that there is a reciprocal relationship between physical environments and people in what Crist (2004, p. 12) calls a "cultivation of receptivity" in which humans can receive meaning from the world through "opening oneself, listening, watching, being within, letting be, or merging into." In this sense, social life and culture will influence place meanings, but the biophysical elements of place are also important.

Stedman (2003) argues that places are co-constructed, drawing on the socially constructed meanings of humans, but also being influenced by the material reality of the biophysical world. He suggests that symbolic and value aspects of place are partially based on some form of material reality. This tension between the material and the subjective leads Malpas (2008, p. 204) to suggest that "culture and heritage are always configured in relation to the 'material', and that there can be no clear or sharp distinction between the natural and the non-natural, the tangible and the intangible." Thus, to understand the cultural values that emerge from marine fisheries there needs to be consideration of the inherent interconnectedness of both material and subjective dimensions. This simple insight into the relational importance of the material and the subjective begins to open up new possibilities for imagining how marine fishing exerts influence in terrestrial locations and how sense of place might be used to capture those relationships.

Although there is perhaps a tendency in popular media not to differentiate between different types of fishing (e.g. inshore/offshore), fisheries activity can give rise to a diversity of relational associations and place characters; for instance large industrial ports to small coves with just a few boats. Sense of place can provide a conceptual framework for understanding this diversity and how people form attachments to and identify with different environmental settings. These relations give rise to a range of material and non-material values associated with fisheries. There is clearly increasing interest in understanding the social and cultural dimensions of fisheries but it is rare to find the idea of sense of place explicitly used even though the idea is often implied. For instance, in the run up to the reform of the CFP, EU Fisheries Commissioner Maria Damanaki stated in a meeting with artisanal and small-scale fishers that "… small-scale fisheries are very important for the survival of coastal communities, for their identity, culture, history and way of life" (ICSF 2010). There seems to be an implicit reference to the idea of sense of place in this comment even though the concept is not directly addressed. Sense of place can be used to explore the elements that Commissioner Damanaki mentions alongside other place-based issues like attachment, identity, dependence and belonging. An important aspect of sense of place is trying to understand how people value their environment, and identifying elements of the environment that are important to them (Schofield and Szymanski 2011).

While fisheries-related studies have not explicitly used sense of place there are calls for the concept to have a larger role in other areas of natural resource management. Williams and Stewart (1998) contend that sense of place allows resource managers to identify and respond to the emotional and spiritual bonds people form with localities. They suggest that sense of place is a way of integrating people into the idea of ecosystem management. Indeed, sense of place is now appearing in areas relating to ecosystem assessment. Church et al. (2011) suggest that a distinctive sense of place can contribute to a range of human value needs. However, sense of place can also help to draw out some of the cultural services provided by ecosystems, for instance, cultural identity, heritage values, spiritual services, inspiration, aesthetic appreciation, recreation and tourism.

The following section presents a case example of marine fishing along the English Channel to illustrate how the concept of sense of place can be a useful approach for unpacking some of the material/subjective social and cultural values of marine fishing. Some background to fishing activity and communities in the Channel is presented first; this is followed by an exploration of how sense of place can be used to reveal some of the tapestry of social and cultural relations that emerge as a result of fishing activity.

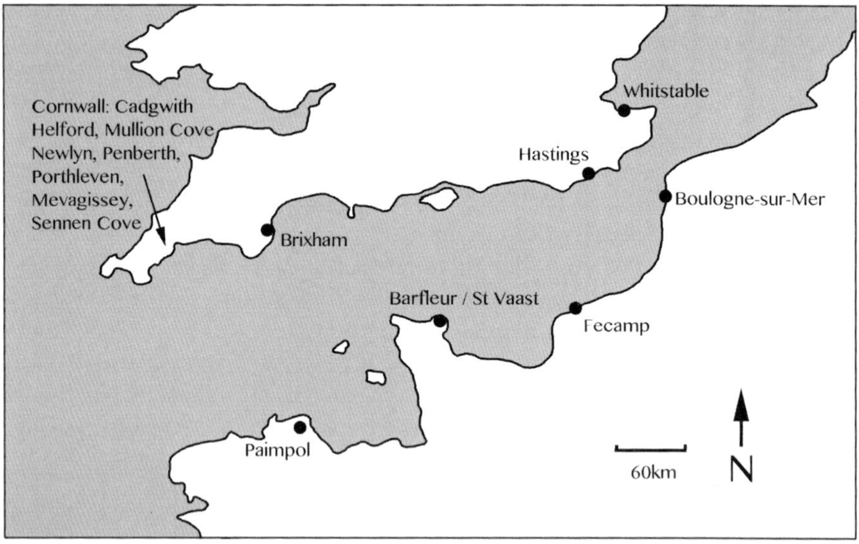

Fig. 14.1 Sense of place study sites

14.3 Fishing Communities Along the English Channel

The English Channel separates southern England from northern France and joins the North Atlantic Ocean to the west and the North Sea to the east. It is approximately 560 km long and varies in width from 240 km at its widest, to 34 km in the Strait of Dover. The Channel covers an area of around 75,000 km^2 and is the world's busiest international seaway, used by over 500 commercial vessels per day (BMT 2009). As part of an Interreg IVa project CHARM III, a study was conducted on the sense of place of fishing communities along the English Channel in southern England and northern France (Acott and Urquhart 2012) (Fig. 14.1). Places were identified to ensure a broad range of sites in terms of the size of port, location, diversity of industries, extent and type of fishing activity and tourism (Box 1).

14.4 Sense of Place as Conceptual Framework

To date fisheries-related studies have not explicitly made sustained use of the idea of sense of place. The present study demonstrates the potential of sense of place to provide a conceptual approach for understanding the complex multidimensional human-environment relationships of marine fishing by considering place as an outcome of tangled material/subjective relations. These relations give rise to a range of values associated with fisheries including heritage values, spiritual values, identity values etc. However, for sense of place to be a genuinely useful approach for policy

Box 1: Case study sites

Case study sites in England were Whitstable (Kent), Hastings (East Sussex), Brixham (Devon) and a range of sites in Cornwall. **Whitstable** is a seaside town and port in northeast Kent well-known for its oysters, which have been collected here since at least Roman times. **Hastings** is a town and borough on the coast of East Sussex and is one of Britain's oldest fishing ports, with boats working from the shingle beach in front of the Old Town for over 1,000 years and it was one of the medieval Cinque Ports. **Brixham** is a small fishing town in Devon, in the south west of England. Fishing and tourism are its main industries and it has been an important fishing port for centuries and it was the largest fishing port in the south-west in the Middle Ages. **Cornwall** is onc of the UK's most important regions for marine fishing, with fishing places ranging from the busy fishing port of Newlyn, with around 150 fishing vessels working out of the harbour, to small fishing coves such as Penberth, with a handful of small open day boats pulled up on the shore.

In France, the case study sites were Boulogne-sur-mer (Nord-pas-de-Calais), Fecamp (Upper Normandy), Barfleur & Saint Vaast (Lower Normandy) and Paimpol (Brittany). **Boulogne-sur-mer** is a city in the north of France in the Nord-pas-de-Calais region and is the most important fishing port in France, with over 7,000 people earning a living in the fisheries sector (in the catch sector and processing). Over 380,000 t of fish are processed annually in the port's fish processing district Capécure, making it Europe's largest fish processing centre. Only around 10 % of the fish is landed in Boulogne is from local boats, the remainder is transported by road from other ports in France and the UK. **Fecamp** is a town in the Seine-Maritime department in the Haute-Normandie. Fishing has been important in Fecamp for centuries, with salt-herrings dating from the tenth century and smoked herrings from the thirteenth century. Its recent history centres around the Newfoundland cod fishing throughout the nineteenth century and up to the 1970s. After the collapse of the cod fishery, only a small inshore fleet remains and pleasure boats dominate the harbour. **Barfleur** is a small harbour town in the Basse-Normandie region of France, located 25 km east of Cherbourg on the eastern coast of the Cotenin peninsula. It is home to a small fishing fleet and is an important site for the harvesting of Barfleur Mussels and oysters, although 700 years ago it was the largest fishing port in Normandy. **Saint-Vaast-la-Hougue** is located about 7 miles south of Barfleur in the Basse-Normandie region of France. Saint-Vaast has extensive oyster beds, where oysters are cultivated for consumption and renowned for their high quality. **Paimpol** is a harbour town in the Côtes-d'Armor department in Brittany and is a popular tourist destination and hosts a bi-annual Sea Shanty festival. The town was important for the Icelandic and Newfoundland cod fishing. Although this extensive fleet has now gone, a small fishing fleet still exists in Paimpol, and there are also oyster beds.

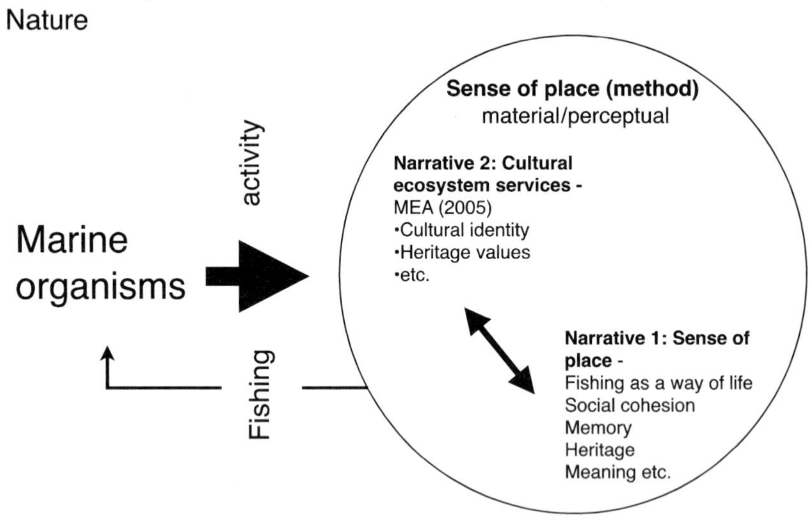

Fig. 14.2 Sense of place conceptual framework

makers and practitioners there needs to be a broader consideration of the epistemological foundation within which such studies are presented. In the UK, and more broadly, there is increasing interest in understanding the value of ecosystem services (Garcia et al. 2003; MEA 2005b; NEA 2011) an important element of which asks how does the natural environment contribute to cultural value.

The relationship between marine organisms, fishing activity, sense of place (method) and emerging narratives is depicted in Fig. 14.2. Fishing activity is a driver that translates marine organisms into a range of socio-cultural effects. Sense of place was used as an approach in our study to view the material/ perceptual dimension of fishing activity and resulted in a range of outcomes being identified. We used sense of place as a methodological approach, but also expressed the results within a narrative framework consistent with this tradition (Fig. 14.2: narrative 1). In addition to fishing activity being a driver of socio-cultural outcomes, there is a feedback loop from those outcomes back to the marine organisms and the fishing activity. If the range of effects spilling out from the fishing activity are not recognised (or valued) that has the potential to modify the fishing effort that will in turn impact on marine ecology.

A multi-method approach was adopted to explore sense of place in the case study sites (see Acott and Urquhart 2012) including a desk study of tourism representation and fishing, scoping visits, photographic surveys of material culture, townscape assessments and a total of 112 semi-structured interviews conducted with a range of stakeholders including fishermen, representatives of fishing communities, fishermen's organisations, heritage providers (e.g. museum curators), tourism providers (e.g. hotel/restaurant owners, tourism office), and artists and galleries. Participants were recruited by identifying the key stakeholders in each case study site through a web-based search, along with snowballing (Babbie 2010), where participants recommended other potential participants.

14.5 Place Identity and Character

Marine fishing can exert a powerful force in shaping the material environment of terrestrial communities and the identity of people that live in and around fishing places. The extent of the impact depends on a range of factors including size of the community, range of other activities in a location, type of fishing operation, and the nature of the physical environment. Large population centres have the ability to dilute the influence of fishing in ways that a small village does not. Similarly, if there is a range of other activities (e.g. tourist attractions, industrial centres, leisure facilities etc.) in an area, for instance Boulogne-sur-mer (France), fishing becomes just one dimension that contributes to place character. However, in a small coastal village such as Cadgwith in Cornwall (England), fishing is the dominant activity and forms the heart and soul of the community. In a range of locations fishing plays an important but different role in the creation of a place. In many cases the central focus of the town is the harbour and fishing is deeply embedded in the character of the place. A participant from a small fishing community, Mevagissey commented:

> The whole place still revolves around the fishing industry really … If you think Mevagissey you think working fishing harbour (Christine[1], fishing community representative, Mevagissey)

While another from a larger fishing port, Brixham, explained:

> It's Brixham, isn't it, the fishing industry is Brixham so, you know, it is just part and parcel of the town (Simon, tourism provider, Brixham)

The material dimension of sense of place includes not just the marine ecosystem and the coastal landscape but also those objects and buildings created by people. In fishing communities, the physical objects associated with fishing contribute to place identity and are markers of identity. These material objects may include fishing boats (Fig. 14.3), nets, pots, street decoration, buildings etc. and the selective (re)production of these material objects can strongly influence the physical character of place. Such objects constitute an important element of the materiality upon which ideas of sense of place can be constructed and interpreted from an aesthetic perspective. As an example, there was overwhelming agreement that harbours look better with fishing boats, rather than other vessels, such as yachts. Participants felt that fishing boats, especially the smaller inshore boats, have a certain charm that contributes to the character of a harbour, perhaps reflecting a sense of authenticity and contemporary working activity.

[1] The names of participants have been changed throughout to preserve anonymity.

Fig. 14.3 Beach launched boats on the Stade at Hastings

Another important element shaping the physical character of fishing places is the abundance of fishing gear (nets, floats, storage boxes etc.). In some places the fishing gear is stored in racks or crates in order to keep it tidy. In other places gears are laid out on quaysides or beaches. Arguably the presence of fishing gear adds authenticity in that contemporary activities of fishermen are contributing and adding to the environment.

They [tourists] love seeing pots and nets on the quayside. (Leon, artist, Barfleur)

It's like crab pots, every photograph of any Cornish fishing village there's a crab pot in the middle of it. (Victor, fisherman, Sennen Cove)

14.6 Authenticity and Fishing Heritage

There were concerns about the authenticity of fishing heritage, with fears of fishing harbours becoming like theme parks if the fishing stopped, with fishermen paid to mend their nets on the harbour-side, wearing sou'westers and talking to tourists about the days when they used to fish. Although it might be possible to recreate the idea of fishing in a staged inauthentic way, this would create a different sense of place to that being created as part of a contemporary working industry. Part of the cultural value created by marine fishing is its active and dynamic contribution to place character.

Fig. 14.4 Preserved capstan wheel and contemporary fishing boats in Penberth, Cornwall

> I mean economically Saint Vaast and Barfleur it's really, I mean fishing it really contributes to them and without fishing or without oyster farming it wouldn't be the same, it wouldn't be Saint Vaast anymore. (Nicole, heritage provider, Saint-Vaast)

> Without the fishing there is no character in Hastings... It's just part of the landscape here, it's part of what Hastings is, it's part of it's cultural identity. (Bob, Fisherman, Hastings)

> And there is an attitude here, in Whitstable, we like the mess. We like the noise. We like the activity, because you can come through here in the middle of winter, on a cold, wet, rainy day, and there will be activity here. There will be people around, life, activity, people going about their day to day jobs. (Adam, harbour master, Whitstable)

In addition to sense of place being able to help construct a picture of contemporary environments, there is also a temporal dimension that is expressed through the idea of heritage. Heritage is part of the narrative in understanding sustainable community development and is important for quality of life and providing a sense of belonging that is central to issues of cultural identity (Tweed and Sutherland 2007). Vileniske (2008) suggests that built heritage plays a role in the creation of local identity, cultural diversity and social cohesion. In addition to providing part of the fabric of the built environment, heritage can also contribute to human needs in providing symbolic meanings. A theme emerged concerning heritage and the importance of fishing as both a past and contemporary activity. Fishing in the past is memorialised in the landscape through the buildings (net huts, fish cellars, warehouses etc.) and infrastructure (e.g. capstan wheels) that remain today (Fig. 14.4). In addition to these tangible elements of heritage there are intangible aspects, including traditions, stories, skills, culture and memories.

Fig. 14.5. Mending nets in Barfleur

Most of the respondents felt it important to preserve this long history and cultural heritage of fishing:

> I think we are interested in activities which form part of who we are and whether it's fishing or whether it's another activity it's, you know, how does that activity define who people are and from that point if you think that yes it is important to preserve the cultural history of an activity, in that case I would say well yes because fishing is part of, it's all part of that. (Aimée, heritage provider, Fécamp)

> Fishing is part of the heritage of Boulogne, it's definitely a part of the heritage of the town. (Pierre, tourism provider, Boulogne)

Having a strong link to the past can help root a community in its locale and give it a sense of identity and strength to adapt to a changing world. There is a connection to the past through the skills that have been passed down through generations, skills that cannot be learnt in a textbook, such as how to get the boats up and down the beach, how to mend nets (Fig. 14.5) and how to prepare the fish. Boats, gear, knowledge and skills have been passed on from father to son for generations and there is often a sense of pride in coming from a local fishing family.

Thus, the past often plays an important role in the construction of collective identities as it can help make sense of and confront future challenges (Dalby and Mackenzie 1997). This is demonstrated by one participant:

> It roots you in something, whether you choose to reject it or embrace it, if you know where you come from perhaps it gives you more confidence to go on. Perhaps you've got more of a chance to make choices if you know where you come from. (Gemma, borough council representative, Hastings)

The way the past is remembered is subject to interpretation and representation. This might be through the way material objects are displayed and interpreted in a museum, interpretation boards on the quay, stories that have been passed down or through the memories of those who were engaged in fishing or fish-related activities. A number of participants spoke about how it is important not to romanticise past fishing activity, but that it was often a difficult and dangerous way of life. The relationship between the heritage of fishing and contemporary practice was important. It was the time-deepened tradition of fishing that provided a cultural rootedness, but this was tied into, and not separate from, the contemporary continuation of fishing. In some of the small fishing places, like Cadgwith, there was a clear concern that if the fishing disappeared the heart would be taken from the community. An understanding of heritage is central to a dialogue on sustainable communities and is closely bound up with the creation of distinctive place characters.

14.7 Art, Fishing and a Sense of Place

The character of a place is subject to many influences, one of which is the objectification of marine fishing through the activities of artists. As Tuan (1976, p. 267) comments: "In artworks people's experiences of life and the world are vividly objectified." The coast has long been a source of inspiration and an attraction for artists, due to the particular light and environment.

> I like that moment, that sort of in-between moment, of the tide, the fluctuations between the high tide and low tide and the whole kind of lights and how the light changes, and at the same time, how the daily life of fishermen changes as well. (Philippe, Artist, Paimpol)

The activity of fishing, fishermen and fishing boats often appear in paintings and influence other artworks. As a creative resource, marine fishing is incorporated into the material environment through the actions of sculptors, artists and writers. Artistic activity is carried out for numerous reasons, including the revenue created by a lucrative tourism market. People are attracted to iconic images as representations of places visited. However, relationships between art and marine fishing are more complex than a single economic dimension might infer. In addition to the 'creation' of artworks, the 'process' of painting or sculpture is related to the negotiation of livelihood opportunities and personal reflection.

In one village an art gallery was providing a source of income for a family where the fisherman had retired from fishing due to ill health. The gallery exhibited a wide range of 'art' products, many associated with marine fishing. A series of paintings, created by the fisherman, reflected his recollections of what happened to him over the years while fishing. Another member of the family had learned to sculpt and was now making models of fishing boats that were for sale in the shop. The gallery was the locus of relationships that tied together fish, fishing, inspiration, reflection, skills (painting and sculpture), family relationships and income. The physical presence of the gallery combined with the production and selling of artworks formed a web of relationships, not just for the family involved but also for visitors and potential customers to the area.

Perhaps one of the most well-known and extensive records of fishing in art is the Newlyn School of Painters, a colony of artists that lived and painted in and around southwest Cornwall in the late eighteenth and early nineteenth centuries. Many of the artists had spent time painting the pastoral lives of people in Brittany in France, and found a similar source of inspiration in Cornwall in the fishing communities. They saw a way of life that seemed untouched by the Industrial Revolution and wanted to capture that on canvas before it disappeared.

14.8 Personal and Community Identity

In addition to exploring the influence of marine fishing on place character, a sense of place approach allowed feelings of personal and community identity to be expressed by participants. Being a fisherman gave fishers and fishing families a rootedness in their community and in the place and provided a ‘marker’ of identity. For these reasons, fishing is more than a means of earning a living and van Ginkel (2001) suggests that fishermen may strive to continue fishing even when it is no longer economically viable to do so, as it defines their identity as individuals, households and communities. For example, one inshore fisherman from Cadgwith spoke of his love for fishing and the way of life it offered:

> You’ve got to want to do the job, it’s not a job you do for the money, you wouldn’t do it in all honesty. You’ve got to love the job, want to do the job and then the money’s secondary you know. That’s the way I see it.... fishing is a way of life, a completely different way of life. (Craig, fisherman, Cadgwith)

Fishers’ attachments to fishing also revolved around the harbour or beach as a physical setting where the fishing activity took place. Harvey (1996) suggests that places are often seen as the ‘locus of collective memory’ where group identity is created through the construction of group memories. Harbours and beaches were important in the construction of both individual and collective identities. Working in a particular environment and use of that space created a sense of belonging to that environment. For fishing communities, as in other rural areas, fishing is part of a cultural process that is collectively constructed and defined. The role of fishing in community life and social cohesion was seen as important by most of the participants, not just the fishermen. This was illustrated by one participant, who came from a fishing family:

> I mean we got a good community here and we’ve got a nice school, lots and lots of different things going on in the village. And I think the fishing industry plays a huge important, well, keeping it all going. (Christine, fishing community, Mevagissey)

These findings support Brookfield et al. (2005, p. 56) who assert that: “the fishing industry is seen to be the forum through which community bonds, values, knowledge, language and traditions are established, confirmed and passed on … For fisheries-dependent communities, fishing is the glue that holds the community together.”

Another reported impact on social structure was the perception of an increase in outsiders moving into communities. Sometimes incomers were seen as diluting community cohesion, and changing the nature of community structure. Incomers and local residents may have different attachments to place, as illustrated by one fisherman from Cornwall. He indicated the potential for conflict between incomers and locals, with each forming different attachments:

> We've been fighting for 14 years to try and get a new quay there so we don't have to put our fish in mud before it goes into the car to bring it over. But all the residents, most of the residents because there are very few permanent residents don't want any development. So which is really hypocritical in a way because they, a lot of those people have bought houses in the village because it's a picturesque fishing village, they don't want any more fishing activity. (Phil, fisherman, Cornwall)

However, this response was balanced by other views that saw the importance of incomers, particularly linked to tourism, as vital for economic revenue.

Fishermen often referred to the physical environment, to the sea, the weather, tides and the coastal landscape. Their identities were co-constructed through a combination of knowledge and relationships to the sea/land-scape together with their individual perspectives, experiences and relationships within the community. The occupational identity of fishers was also mediated by their daily engagement with the natural environment. Although conversations were not necessarily framed around religion and spirituality there was clearly a connection being made with the sea that was deeply emotional and bonding:

> It's for the love of the work of the job and the love of the sea which is why, I mean you don't become a fisherman randomly just like that. There is the love of the sea and of the environment itself which is important which pushes to accept the difficulties of the profession which is actually quite, it is hard work being a fisherman … Well the freedom, the pleasure of fishing, of trying to understand nature – there is quite a few things that, there is quite a few factors which bring you to loving what you do to the kind of work and to the profession and that bring you to practising it. (Théodore, fishermen's organisation, Boulogne)

14.9 Cultural Ecosystem Services

The earlier sections have provided a brief description of the emergent themes relating to sense of place. In discussions with stakeholders and policy makers it is clear there are conceptual difficulties in communicating qualitative sense of place ideas, in particular to those more familiar with an epistemological position emerging from positivism and the harder natural and physical sciences.

For studies to have impact and applied meaning beyond life in academic books and journals, thought must be given to the salience of the results for other stakeholders and the broader policy-making community. Currently there is increasing importance being given to the idea of ecosystem services (Fletcher et al. 2011; Lundy and Wade 2011; Chan et al. 2012; Mace et al. 2012; Robertson 2012) and the ecosystem approach to fisheries (Garcia et al. 2003). However, the valuation of cultural ecosystem services is a particularly difficult area as identified in the UK

Table 14.1 Sense of place and cultural ecosystem services of marine fisheries

Sense of place themes	Cultural ecosystem services
Fishing as 'a way of life' Contribution of fishing to social cohesion and community identity Fishing as 'iconic' marker of identity	*Cultural identity*
Fishing is represented through 'memories' in the landscape: harbours, capstan houses, net huts etc. Memory of past fishing activity influences identity	*Heritage values*
Fishers feel a deep connection to the sea through their daily engagement with it	*Spiritual services*
Fishing and the marine environment provide inspiration for artists, music and literature	*Inspiration*
Fishing influences place character through its material presence (contemporary and historic), e.g. boats, fishing gear, buildings, street decoration etc.	*Aesthetic appreciation*
'Fishing culture' contributes to the appeal of places for tourism through the presence of fishing fleets, heritage and fish as cuisine Tourist boat trips, buying fresh fish etc.	*Recreation & tourism*

National Ecosystem Assessment (Church et al. 2011, p. 63): "One important challenge is how to develop a conceptual and/or methodological approach which allows the humanities and more interpretive social science disciplines to make their distinctive contributions to the assessment in such a way as to strengthen the integration of scientific, economic, cultural and socio-political evidence for policy." This perspective is echoed by Chan et al. (2012) who suggest that methods from diverse social sciences should be used to help characterise cultural services. In this regard, sense of place may offer a conceptual and methodological approach to investigate socio-cultural values associated with marine fishing.

Aligning sense of place with cultural ecosystem services has the potential to provide a policy-relevant context within which the contribution of marine fisheries to sustainable community development can be articulated. Therefore, drawing on the framework provided by the MEA (2005), we interpreted our results in the context of cultural ecosystem services (Table 14.1; see also Fig. 14.2: narrative 2). As Table 14.1 illustrates, many of the sense of place themes identified in our study can be understood as cultural ecosystem services.

However, some caution does need to be taken in pursuing this goal. Chapter 16 of the UK National Ecosystem Assessment is a discussion of cultural ecosystem services (Church et al. 2011). This primarily deals with the way particular 'environmental settings' can provide ecosystem services. Through the lens of sense of place, we have revealed how an environmental setting is a result of a complex entanglement of nature and society where marine fishing is acting as a process of identity creation that connects the often unseen undersea world with places on land.

This might be through direct experience of the fishermen, or indirectly through the way fishing activities spill out into coastal locations (place character, community identity, personal identity).

There are two important conceptual points to note about using sense of place and ecosystem services in this way. First, sense of place is understood with reference to relational associations of activities (fishing and non-fishing), events and 'things' that connect the undersea world with terrestrial places. The focus is, therefore, to understand place-making as processes that span different environmental settings of the sea and land. In using sense of place as a lens to understand these processes, we are exploring sets of relationships being created and in part driven by marine fishing. Part of the marine ecosystem service is a direct experience of fishing and being at sea, but further complications emerge as the idea of fishing gets incorporated into the socio-cultural mix of different places. Mechanisms should be found to incorporate this process-driven, relational understanding into fisheries policy, coastal zone management and sustainable community development more broadly. It is possible that the popular categories of supporting, provisioning, regulating and cultural services might prove conceptually limiting and new approaches to thinking about relationships between the material/subjective worlds might be needed. Possible conceptual frameworks for building this approach include actor network theory (Latour 2005), hybrid geography (Whatmore 2002) and non-representational theory (Thrift 2007). Actor network theory (ANT), as an example of a non-dualistic approach, offers the potential to explore processes and relations that are driven by the activity of marine fishing. ANT is co-constructionist and can be used to identify "how relations and entities come into being together" (Murdoch 2001, p. 111). In addition to conceptual ideas like ANT there is a broad area of research examining the inter-connection of social and ecological components within marine systems (Perry and Rosemary 2010). A special issue of Marine Policy (Vol. 37, 2013) reports on approaches to help bridge the social-ecological divide while developing perspectives on governance strategies.

Second, we adopted a narrative approach where the voices of the participants were allowed to emerge and tell the story of the importance of marine fishing in their lives and the places in which they live. While economic valuation in ecosystem services is fundamentally important, the relationships between land and sea, human and ecosystem, need to be understood through narrative as well as economic valuation. In order to plan for a sustainable future for fishing communities, there needs to be broader understanding of the multiple cultural ecosystem values that result and spill out from the act of fishing. The economic value of fishing, and associated activities, is ultimately only one type of value in a broader network of relationships that emerge where the activity of fishing occurs. These values are depicted in Table 4.1 as relating to various dimensions of identity, heritage, spirituality, inspiration, aesthetic and recreation. Chan et al. (2012, p. 16) call for a "new research community and program at the nexus of ecological–economic analysis and the social sciences of decision- making, a program dramatically different from the existing ES [ecosystems services] research program."

14.10 Conclusion

The focus of this chapter has been to generalise the contribution that marine fishing makes to sense of place across the different study locations. However, the qualitative research undertaken provides a rich source of data that describes material and subjective associations with marine fishing in particular localities. Although beyond the boundaries of this chapter there is considerable scope to mine the data to provide local contextual accounts of sense of place related to particular villages, towns or regions (for examples see Urquhart and Acott (2013) and Urquhart and Acott (2013)). The cultural ecosystem services identified during our research form part of the broader tapestry of issues important in sustainable community development. Tweed and Sutherland (2007) discuss cultural heritage for sustainable urban development while Roseland (2000) considers the broader necessity of integrating environmental, economic and social objectives into sustainable community development. This theme is taken further by Britto (2011) who suggests a multi-modal systems method for capturing the broad dimensionality associated with sustainable development. We believe that sense of place allied to a cultural ecosystems services perspective can produce a policy-relevant account of the importance of marine fishing. The challenge will then be to embed that perspective into local and regional planning frameworks so that the broader cultural contribution of marine fishing to sustainable community development can be evaluated.

Brookfield et al. (2005) contend that fishing is the glue that holds the community together. Perhaps rather than thinking about it as 'sticking' together pre-existing entities, fishing can be considered as an activity that drives the process of place-making. Particular places emerge with fishing playing an integral part in the formation of place character and identity. If policy makers involved with fishing places are to plan for a sustainable future there needs to be a clear articulation and understanding of the myriad social, economic and environmental relationships that emerge as a result of marine fishing. Sense of place provides a starting point for exploring these relationships. Without such a perspective, social and cultural values can be overlooked or only partially addressed in the decision-making process. We argue that sense of place studies can deliver a policy-relevant account of cultural ecosystem values that emerge when marine and terrestrial environments are drawn together by the activity of fishing. Such analysis needs to be combined with ecological, environmental and economic valuations to understand how marine fishing can contribute to a sustainable future for inshore fishing communities.

Acknowledgements Research for this chapter was conducted as part of the project: CHARM 3 (CHannel integrated Approach to marine Resource Management), selected within the scope of the INTERREG IVA France (Channel) – England cross-border European cooperation programme, co-financed by the ERDF.

References

Acott, T., & Urquhart, J. (2012). *Marine Fisheries and Sense of Place in Coastal Communities of the English Channel/La Manche.* Final report of Action 6.2 of the CHARM III Interreg 4a project. Chatham: University of Greenwich.

Agyeman, J., & Evans, B. (2004). 'Just sustainability': The emerging discourse of environmental justice in Britain? *The Geographical Journal, 170*(2), 155–164.

Altman, I. A., & Low, S. M. (Eds.). (1992). *Place attachment*. New York: Plenum.

Babbie, E. (2010). *The basics of social research*. Belmont: Wadsworth Cengage Learning.

Blay-Palmer, A. (2011). Sustainable communities, an introduction. *Local Environment, 16*(8), 747–752.

BMT. (2009). Reducing Risk in the English Channel/La Manche Traffic Separation Schemes. final report for the Maritime and Coastguard Agency (MCA) and the Ministere de l'Ecologie, de l'Energie, du Developpement durable et de l'Amenagement du territoire, BMT Isis Ltd.

Bradshaw, M., Wood, L. & Williamson, S. (2001). Applying qualitative and quantitative research: A social impact assessment of a fishery. *Applied Geography, 21*, 69–85.

Britto, C. M. (2011). Sustainable community development: A brief introduction to the multi-modal systems method. *Systemic Practice and Action Research, 24*(6): 533–544.

Brookfield, K., Gray, T., & Hatchard, J. (2005). The concept of fisheries-dependent communities. A comparative analysis of four UK case studies: Shetland, Peterhead, North Shields and Lowestoft. *Fisheries Research, 72*, 55–69.

Chan, K.M.A., Satterfield, T., & Goldstein, J. (2012). Rethinking ecosystem services to better address and navigate cultural values. *Ecological Economics, 74*, 8–18.

Church, A., Burgess, J., Ravenscroft, N., Bird, W., Blackstock, K., Brady, E., Crang, M., Fish, R., Gruffudd, P., Mourato, S., Pretty, J., Tolia-Kelly, D., Turner, K., & Winter, M. (2011). Cultural services. UK National Ecosystem Assessment: Technical Report. Cambridge, UNEP–WCMC.

Creswell, T. (2004). *Place: A short introduction*. Oxford: Blackwell Publishing.

Crist, E. (2004). Against the social construction of nature and wilderness. *Environmental Ethics, 26*, 5–24.

Dalby, S., & Mackenzie, F. (1997). Reconceptualising local community: Environment, identity and threat. *Area, 29*(2), 99–108.

EC. (2009). Green paper: Reform of the common fisheries policy. Brussels, European Commission.

Eftec. (2010). Provision of sustainable access to Inshore Fisheries Research Gap Analysis. London, Defra.

Eisenhauer, B.W., Krannich, R.S., & Blahna, D.J. (2000). Attachments to special places on public lands: An analysis of activities, reason for attachments, and community connections. *Society and Natural Resources, 13*, 421–441.

Firn Crichton Roberts Ltd & Graduate School of Environmental Studies, University of Strathclyde (2000). An assessment of the socio-economic costs & benefits of integrated coastal zone management. Final report to the European Commission.

Fletcher, S., Saunders, J., & Herbert, R.J.H. (2011). A review of the ecosystem services provided by broad-scale marine habitats in England's MPA network. *Journal of Coastal Research, 64*(1): 378–383.

Forst, M. F. (2009). The convergence of Integrated Coastal Zone Management and the ecosystems approach. *Ocean & Coastal Management, 52*(6), 294–306.

Garcia, S. M., Zerbi, A., Aliaume, C., Do Chi, T., & Lasserre, G. (2003). The ecosystem approach to fisheries. Issues, terminology, principles, institutional foundations, implementation and outlook. FAO Fisheries Technical Paper. Rome, FAO: 71.

Harvey, D. C. (1996). *Justice, nature and the geography of difference*. Cambridge: Blackwell.

Holloway, L., & Hubbard, P. (2001). *People and place: The extraordinary geographies of everyday life*. Harlow: Pearson Education Ltd.

Hummon, D. M. (1992). Community attachment: Local sentiment and sense of place. In I. Altman & S. M. Low (Eds.), *Place attachment*. New York: Plenum Press.

ICSF. (2010). ITQs a User Right, not a Property Right, confirms EU Commissioner for Fisheries and Maritime Affairs. I. C. i. S. o. Fishworkers.
Jacob, S., Farmer, F., Jepson, M., & Adams, C. (2001). Landing a definition of fishing dependent communities. *Fisheries, 26*(10), 16–22.
Jacquet, J. (2009). Silent water: A brief examination of the marine fisheries crises. *Environment Development and Sustainability, 11*, 255–263.
Jentoft, S. (2000). The community: A missing link of fisheries management. *Marine Policy, 24*, 53–59.
Kaltenborn, B. P. (1998). Effects of sense of place on responses to environmental impacts. *Applied Geography, 18*(2), 169–189.
Latour, B. (2005). *Reassembling the social: an introduction to Actor-Network-Theory*. Oxford: Oxford University Press.
Low, S. M., & Altman, I. I. (1992). Place attachment: A conceptual inquiry. In I. Altman & S. M. Low (Eds.), *Place attachment*. New York: Plenum Press.
Lundy, L., & Wade R. (2011). Integrating sciences to sustain urban ecosystem services. *Progress in Physical Geography, 35*(5), 653–669.
Mace, G. M., & Bateman, I. (2011). Conceptual framework and methodology. The UK National Ecosystem Assessment: Technical Report. Cambridge, UNEP–WCMC.
Mace, G.M., Norris, K., & Fitter, A.H. (2012). Biodiversity and ecosystem services: a multilayered relationship. *Trends Ecol Evol, 27*(1), 19–26.
Malpas, J. (2008). New media, cultural heritage and the sense of place: Mapping the conceptual ground. *International Journal of Heritage Studies, 14*(3), 197–209.
Marsden, T. & Hines, F. (2008). Unpacking the quest for community: some conceptual parameters. In T. Marsden (Ed.), *Sustainable Communities: new spaces for planning, participation and engagement*. Oxford: Elsevier.
Massey, D., & Jess, P. (Eds.). (1995). *A place in the world?* Oxford: Oxford University Press.
McClanahan, T.R., Castilla, J.C., White, A.T., & Defeo, O. (2009). Healing small-scale fisheries by facilitating complex socio-ecological systems. *Rev Fish Biol Fisheries, 19*, 33–47.
MEA. (2005a). Ecosystems and human well-Being, Synthesis Report. Washington, D.C.: Island Press.
MEA. (2005b). Millennium ecosystem assessment: Current state and trends assessment. Washington D.C.: Island Press, Millennium Ecosystem Assessment.
Murdoch, J. (2001). Ecologising sociology: Actor-network theory, co-construction and the problem of human exemptionalism. *Sociology, 35*(1), 111–133.
Nadel-Klein, J. (2000). Granny baited the lines: perpetual crisis and the changing role of women in Scottish fishing communities. *Women's Studies International Forum, 23*(3), 363–372.
Nadel-Klein, J., & Davis, D. L. (1988). To work and to weep. St John's, Newfoundland, Institute of Social and Economic Research, Memorial University of Newfoundland.
NEA. (2011). UK National Ecosystem Assessment: Understanding Nature's Value to Society. Synthesis of the Key Findings. Cambridge, UNEP–WCMC.
Nuttall, M. (2000). Crisis, risk and deskillment in North-east Scotland's fishing industry. In D. Symes (Ed.), *Fisheries dependent regions*. Oxford: Blackwell Science.
Perry, I. R., & Rosemary, O. E. (2010). Introduction: Coping with global change in marine social-ecological systems. *Marine Policy, 34*(4), 739–741.
Pollnac, R.B., Abbott-Jamieson, S., Smith, C., Miller, M.L., Clay, P.M., & Oles, B. (2006). Toward a model for fisheries social impact assessment. *Marine Fisheries Review, 68*(1–4), 1–18.
Proshansky, H.M., Fabian, A.K., & Kaminoff, R. (1983). Place-identity. *Journal of Environmental Psychology, 3*, 57–83.
Relph, E. (1976). *Place and placelessness*. London: Pion.
Robertson, M. (2012). Measurement and alienation: making a world of ecosystem services. *Transactions of the Institute of British Geographers, 37*, 386–401.
Roseland, M. (2000). Sustainable community development: integrating environmental, economic, and social objectives. *Progress in Planning, 54*, 73–132.

Schofield, J., & Szymanski, R. (2011). Sense of place in a changing world. In J. Schofield & R. Szymanski (Eds.), *Local heritiage, global context: Cultural perspectives on sense of place*. Farnham: Ashgate.

Shumaker, S. A., & Taylor, R. B. (1983). Toward a clarification of people-place relationships: A model of attachment to place. In N. R. Feimer & E. S. Geller (Eds.), *Environmental psychology: Directions and perspectives* (pp. 219–251). New York: Praeger.

Stedman, R. C. (2003). Is it really just a social construction?: The contribution of the physical environment to sense of place. *Society and Natural Resources*, *16*(8): 671–685.

Steelman, T. A., & Wallace, R. L. (2001). Propety rights and property wrongs: Why context matters in fisheries management. *Policy Sciences*, *34*, 357–379.

Symes, D. (2005). Altering course: Future directions for Europe's fisheries policy. *Fisheries Research*, *71*(3): 259–265.

Symes, D., & Frangoudes, K. (2001). The social organisation and reproduction of inshore fishing. In D. Symes & J. Phillipson (Eds.), *Inshore fisheries management*. Dordrecht: Kluwer.

Symes, D., & Phillipson, J. (2009). Whatever became of social objectives in fisheries policy? *Fisheries Research*, *95*(1), 1–5.

Thrift, N. (2007). *Non-representational theory: Space, politics, affect*. Oxon: Routledge.

Tuan, Y.-F. (1974). *Topophilia: A study of environmental perceptions, attitudes and values*. Prentice Hall.

Tuan, Y.-F. (1977). *Space and place: The perspectives of experiance*. Minneapolis University of Minnesota Press.

Tweed, C., & Sutherland, M. (2007). Built cultural heritage and sustainable urban development. *Landscape and Urban Planning*, *83*(1), 62–69.

Urquhart, J., & Acott, T. (2013). Constructing 'The Stade': Fishers' and non-fishers' identity and place attachment in Hastings, south-east England. *Marine Policy*, *37*, 45–54.

Urquhart, J., & Acott, T. (2014). A sense of place in cultural ecosystem services: The case of Cornish fishing communities. *Society and Natural Resources*. 27(1), 3–19.

Urquhart, J., Acott, T., Reed, M., & Courtney, P. (2011). Setting an agenda for social science research in fisheries policy in Northern Europe. *Fisheries Research*, *108*, 240–247.

van Ginkel, R. (2001). Inshore fishermen: Cultural dimensions of a maritime occupation. In D. Symes & J. Phillipson (Eds.), *Inshore Fisheries Management*. Dordrecht: Kluwer.

Vileniske, I. G. (2008). Influence of built heritage on sustainable development of landscape. *Landscape Research*, *33*(4), 425–437.

Whatmore, S. (2002). *Hybrid geographies: Natures, cultures and spaces*. London: Routledge.

Williams, R. (2008). *Changing constructions of identity: Fisher households and industry restructuring*. PhD, Newcastle University.

Williams, C. C., & Millington, A. C. (2004). The diverse and contested meanings of sustainable development. *The Geographical Journal*, *170*(2), 99–104.

Williams, D. R., & Stewart, S. I. (1998). Sense of place: An elusive concept that is finding a home in ecosystem management. *Journal of Forestry*, *96*, 18–23.

Chapter 15
Heritage, Skills and Livelihood: Reconstruction and Regeneration in a Cornish Fishing Port

Tim Martindale

Keywords Fishing heritage • Fishing community • Sustainable coastal livelihoods • Critical nostalgia • Cornwall • Transmission • Recovery of skills • Newlyn, arts and craft

15.1 Introduction

> Every rock and cove in the parish has a name. Porth & Gear, The Nancy, Rubble Cove, Buttercove, Porthmear, Tottycove, Trescore, The Turtle—These names show how much this coast was once used but they are nearly forgotten. If that happens their history will be gone forever, and if a community loses its past it's in danger of losing its way. (*Nick Darke, The Wrecking Season*)[1]

This quotation from the Cornish playwright, fisherman and 'wrecker'[2], Nick Darke, points to connections between knowledge, memory, community and place. Intimate knowledge of shoreline features once common amongst coastal dwellers was certainly an unwritten, informal knowledge and language shared between generations. However its transmission is dependent on how the coast is used, that is on practice and labour. Nick Darke was speaking in the context of his fears about the loss of traditional livelihoods in the small farming and fishing community where he lived and was born, and the encroachment of tourism and property speculation. In his view tourism offered little remuneration and less dignity to its workforce:

> Trade is usually what happens after a product has been manufactured, extracted, harvested, or caught. Now that trade is all that is left any sense of community and common purpose once provided by those activities is lost. Culture is debased and everything, including history, becomes a commodity (Darke 1999, p. xiii)

[1] The Wrecking Season (2005) directed by Jane Darke and produced by Boatshed Films.

[2] Wrecking: the traditional term (and the term Darke preferred) for salvaging wrecks or other flotsam and jetsam washed up on the shoreline,(otherwise known as 'beachcombing').

T. Martindale (✉)
Goldsmiths, University of London, 222 Croxted Road,
London, SE24 9DG, UK
e-mail: tim_martindale@hotmail.com

J. Urquhart et al. (eds.), *Social Issues in Sustainable Fisheries Management*,
MARE Publication Series 9, DOI 10.1007/978-94-007-7911-2_15,

Darke is critiquing the loss of livelihoods, related to the production of tangible things, which for him is the basis of community. He seems even to deplore the production of heritage as one more form of extraction without return. Whilst acknowledging issues of conflict, class and complexity around heritage, in this chapter I put forward an argument that heritage can contribute to sustaining livelihoods in fishing communities. I underline the importance of coastal use and knowledge, discussed through the example of the reconstruction of a sailing vessel of the kind once commonly used in fishing off Britain's coasts. This example leads me to emphasise non-essentialist meanings of heritage (Harvey 2001; Howard and Pindar 2003) and practice-based and labour-centred views on place and identity (in the manner of Ingold 2000; Gray 2000; Howard 2012).

Although there seems to be wide consensus about the necessity of working towards sustainability in fisheries there is also a healthy element of scepticism amongst scientists, ranging from cautiousness (Pauly et al. 2002) to doubt and criticism of some of the assumptions of the discourse (Longhurst 2006). This usually rests on consideration of the unsustainability of fisheries in the long, historical view and the complexity of understanding and regulating ecosystems and human impact. 'Restoration ecology' advocates have called for 'reconstructing the past to salvage the future' (Pitcher 2001, p. 601) referring to building datasets of past, 'pristine' conditions to inform present-day management models. However, this should not be misinterpreted as a prescription for anything as simple as 'going back' to pre-industrial technologies or social formations. As Longhurst (2006) and Thurstan et al. argue (2010), serious depletion of stocks in the North East Atlantic and North Sea had already occurred before the transition to steam and diesel powered boats. My argument for the relevance of fisheries heritage to sustainability (in the broad, multi-faceted sense of the latter term) rests on three considerations: the need for alternative livelihoods and diversification in the wake of fleet reduction, declining incomes, rising costs and restricted access (Symes and Phillipson 2009; Urquhart and Acott 2013); the potential contribution to promoting and strengthening the links between 'the catch and the locality' (Reed et al. 2013); and its role as a source of 'critical nostalgia' (Clifford 1986)—allegorical as well as practical instruction in local and regional resourcefulness (Macdonald 2002).

15.2 Cultural Loss and Salvage: Heritage Meanings and Practices

On the subject of heritage and sustainability in the 'coastal zone' of South West England, Howard and Pindar (2003) articulate two concerns: to outline a perspective on 'fields of heritage' and to question the validity of any rigid distinction between 'natural' and 'cultural' heritage; and to consider the implications of 'modes of cultural heritage consumption' (p. 57). Whilst not spelling out the way they define 'heritage', it is clear from their discussion that heritage involves a concern, intention or practice to *conserve*, whether the object is a building, species, language, or skill. In most definitions of heritage the term is confusingly conflated with 'culture', especially since in recent times notions of 'intangible', or 'non-material' heritage have become more commonplace in official and policy-led definitions, alongside the previous emphasis on the historic (built) environment (Ahmad 2006). It is important, however, to distinguish 'heritage' from a more generic notion of 'culture' because otherwise we lose sight of the fact that heritage entails a selective and explicit attention to particular manifestations of culture that are deemed worthy, or in need, of preservation and maintenance.

Although heritage is concerned with articulating a view of what is traditional, this does not mean it is necessarily conservative, as heritage critics such as Hewison (1987) have argued (neither in the sense that it must be against innovation or change, nor in the sense that it is always elitist). Heritage derives from processes of social change, especially declining and/or shifting patterns of labour and livelihood and the accompanying movements of people and social encounters through which different places and classes interact. Heritage is not only a preoccupation of the middle and upper classes, but also derives from the experience of the working class in interaction with other classes within ongoing processes of social transformation. Rather than seeing heritage as merely a thing or an essence then, I follow Harvey (2001, p. 327) in taking a relational view on heritage as "a process, or a verb, related to human action and agency, and an instrument of cultural power… a contemporary product shaped from history."

Howard and Pindar's six fields of heritage (*ibid*) are: landscapes (including seascapes), monuments, sites, artefacts, activities and ways of life and finally, people. Some landscapes, places or practices associated with heritage might incorporate multiple fields. One of Howard and Pindar's examples, the South West Coast Path (a National Trail) incorporates a mixture of protected and unprotected, natural and cultural features such as cliffs, coves and fishing villages. These may include sites of remembrance, including tragedies such as the loss of the Penlee lifeboat in 1981 when it went to the aid of a stricken ship, *The Union Star*. The old Penlee lifeboat house near Newlyn is now a monument for the crew lost in that incident—mostly fishermen from the village of Mousehole—and their families. An active fishing village is also the base for the production of a range of artefacts linked to activities and ways of life.

As Howard and Pindar (2003) observe "the heritage of the coastal zone includes the entire culture of how to use it… Heritage is not only the material lobster pot, but also the ability to make one" (p. 61). They also cite the examples of individuals or societies that preserve authentic old ships in order to sail them, and the recent revival of racing pilot gigs. The Cornishman and master craftsman Ralph Bird who made 29 of the 141 registered gigs in use today, described the pilot-gig as formerly being the 'white van' of maritime Cornwall—an all-purpose workhorse used to ferry pilots out to ships as well as involved in salvage and rescue operations[3]. There is a growing number of racing clubs mainly based in active and former fishing communities, and as Howard and Pindar put it, the sport is "one example of a

[3] Obituaries, The Times (November 14 2009, p. 115).

determination to conserve the activity as well as the artefact. It is more interested with rowing than carpentry" (p. 61).

Laurier (1998) has studied formal and informal projects of ship replication and restoration, highlighting the meanings and skills involved for participants and audiences engaged in projects that involve craft production. Laurier's sense of the term 'craft' recalls a 'pre-Fordist' era of connection between maker and product that lies counter to a wider trend of capitalist alienation (Greenhalgh 1997). However, 'craft' is also relevant here in Sennett's (2009) sense of the term as technique or expertise that calls upon both manual dexterity and intellect, the problem-solving abilities of 'hand' and 'head' combined. Laurier notes the significant amount of historical research that both expert and amateur boatbuilders undertake—a dynamic process involving embodied knowledge, a makeshift approach to old and new, and trial and error. The informal boat restorers in particular work like genealogists contacting families to trace the biographies of previous owners or sailors and investigating archives. However, Laurier (1998) concludes that "the vital part of restoration is the reacquisition of skills and this forms a final embodied link to the past" (p. 47) underlining the importance in this context of an informal, 'learning by doing' approach. Similarly, Easthope (2001) also distinguishes 'kinaesthetic' from 'intellectual' engagements with maritime heritage. In his seminal work on livelihood, dwelling and skill, Ingold (2000) contrasts 'skill' or 'technique' with 'technology'. Associating the latter with formal, epistemic knowledge, skill by contrast is "tacit, subjective, context-dependent, practical 'knowledge how', typically acquired through observation and imitation rather than verbal instruction" (p. 316) He is also concerned with showing the links between place, practical knowledge and the human subject (the fundamental emplacement of such knowledge and its ties to the body, personhood and social relationships). This perspective enriches our understanding of the relationship between heritage practices and place.

Crang argues (1994, p. 151) that "each [*heritage*] practice has as its effect a different space for the past." Many of the conflicts and tensions surrounding heritage production relate to how practices (such as replicating or restoring a boat) are incorporated into the redevelopment of space (such as waterfronts). Steinberg (1999, p. 41) observes that an image of the ocean as a nostalgic space finds contemporary salience in the 'postmodern urban waterfront', examples being the festival market places, high income housing and maritime museums of Boston, Baltimore, Bristol, Cape Town, Lisbon and Sydney. "Here, the sea is referenced as a crucial source for folk culture and past economic glory, but the role of the ocean in contemporary political economy is reduced to that of a provider of images to be consumed" (p. 407). Steinberg quotes Sckula (1995, p. 12): "The old harbour front, its links to a common culture shattered by unemployment, is now reclaimed for a bourgeois reverie on the mercantilist past."

Observing the recent movement towards 'vernacular' modes of heritage, a fascination with the mundane and growth in 'interactive' and local heritage museums, Day and Lunn (2003, p. 296) consider whether "nostalgia is indicative of a more participatory and multilayered sense of the past?" Or if, "what generally passes for nostalgia-driven heritage is in fact a version of a past which is romanticized and distanced from the everyday experiences of most people"—the sights, sounds, smells and dangers? A cautionary tale is told by Atkinson et al. (2002). In the place marketing and redevelopment of the city of Hull, a former distant-water fishing port, efforts have been made to 'exorcise' fishing (including the smell of fish) from the 'civic image'. Illustrating the inherent selectivity of heritage, the city's maritime heritage is referenced in terms of the romanticised, historic age of sail. Meanwhile there has been contestation about the redevelopment of the dock that challenges any simplistic counter-narrative about working class community. The dock includes a site where an annual memorial event is held to the 8,000 trawlermen lost at sea, showing how the built environment acts as a repository of collective place-memory. However, Atkinson et al. (2002) point out that not all of Hull's fishing community would want the trawler-owners building to be preserved—a reminder that 'collective' memories may also be formed and informed by contexts of inequality. In a comparative study of small museums in North Carolina mill towns and the 'Time and Tide' project in Great Yarmouth, Wedgwood (2009) has asked whether working classes can also gain from preservation. She noted that "Yarmouth people wanted to turn an empty fish-factory into a museum, while retaining the fishy smell, and a fire-damaged wall", suggesting the importance of personal memory in this context. The tension these examples highlight seems to be one between heritage that presents a homogenised and sanitised version of the past, and one that acknowledges a heterogeneity of local experiences and interests (including inequality) and which enables a more "critical presentation of the past" and its "links with, or contingency on the present" (Walsh 1992, cited by Day and Lunn 2003, p. 297).

Rural and industrial heritage may be a source of belonging and identity long after the labour which it draws on has ceased. The inhabitants of Ferryden in Scotland highly prize their identities as 'fisherfolk' despite the fact that the place no longer has an active fishing industry. Nadel-Klein (2003, p. 8) situates their role in the invention and perpetuation of idealised aspects of the fishing past as a response to the ongoing marginalisation of rural places within a capitalist political economy. In the process there has been a move from 'fishers' material status as primary producers of food to their symbolic status as objects of the 'tourist gaze'. Resentment and resistance towards the prospect of becoming the latter is, however, keenly expressed by fishers who remain active in the industry. This brings us back to the second of Howard and Pindar's (2003) concerns about the implications of modes of cultural heritage consumption.

Cottages and sail lofts in Cornish fishing villages have in a sense been 'preserved' by conversion to holiday lets, second homes and artists' studios. Of course the consequence of this market in desirable locations is that many locals are priced out of property ownership in these villages. The author of one travel article who visited Salcombe in Devon seemed either unaware or uninterested in the area's maritime heritage, other than the most superficial aspects, and more impressed by the "breathtaking prime real estate… which has turned this formerly sleepy fishing village into the Knightsbridge of Devon"[4]. There seems to be a real spectrum

[4] 'Devon Sent' (Evening Standard Magazine, Standard.co.uk/Lifestyle).

of tourist consumption from this example to more informed and sensitive perceptions of visitors keenly interested in signs of a working fishing industry as found by Urquhart and Acott in Hastings (2013). The nuances of class and community in such encounters is discussed by Walton (2000) through the life and work of Stephen Reynolds, an author and fisheries inspector who lived and worked amongst the fishing community of Edwardian Sidmouth for a time. Reynolds felt that the fishermen he came to know were able to identify more with working class visitors who were increasingly arriving on the railway, than with "people of other classes who had lived in the neighbourhood all their lives" (p. 131). Reynolds articulated a sense of the distinctive craft of the inshore fisherman: "local knowledge, coupled with 'pluck' and the practical skills of the seaman" (p. 134) but he also expressed some of the anxieties—although filtered through his own romantic and sensitive preoccupations—regarding the potential loss of manliness, dignity and 'degradation of craft' posed by the opportunity for fishermen to work as pleasure boatmen for the visitors.

The most distinctive and important characteristic of the coast may be, as Walton has argued (2010), that it is an 'informal space'—one that is deeply evocative for personal as well as collective memory, whether as a source for recollection of childhood seaside holidays or one connected to making a livelihood from the sea. Commenting on Casey's (2002, p. 76) argument that to "know a region is also to be able to remember it"—Matsuda (2004, p. 262) says this mnemonic sense of place "defies mere 'representation' because it is not about symbolism, but about finding presence in shifting temporal registers of a lived past." As visitors and diverse local inhabitants and workers attempt to 'find presence' in relation to past and contemporary rural life-ways, there is potential for both connection and disconnection to nature, work, things made, other people. In any case tourism need not be the only target of coastal heritage and Howard and Pindar (2003) seriously question the sustainability of basing coastal economies around tourism. Rather, "if tourism can never be sustainable, then conserving heritage to serve the local population in very different ways might be" (p. 67). The implications of different modes of heritage production may then be as important as modes of consumption. With this thought in mind, I turn now to considering an ethnographic context that illustrates these concerns.

15.3 Research Context

The data presented in this chapter draw on research in Cornwall conducted over the course of 1 year from summer 2008 to 2009 and during subsequent short visits. The project explored connections between fishing livelihood, craft and heritage in Cornwall, with a focus on the port of Newlyn. It has enquired into different forms of knowledge and knowledge transmission and reflects on notions of sustainability, community and place in light of social change in Cornish fishing villages. A variety of methods were employed, including: archival research; formal interviews and casual conversations with fishermen, fishermen-artists, boatbuilders, fish merchants, people working in fisheries management and welfare and a range of non-fishing

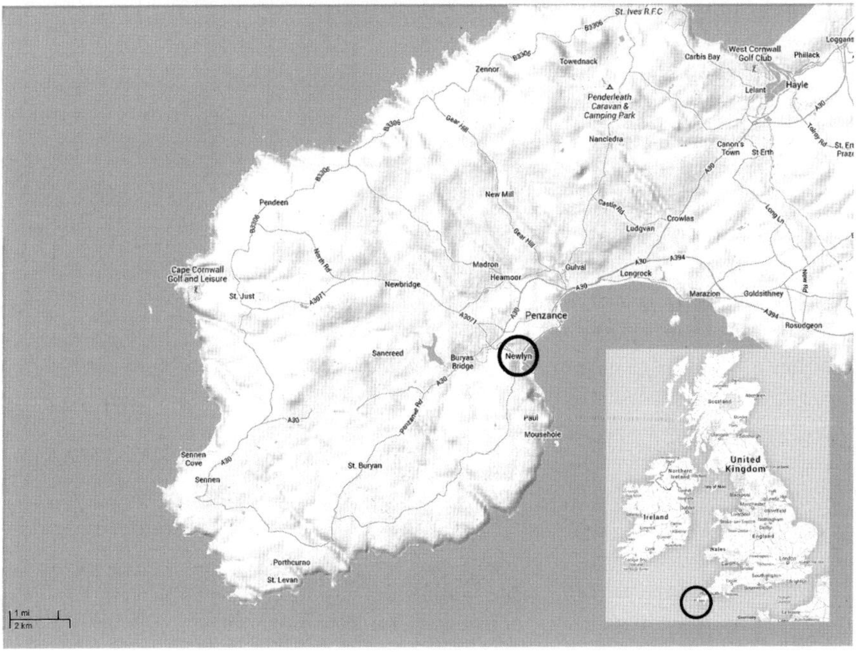

Fig. 15.1 Location of Newlyn, near Penzance, in Cornwall, South-West England

locals; and finally participant observation in a range of activities including fishing trips, formal basic training for fishermen, visiting museums and art exhibitions and learning to sail a heritage fishing boat. This chapter deals in particular with the latter heritage initiative but before describing that, a brief account of the study site is necessary to put it into context[5].

Situated near the south west tip of Cornwall (Fig. 15.1), Newlyn supports a large and diverse fishing fleet including beam trawlers, stern trawlers, ring-netters, gill-netters, crabbers and handline fishing boats. These exploit fisheries primarily in the Celtic Sea, but also the English Channel, the Bristol Channel and the Irish Sea. The various currents that converge off the south-west coast provide a rich and diverse species range and the offshore vessels in particular target premium fish such as monkfish, megrim and sole. In 2011, fish worth £ 22 million were landed into the port (MMO 2012) and in 2009, 156 fishing vessels were operating from Newlyn as a base port employing about 255 fishermen[6]. Census figures from 2001 show that in the local authority area of Penzance South (which includes Newlyn) only 98 people

[5] The names of interviewees quoted in this chapter have not been disclosed to protect anonymity, with the exception of John Lambourn who provided permission to disclose. All images are the author's own unless credit is given.

[6] Data courtesy of Cornwall Sea Fisheries Council (2009).

(5 % of the population) were employed in the fishing industry[7]. This indicates that the majority of fishers operating from Newlyn live outside of the ward. There is a need for local employment far beyond what the fishing industry can provide; however in this region there are limited alternatives. The five largest sectors of employment in 2001 were: wholesale and retail (18.4 %), health and social work (12.3 %), hotels and catering (11.4 %), real estate renting and business activities (10.1), and manufacturing (8.4)[8]. The region is very dependent on tourism as well as public sector work, much of the employment is seasonal and/or low paid and there is high unemployment (4.2 % compared to 2.6 % average for the southwest and 3.4 % average for England)[9].

The Superintendent of the Fishermen's Mission (Newlyn branch)[10], emphasised how the local fishing industry was experiencing problems associated with rising costs, restricted access and declining incomes, including a fall in recruitment of young people and a loss of more experienced fishers. Over the last 6 to 8 years the latter had been commonly the 'less competent ones' he said, but more recently they were losing 'good skippers' leaving for jobs in the North Sea energy industry where they can get a regular wage and a predictable rhythm of 1 month at work and 1 month on leave. Not only were they losing valuable skill-sets that would be difficult to replace once they had gone, but he also felt that young people were turning to local jobs (high street and supermarket retail etc.) that were in his view not of 'high quality', in other words not career jobs. The Superintendent suggested this is linked to the fact that fishing was no longer seen as a 'valued career' and parents in Newlyn were also reluctant for their sons to follow their fathers into fishing. The lack of optimism about fishing futures I encountered at Newlyn contrasted with my experience of fishers in other ports like Mevagissey. One reason for this may be because of contrasting forms of economic organisation, with Newlyn particularly dependent on the fleet of offshore trawlers that go to sea for up to 14 days at a time. Furthermore, most of the beam-trawl vessels are company-owned and a study of how the Newlyn fleet was adapting to rising fuel costs (Abernethy et al. 2010) found that not only was this section of the fleet particularly heavy on fuel consumption but there was also less optimism and sense of economic security amongst the company skippers then the independent owner-skippers.

Newlyn has a diverse population that reflects both its industrial past and its history as a centre for arts and crafts going back to the Newlyn School of artists that emerged in the 1880s, who were primarily drawn to Newlyn by the fishing industry. Despite the fact that fishing only employs a minority of the local population it dominates the village physically and socially and residents draw meaning and a sense of place from it. A number of non-fishing locals described the attraction

[7] 2001 Census for England and Wales. Source: Office for National Statistics licensed under the Open Government Licence v.1.0.

[8] Ibid.

[9] Ibid.

[10] The Superintendent emphasised that the views expressed were his own and that he was not speaking on behalf of the Fisherman's Mission.

of it as something that felt 'vital' and 'real'. Although Newlyn includes multiple, overlapping communities, with various factions, tensions and solidarities, fishing and non-fishing locals take pride in the fact that it is a 'rough and ready', 'working sea port'. It has a wider reputation as having a roguish element and Newlyners seemed to actively cultivate this image whether or not they felt it related directly to them. Unlike most other fishing villages in Cornwall, Newlyn only has a marginal tourist industry. It draws in a small number of visitors from the much larger number that visit the neighbouring picturesque fishing village of Mousehole, which now harbours a tiny active fishing fleet for part of the year and where a large proportion of the cottages are now holiday lets. It was not uncommon for people I spoke to in Newlyn to draw a contrast with Mousehole with a mixture of pride and also anxiety about the prospect of Newlyn sharing the same future.

Against this background of concern about loss of skills and loss of jobs, I encountered a heritage initiative that was attempting to revive old skills and create new jobs. In 2004, *Ripple* was salvaged from a muddy grave in a river estuary in Cornwall. Built in 1886, *Ripple* was the remains of a Cornish sail fishing boat, known as a *lugger*. The boat was brought back to Newlyn where it had docked prior to 1933 when it began its life as a leisure yacht and houseboat. During this latest reincarnation, retired ship captain and civil servant John Lambourn spent the next 5 years restoring her. *Ripple* was re-launched in 2007, hauled to the sea by a local rugby team watched by a crowd of several hundred, and given a blessing by the local parish minister. The sight of *Ripple* moored in the harbour drew my curiosity, incongruous alongside the other boats but also somehow fitting in the wider land/seascape. It begged the question whether this heritage project could have something to say and to contribute to the problems in the fishing industry or whether they were operating in two different social and economic domains—separate and even in antagonism.

Following Cornwall's designation as an Objective One area for the 2000–2007 EU funding programme a fisheries task force was set up consisting of fisheries regulators, port managers, fish merchants and processors, agents and fishermen's representatives to plan and implement how the money potentially available for fisheries was to be spent. A sub-group—the Newlyn Fishing Industry Forum (NFIF)—was given the task of studying the potential for regeneration in the port through developing fisheries infrastructure and identifying and capitalising on "opportunities that can be gained by combining aspects of tourism, leisure and fishing industries and to encourage all sectors of the fishing industry to be more accessible to the public"[11]. Proposals were not only informed by top-down policy directives; rather in public discourse surrounding regeneration in Newlyn, in online blogs and in my interviews, there was talk of the need for change and modernisation and of addressing patterns of stagnation and narrow development. One of the members of the NFIF was the Methodist minister for Newlyn. Although recognising the complex and longstanding ties between Newlyn and the firm that own the majority of the beam trawl fleet, he questioned the public benefit of the harbour having "all its

[11] Objective One Partnership for Cornwall and Scilly (n.d.).

Fig. 15.2 *Ripple,* Looe Lugger Festival 2009. Author's own photograph

resources tied up in boats that were unable to go to sea" (referring to the impact of the fuel cost and recruitment issues). A harbour is about more than those who go to sea, he added: "You cannot separate a harbour from the people that live around it."

John Lambourn was also a member of the NFIF and envisaged the boat contributing to its heritage and regeneration goals, but essentially the idea, the finance for the project, and a lot of the restoration work, was all his own. In fact it took 5 years of hard work before she was seaworthy. John grew up close to Newlyn and was the son of an artist. As a young man he left to join the merchant navy, and became a ship's captain ferrying cargo and passengers all over the world. This led to a position as assistant harbour master and civil servant in the Marine Department of the port of Hong Kong. On retirement he returned to live in Newlyn where his brother is a fisherman. Acknowledging the unique character of Newlyn, John has said that a 'too tidy approach' to promoting the area's heritage would not sit well. He intended for *Ripple* to be a working boat and to have a 'rural' rather than an 'academic' or 'sacred' function. His vision was to set up a sailing school that would give young people as well as paying tourists a practical educational experience. He envisaged that the learning of seamanship skills through luggers would not only be a means for personal development and life skills amongst young people but would also stimulate a growing interest in traditional boatbuilding in Cornwall. Furthermore, re-registering it under its original fishing vessel number he hoped to use it to demonstrate fishing techniques and land fish to the market (Fig. 15.2).

15.4 Reconstruction

> No one knows when the first boat was built, or where, or by whom, or why. Boats began before history; boats are part of our cultural memories. Why else do people gather at the water's edge when tall ships appear? *Dick Wagner, founding director, The Centre for Wooden Boats, Seattle* (Hendrickson 2012, p. 21)

Ripple constitutes a particular form of heritage but it also signifies and embodies a range of other relatively longstanding heritages—the influence of the Newlyn School of artists that drew John's father to the area and which, alongside photographs, provide visual referents, in the absence of living memory, of the days when luggers were in common use; its biography tells a story drawing on a tradition of folk tales centred around boats and their journeys; and finally it embodies a range of craft skills, which some have even considered to be 'arts' of their own. A former fisherman and artist remarked:

> A lot is said about art in Cornwall, but hammer and chisel art, the art of bending and avoiding splitting, the art of each fastening being driven in and making up the overall strength of the *Ripple* seem to be John Lambourn's art. What a beauty the *Ripple* is after so much work. I get the feeling John could see her finished before he started.

Luggers are heavy-framed, beamy[12] craft built for fishing that are easily identifiable by two perpendicular sails located fore and aft[13]. They are carvel-built, which means that once the keel[14] is laid and the ribs are in place, the planks of the hull are pegged flush against one another, rather than overlapping, as is the clinker tradition. Clinker design was used by the Vikings to build their longboats whilst carvel techniques were common in the ancient eastern Mediterranean. Maritime historians are not clear on how these influences spread, or whether techniques in different areas developed independently (Oliver 1971). The Cornish lugger that evolved was very particular to the region, whilst drawing in influences such as Breton vessels encountered during smuggling expeditions. The design continued to develop even after the arrival of the railway into Cornwall, as the need to be first back to market to get a good price for the catch intensified. There were differences in design between East Cornwall and West Cornwall and even between ports such as St Ives and Mount's Bay. There is no one alive now who made a living from sailing these boats, and few people who have the knowledge to build them. However, there is a rich variety of historical sources that John could go to.

To begin with there were the technical drawings and writings of maritime historians like Philip Oak and Edgar March, commissioned between the 1930s and 1950s by the National Maritime Museum to travel the length and breadth of the British Isles recording both the design of traditional craft and the memories of the boatbuilders and mariners, as these craft were being replaced by engine powered boats and steel hulls. There were also 'hand me down stories' (in John's words) and family archives. Once the restoration had begun, descendents of her former owners began rummaging around in attics and producing photographs and records that revealed *Ripple's* biography. There are a great many photographs as well as paintings depicting luggers in the late nineteenth century especially. This was the moment shortly after the arrival of the railway, when the lugger fishing industry was at its peak and

[12] Beamy: a terms used to describe a vessel that is broad (i.e. the proportion of its 'beam' or breadth relative to length).

[13] Fore and aft: referring to the front and rear sections of a vessel, or towards the 'stern' and the 'bow' respectively.

[14] Keel: A lengthwise structure along the base of a ship.

artists' communities as well as early tourism were beginning to flourish, especially at Newlyn and St Ives. John had never built a boat before, let alone a lugger and yet the small details such as the rigging match these visual representations perfectly. Finally, there is a huge amount of what John has called 'the social history of Cornwall'—the historic documents in public archives that record who built the boats, who owned and had shares in the boats, who skippered them and how much they caught.

The construction of *Ripple* and the material networks it embodies is mirrored in the social and economic relationships evident in the archives. Census data for St Ives in the late nineteenth century shows that the boatbuilders, sail-makers, blacksmiths, rope-makers, coopers etc. all lived alongside the fishers and mariners. The legal ownership of each boat consisted of 64 shares and the Merchant Shipping Records show that these trades people as well as widows frequently held shares in the boats. The profits of each catch were divided between a boat-share (which was divided again between the owners), a body-share which able fishermen received (boys received a half-share) and a finally a net-share for those crew members that owned a net or a piece of a net[15]. The wives of the fishermen often made and mended the nets, especially as fishing industrialised and the boats were away for up to 3 months at a time chasing the shoals of herring in an annual circumnavigation of Britain.

The way that *Ripple* is a conduit for the transmission of historical knowledge lies not only in archival repositories of social history and memory, but also in the skills and insight acquired through learning to sail her. I experienced this first-hand as a member of the crew sailing her for the first time since the 1930s. The ability to sail a lugger, as to build one, was also a skill that had to be recovered and relearned—and the only way to do this was through practice. It was a tough, very physical challenge that gave us a direct connection to a bygone way of life. There were moments of exhilaration when body and limb, wind and sailing rig finally worked in tandem, and boat and crew achieved momentary gracefulness. After a race at a lugger regatta, we rowed into the harbour, two men to each massive oar. A crowd was gathered on the piers, and cheered as we passed through the gaps. However, for the most part the experience was punishingly hard and sometimes frightening. With a dipping lug rig, every time it was necessary for the boat to tack, the foresail (about 700 sq feet of canvas and a heavy wooden spar) had to be quickly lowered, passed around the mast and re-hoisted, without losing the wind or getting things tangled up. It was a difficult procedure for a bunch of novices. Due to John's commitment to authenticity, the sails were held in place by large iron hooks which passed through a round iron ring or *cleat* in the corner of the sail and were connected to the *sheets* which passed inside the gunwale and up to the *halyards* (ropes on which you pulled or 'let go' to raise or lower the sail). Sometimes when sailing the wind slackened for a moment and the iron cleats would come free. The sail would start to whip and crack like lighting, the iron ring flying dangerously around our heads, until some brave soul caught it and wrestled it back into place. On a failed attempt to make it to the Isles of Scilly in heavy seas, a crew member took a nasty hit to the head, and a lifeboat was called to tow the boat back to safety.

---

[15] Thanks to Tony Pawlyn, maritime historian (personal communication) for information regarding the social history of the Cornish fishing industry in the lugger era.

Fig. 15.3 The crew of *Ripple* working together to furl and cover a sail, Mount's Bay 2009. Author's own photograph

Through these experiences we were given an insight into how tough the men must have been that sailed these boats for a living and we learnt that fishing and sailing, in the era before fishing boats were mechanised, were interdependent and advanced whole-body skills, to which a practical education from a young age would have been a great advantage. We also had to struggle with the nautical terminology and phrases that John insisted on using, as if to show that *Ripple* was part of a much broader maritime tradition. About 30 original Cornish luggers and 3 replicas are still sailing today. There is an element of performance and spectacle about these, which is romantic particularly for bystanders and onlookers; as crew-members we were only too aware of the dangers. However, when the festivals and regattas are in full sway they are a thrilling experience for crews and spectators because they animate seascapes in a way otherwise rarely seen today which perhaps resonates with a deep, subconscious memory of past eras. As fishing and other coastal industries, such as shipping and shipbuilding, have industrialised, specialised and in many areas declined (Smith 1999; Starkey 1998), these tangible links with working seascapes have also been lost (Figs. 15.3 and 15.4).

Just as John's project is given meaning by local repositories of history and memory, this last aspect brings into view a body of academic work in archaeology and historical geography. Braudel's approach to Mediterranean history set a precedent (1995 [1972[16]])—an influence evident in some subsequent scholarly treatments of Atlantic history, for example, Bowen's *Britain and the Western Seaways* (1972) and Cunliffe's *Facing the Ocean: the Atlantic and its Peoples* (2001). These identify the Atlantic coastal routes of the western fringes of Europe as constituting a distinct realm of cultural contact and exchange, fostering over millennia littoral cultures with similarities in technology, language, religion and other aspects of culture. Whilst the *Ripple* restoration is somewhat particular given the local context,

[16] First published in France under the title La Méditerranée et la Monde Méditerranéen á l'Époque de Philippe II, 1949.

Fig. 15.4 *Ripple* chasing the pack, Looe Lugger Festival 2009. Author's own photograph

it is also one of a growing network of maritime heritage projects spanning Atlantic North West Europe explicitly celebrating a common heritage which some see as part of an ongoing 'Celtic' identity. This new maritime heritage has echoes of a deep past in which, from one angle, shared marginality drove a common seafaring culture on the fringes of Europe. From another angle such cultures were part of a cosmopolitan oceanic world, which from prehistory to the middle and early modern ages, was at the centre of trade, migration and innovation.

To recall my first encounter with John and the *Ripple*, I had asked him how he had the skills to build such a boat and he replied: "Oh, when you grow up, in an environment where things are made, you just pick it up, like you do when you work on your own house… That's what they should be teaching young people. It gives you"… (he searched for the right words). "Freedom!" his friend, a fisherman, put in. "Freedom, yes", John continued "and also a sort of 'can do' attitude—if you have a dream and you can do the work yourself, well then that can make the difference between achieving something and never even beginning." Macdonald (2002) has said that displays of vernacular material culture represent a critical commentary on resourcefulness that is expressive both of a locality and a way of life that is broader than the locality. Laviolette (2006) makes a similar argument about contemporary maritime art in Cornwall that makes use of recycled and salvaged material recalling the work of the St Ives fisherman-artist Alfred Wallace (1855–1942).

Ripple represents a technological tradition that is unique to west Cornwall and simultaneously it can also be interpreted as representative of ways of life collectively associated with the broad historical-geography of maritime regions and more specifically, with fishing. The emphasis John placed on reviving a sense of local resourcefulness and independence is pertinent in the context of the fishing industry in an era where entry costs are increasingly prohibitive for young people.

15.5 Regeneration

Whilst *Ripple* can be seen as an example of 'critical nostalgia' (Clifford 1986) it is arguably also a pragmatic and forward-looking enterprise. John had said one of his aims was "to open people's eyes to the lessons of a 100 years ago, when there was no oil and only wind." Recently one of the last of the Westcountry ketches[17], *Irene*, sailed for Brazil, via the Mediterranean. It was transporting and trading in ethical and organic food produce between ports on route, and is an imaginative attempt to explore a market for low carbon cargo. Such ventures are also being explored on a commercial level by companies such as B9 Shipping Company. Given the pressures in the fishing fleet owing to rising fuel costs[18], *Ripple* provides an allegory about the need to explore alternative technologies and the role for the past as a resource and stimulus for future innovation. Several vessels at Newlyn have already experimented with incorporating sail power to make them less reliant on diesel. With multiple major redevelopment plans, public and private, having been discussed for Newlyn and the surrounding vicinity, John could also see an opportunity for revived boatbuilding and servicing yards. These could not only provide alternative jobs but also potentially be a stimulus for technological innovation in the fisheries sector.

Attitudes towards John's project from members of the local fishing industry have, however, have been mixed. A colourful and provocative character, his vision has been regarded by some as romantic and even a tad eccentric.. Until recently, he was a newly appointed member of the Newlyn Pier and Harbour Commissioners, following a government Harbour Review Order in 2010. John, along with two others, was later voted out by secret ballot. Little information was given to the public as to the reasons for this ousting but it is no secret that there were disagreements regarding harbour redevelopment and regeneration plans, in particular a proposal for a new fish market.

One of the problems identified with the existing market has been is that it is an extremely functional and not aesthetically pleasing building that dominates the seafront and blocks views from the centre of the village to the sea. Most tourists either bypass Newlyn or pass straight through on the way to Mousehole. Ambitious plans were advocated by some people, including John, for redevelopment of the harbour that would include 'visitor friendly' features and a more effective marketing of the heritage of the village[19]. Others, including some fishermen and fish merchant firms, felt that the fishing industry could not afford such plans and were wary of whom it would benefit. Whilst yet another group of port users, including fishermen I spoke to who operated from Newlyn but did not land their fish there, or reside in the locale, felt that any new market would ideally be located not in the village at all but

[17] Ketch: two-masted, fore and aft sailing ship traditionally used for transporting small cargoes.

[18] The overheads created by fuel expenses are huge—for some Newlyn based trawlers £ 10,000 of diesel per trip on average—arguably a significant pressure towards overfishing as well as affecting fisher incomes.

[19] Plans were informed by the Cornwall Archaeological Unit report: Cornwall and Scilly Urban Survey. Historical characterization for regeneration: Newlyn (Russell 2003).

favoured a new central inland market and distribution centre directly linked to one of the main roads, where lorries can get into and out of easily. This could then serve both large fishing ports in the region and other smaller ports that presently overland their fish to existing markets.

Clearly there are widely divergent views represented here about the particular ties between fishing markets, towns/villages, harbours and fleets and how these are to evolve and adapt to changing European and global economic and political conditions. For one prominent fish merchant and multiple fishing vessel owner in Newlyn, the bottom line must be the priorities of the catching sector. They had some sharp but pertinent points about the limitations of John's project, saying, "it doesn't do a lot to the port in terms of economic strength or economic financial benefits, you know, it doesn't employ anybody as such and it doesn't bring in an awful lot of bread and butter, does it?" I asked if they could see a role for the kinds of diversification that John's project might help promote such as traditional boat building and boat servicing yards. They replied that they could not see anything necessarily wrong with it but doubted whether it was "a viable thing bringing in an economy to the harbour in terms of fish landed."

> He hasn't landed a fish yet with the *Ripple* and I don't think he ever will land a fish. If he wants to have the *Ripple* and go sailing, that's fine and have people building luggers and things, I haven't got a problem with that. There is certainly a need for carpenters, but whether it is viable without grants and things like that, that's quite a different issue.

Perhaps naively I then asked my interviewee whether they felt that projects like John's might have a role in informing wider communities about fishing in Cornwall and potentially attract new recruitment. Once again their response was to the point:

> I haven't seen any single person come in yet that's gone commercial fishing, having gone on the *Ripple*. And I mean, maybe it will, but I very much doubt it. There's quite a different set of skills needed to go commercial fishing on some of the steel vessels than the sailing vessel.

The comments of the fish merchant convey a sense of boundaries, social distance and different economic priorities that in her view mark a sharp divide between fishing 'industry' and the kinds of 'heritage' John is promoting. Apprenticeship into fishing is typically by an informal process of experience, observation and practice. Pálsson (1994) has drawn an analogy between acquiring fishing and seafaring skills and going on a journey, finding that Icelandic fishermen spoke of overcoming seasickness as 'getting one's sea-legs', therefore providing a metaphor for the corporeal nature of gaining competence at sea. Along with Ingold (2000) he has used the term 'enskillment' to denote this kind of knowledge acquisition that comes from active engagement with the social and physical environment. Cognitive social learning theory also points to the importance of environments for learning, incorporating such processes as 'legitimate peripheral participation' (Lave and Wenger 1991) and 'communities of practice' (Wenger 1998). One fishermen I interviewed recalled helping his father and uncles out as a boy—cleaning up the boat when it landed, going to sea with them, being allowed to keep and sell crabs and generally learning "small things, like tying your knots".

Whilst *Ripple* is a form of heritage that calls on informal, practical and experiential learning of skills, this process and the skills learnt are comparable to modern fishing but not alike. It can by no means be a replacement for occupational skill-sets in fishing. What is being learnt during an apprenticeship on a boat is not only mechanical skills but also how to fit in to a social system which is both embodied in the habitus of the crew but also transcends the crew as a wider way of life (van Ginkel 2001; Simpson 2006). To 'learn the ropes' as a new recruit into fishing van Ginkel (2001, p. 179) says, is a process:

> ...not limited to the mere performing of tasks; it includes internalizing the norms, values, attitudes, interests, knowledge and skills necessary to become an accepted member of the occupational group, to do the job properly, and to legitimize the work world. Compatibility with the crew's ideology is an important factor.

Fishers' attachment to their way of life is often explained in terms of features of 'occupational community' (Davis 1986; Lummis 1985) such as a strong sense of pride and satisfaction in ones work and identity, specialised knowledge and skills, "an 'egalitarian ideology' combined with rhetoric's and concepts of independence, self-reliance, freedom and so forth" (van Ginkel 2001, p. 178). One aspect of this pride and satisfaction in work and identity, which is surprisingly sometimes overlooked in these analyses, is status as "primary producers of food" (Nadel-Klein 2003, p. 8). In an era where more than half of the world's population live in cities and the majority of people—at least in the industrialised West—are not directly engaged in producing food, the close associations between fisher and fish as a vital and often messy, bloody, smelly life source and the idea of 'putting food on plates' is significant, not least to many fishers themselves. As one fisherman expressed it commenting on his sons following him into fishing:

> They went fishing of their own choice. They didn't come because I made them come. They wanted to go fishing. But if I didn't think fishing had a future, I would have tried to put them off. But I've always believed that fishing's got a good future. Because the way I always look at it, in simple terms is, you got to eat [pointing], and everybody else got to eat on this planet, and there's only so many people producing food.

Whilst, to my knowledge, *Ripple* may not yet have been dirtied with the blood and fish guts of a commercial catch, and whilst the reconstruction was self-funded by John, it nonetheless does have something to contribute directly to the catching sector and to the local economy. By evoking a sense of the past in a tangible way, by recalling and bringing to life scenes depicted in photographs and artworks, *Ripple* makes a link between different 'fields' of heritage—the production and consumption of local history, visual cultures (including art galleries and museums) and fishing in a contemporary working harbour—domains that might otherwise remain separate, disconnected and fragmented spheres to the detriment of all. This has already contributed to generating a 'sense of place' that is fostering stronger links between the 'catch and the locality' (Reed et al. 2013) as in the example of the recent revival of the fishery for pilchards in Cornwall, now rebranded, as the 'Cornish sardine'. One firm is now selling Cornish sardines in tins illustrated with Newlyn School

Fig. 15.5 A tin of 'Cornish sardines' featuring *The Greeting* by Newlyn School artist Walter Langley. Author's own photograph

Fig. 15.6 Artist Bernard Evans painting *Ripple* and other luggers alongside the medieval pier in Newlyn harbour during the 'Painting Party on the Quay' event, British Tourism Week, March 2011. Courtesy of Steven Walker

paintings featuring luggers in Newlyn harbour (Fig. 15.5). Now that real luggers can once again be seen alongside the medieval 'old quay' in Newlyn this marketing has been used to good effect. The pilchard fishery is not regulated by quotas and is being promoted by catchers and merchants as both indigenous and sustainable. This marketable 'sense of place' as food provenance is not to be mistaken for the diverse and grittier everyday experience and place attachments of fishers and other workers and residents in places like Newlyn. Nonetheless it is an important one when the sustainability (social, economic, environmental) of forms of fish production reliant on bulk overseas export is questionable (Fig. 15.6).

15.6 Conclusions

The example of fisheries heritage considered in this chapter challenges notions of 'industry' and 'heritage' as being separate and opposed domains. It also demonstrates an important role for heritage that exists outside of museum contexts and which incorporates informal learning, and production and use of material artefacts including craft skills. This can be a source of alternative and diversified fishery-linked livelihoods, a factor in strengthening and promoting links between catch and locality, and a powerful source of critical nostalgia to stimulate imagination and innovation. Alternative forms of heritage production have implications for alternative forms of consumption (including tourism) and even alternative forms of fishing. Dependency of communities on harbours and on the sea in places with a history of fishing is broader and more complex than merely landing of fish. The current period of economic recession and rising unemployment has consequences for young people in rural maritime regions, across class, occupational and family backgrounds. In a context of frequent anxiety about the loss of 'real' and 'tangible' jobs (Crow et al. 2009) and growing disparities between 'financially rich' centres and 'heritage rich' peripheries (Howard and Pindar 2003, p. 65), heritage initiatives that can strengthen regions, livelihoods and diversity of skill-base are to be supported.

Nonetheless, maintaining existing fishing harbours and beaches as bases for catching fish, remains the highest priority for the sustainability of coastal economies and the integrity of coastal places. Needless to say this should be complemented and stimulated by heritage initiatives, rather than replaced. This entails a historically informed conception of 'fishing communities'—both in the traditional and occupational sense—as mixed economies, which are always changing and evolving. Walton (2000) quotes a Lowerstoft man born in 1902 speaking of what made community in his local fishing context (p. 128), and concludes that the basis of community in this view was "commitment to an industry, not necessarily entailing actually going to sea, but being part of a network of shared interests and concerns that surrounded the fishing." A similar case is made in the context of contemporary Scotland by Ross (2013). Academics can inform policymakers, local authorities and non-state actors about how to support these kinds of communities by observing the connections and disconnections between the various practices, politics and priorities of their different sectors. Ultimately neither state, industry nor community models of development will be sufficient alone to articulate and manage their complex ties.

Acknowledgments This chapter is based on doctoral research funded by the Economic and Social Research Council. The author is grateful for the support of supervisors Frances Pine, Mao Mollona and Rebecca Cassidy and for the time and good faith given by research participants. Early versions were presented to seminars at Goldsmiths, University of London. It has benefited from the helpful feedback of colleagues and peer reviewers to who thanks are also extended.

References

Abernethy, K., Trebilcock, P., Kebede, B., Allison, E., & Dulvey, N. (2010). Fueling the decline in UK fishing communities? *ICES Journal of Marine Science, 67*.

Ahmad, Y. (2006). The scope and definitions of Heritage: From tangible to intangible. *International Journal of Heritage Studies, 12*(3), 292–300.

Atkinson, D., Cooke, S., & Spooner, D. (2002). Tales from the Riverbank: Place-marketing and maritime heritages. *International Journal of Heritage Studies, 8*(1), 25–40.

Bowen, E. G. (1972). *Britain and the Western seaways*. Thames and Hudson Limited.

Braudel, F. (1995). *The mediterranean and the mediterranean world in the age of philip II* (Vol. I). Berkeley: University of California Press. (Original edition, 1972).

Casey, E. S. (2002). *Representing place: Landscape painting and maps*. University of Minnesota Press.

Clifford, J. (1986). On ethnographic allegory. In J. M. Clifford & G. E. Marcus (Ed.), *Writing culture: The poetics and politics of ethnography*. University of California Press.

Crang, M. (1994). On the heritage trail: Maps of and journeys to olde Englande. *Environment and Planning D: Society and Space, 12*(3), 341–355.

Crow, G., Hatton, P., Lyon, D., & Strangleman, T. (2009). New divisions of labour? Comparative thoughts on the current recession. *Sociological Research Online, 14*(2), 10.

Cunliffe, B. (2001). Facing the ocean: The Atlantic and its peoples 8000 BC—AD 1500. Oxford University Press.

Darke, N. (1999). *Plays, 1*. London: Methuen Drama.

Davis, D. L. (1986). Occupational community and Fishermen's wives in a Newfoundland fishing village. *Anthropological Quarterly, 59*(3), 129–142.

Day, A., & Lunn, K. (2003). British maritime heritage: Carried along by the currents? *International Journal of Heritage Studies, 9*(4), 289–305.

Easthope, G. (2001). Heritage sailing in Australia: A preliminary schema. *International Journal of Heritage Studies, 7*(2), 185–190.

Gray, J. N. (2000). *At home in the hills: Sense of place in the Scottish borders*. New York: Berghahn Books.

Greenhalgh, P. (1997). The history of craft. In P. Dormer (Ed.), *The culture of craft*. Manchester: Manchester University Press.

Harvey, D. C. (2001). Heritage pasts and heritage presents: Temporality, meaning and the scope of heritage studies. *International Journal of Heritage Studies, 7*(4), 319–338.

Hendrickson, P. (2012). *Hemingway's boat: Everything he loved in life, and lost, 1934–1961*. Random House.

Hewison, R. (1987). *The heritage industry: Britain in a climate of decline*. London: Methuen London.

Howard, P. M. (2012). Workplace cosmopolitanization and the power and pain of class relations at sea. *Focaal*, (62), 55–69.

Howard, P., & Pindar, D. (2003). Cultural heritage and sustainability in the coastal zone: experiences in south west England. *Journal of Cultural Heritage, 4*, 57–68.

Ingold, T. (2000). *The perception of the environment: essays on livelihood, dwelling and skill*. London: Routledge.

Laurier, E. (1998). Replication and restoration: Ways of making maritime heritage. *Journal of Material Culture, 3*(1), 21–50.

Lave, J., & Wenger, E. (1991). *Situated learning: Legitimate peripheral participation*. Cambridge: Cambridge University Press.

Laviolette, P. (2006). Ships of relations: Navigating through local cornish maritime art. *International Journal of Heritage Studies, 12*(1), 69–92.

Longhurst, A. (2006). The sustainability myth. *Fisheries Research, 81*(2–3), 107–112.

Lummis, T. (1985). Occupation and society: the East Anglian fisherman 1880–1914. Cambridge: Cambridge University Press.

MacDonald, S. (2002). On 'Old Things': The fetishization of past everyday life. In N. Rapport (Ed.), *British subjects: An anthropology of Britain*. Oxford: Berg.

Matsuda, M. (2004). A place for history. *History and Theory, 43*(2), 260–271.

MMO. (2012). *The UK fishing industry in 2011: Landings*. Marine Management Organization.

Nadel-Klein, J. (2003). *Fishing for heritage: Modernity and loss along the Scottish coast*. Oxford: Berg.

Oliver, A. S. (1971). *Boats and boatbuilding in West Cornwall: A study of traditional cornish fishing craft*. Barton.

Objective One Partnership for Cornwall and Scilly (n.d.). 'Cornwall and Isles of Scilly Fishing Industry Task Force Strategy 2000–2010'. (http://www.objectiveone.com/ob1/pdfs/FTF%20STRATEGY%20final.pdf)

Pálsson, G. (1994). Enskilment at sea. *Man, 29*(4), 901–927.

Pauly, D., Christensen, V., Guenette, S., Pitcher, T. J., Sumaila, U. R., & Walters, C. J., et al. (2002). Towards sustainability in world fisheries. *Nature, 418*(6898), 689–695.

Pitcher, T. J. (2001). Fisheries managed to rebuild ecosystems? Reconstructing the past to salvage the future. *Ecological Applications, 11*(2), 601–617.

Reed, M., Courtney, P., Urquhart, J., & Ross, N. (2013). Beyond fish as commodities: Understanding the socio-cultural role of inshore fisheries in England. *Marine Policy, 37*, 62–68.

Ross, N. (2013). Exploring concepts of fisheries 'dependency' and 'community' in Scotland. *Marine Policy 37*, 55–61.

Russell, S. (2003). Cornwall and Scilly urban survey. Historical characterization for regeneration. Newlyn, Cornwall Archaeological Unit.

Sckula, A. (1995). *Fish story*. Dusseldorf: Richter.

Sennett, R. (2009). *The craftsman*. Penguin Books.

Simpson, E. (2006). *Islam and society in the Indian Ocean: Seafarers of Kachchh*. London: Routledge.

Smith, H. (1999). The historical geography of the seas around the British Isles. *Marine Policy, 23*(4–5), 275–287.

Starkey, D. J. (1998). Growth and transition in Britain's maritime economy, 1870–1914: The case of South-West England. In D. J. Starkey & A. G. Jamieson (Eds.), *Exploiting the sea: Aspects of Britain's maritime economy since 1870*. Exeter: University of Exeter Press.

Steinberg, P. E. (1999). The maritime mystique: Sustainable development, capital mobility, and nostalgia in the world ocean. *Environment and Planning D: Society and Space, 17*(4), 403–426.

Symes, D., & Phillipson, J. (2009). Whatever became of social objectives in fisheries policy? *Fisheries Research, 95*(1), 1–5.

Thurstan, R. H., Brockington, S., & Roberts, C. M. (2010). The effects of 118 years of industrial fishing on UK bottom trawl fisheries. *Nature Communications, 1*, 15.

Urquhart, J., & Acott, T. (2013). Constructing 'the stade': Fishers' and non-fishers' identity and place attachment in Hastings, South-East England. *Marine Policy* 37, 45–54.

van Ginkel, R. (2001). Inshore fishermen: Cultural dimensions of a maritime occupation. In D. Symes, & Phillipson, J. S. (Eds.), *Inshore fisheries management*. Kluwer: Springer.

Walsh, K. (1992). *The representation of the past: Museums and heritage in the post-modern world*. Taylor & Francis.

Walton, J. K. (2000). Fishing communities, 1850–1950. In D. J. Starkey, C. Reid, & N. Ashcroft (Eds.), *England's sea fisheries: The commercial sea fisheries of England and Wales since 1300* (pp. 127–137). London: Chatham Publishing.

Walton, J. K. (2010). *English seaside towns: Past, present and future*. Coastal Communities Alliance.

Wedgwood, T. (2009). History in two dimensions or three? Working class responses to history. *International Journal of Heritage Studies, 15*(4), 277–297.

Wenger, E. (1998). Communities of practice. Learning as a social system. *Systems Thinker*.

Chapter 16
The Socio-Cultural Impact of Industry Restructuring: Fishing Identities in Northeast Scotland

Ruth Williams

Keywords Identity • Restructuring • Fishing community • Place • Symbols • Scotland

16.1 Introduction

Fishing as an occupation provides more than a way of earning a living (Pettersen 1996; van Ginkel 2001). Its traditions, structures and dynamics influence all aspects of the lives of individuals and households who work within and live alongside the industry. However, the Scottish fishing industry is experiencing a period of major restructuring, driven by an over capacity in the fleet (Royal Society of Edinburgh 2004) and reflecting global problems in key commercial stocks (Symes 2001). The impacts of this restructuring have repercussions for the individuals and households who depend on the fishing industry for their economic and socio-cultural resources. Although fisheries policy has begun to recognise the socio-cultural dimensions of the industry, there is little clarity over what these might be, how they can be assessed and what they may mean for fisheries policy. In this chapter the concept of identity is used to encapsulate and explore the socio-cultural dimensions of the fishing industry and the impacts of restructuring in northeast Scotland and the fishing settlements of Lossiemouth, Cullen, Buckie and Fraserburgh in particular. The aim is to explore the role of fishing communities in the construction and performance of fishing identities and the changes to this brought about by restructuring. Qualitative data from interviews with fishermen, former fishermen and their wives help to unpack what fishing communities mean to people and the roles they play in the construction and performance of fishing identities.

R. Williams (✉)
Centre for Rural Economy, School of Agriculture, Food and Rural Development, Newcastle University, Newcastle upon Tyne, NE1 7RU, UK
e-mail: ruthwilliams@outlook.com

J. Urquhart et al. (eds.), *Social Issues in Sustainable Fisheries Management*, MARE Publication Series 9, DOI 10.1007/978-94-007-7911-2_16,

16.2 Fishing Industry Restructuring in Northeast Scotland

The UK fishing industry is highly concentrated in particular locations, with northeast Scotland an important area within both the Scottish and UK contexts. In 2005, 491 vessels were registered in northeast Scotland (Aberdeenshire Council 2007). For over 10 m vessels, northeast Scotland accounted for 35 % of the Scottish fleet and 17 % of the UK fleet. Employment in the catching sector of northeast Scotland was 1,448 in 2005, accounting for 30 % of Scotland's and 11 % of the UK's employment in fishing. In the same year, 52 % of the Scottish catch by volume and 50 % by value, and 39 % of the UK catch by volume and 34 % by value, was landed into northeast Scotland's ports (Aberdeenshire Council 2007). Despite contraction within northeast Scotland's fishing industry following two phases of decommissioning, the region has maintained its dominance within the UK in terms of both landings and employment (Scottish Government 2010). The analysis focuses on the circumstances surrounding the demersal sector which targets mixed whitefish species that live on or near the seabed such as cod, haddock, whiting, monkfish and also Norwegian lobster (*Nephrops*) and landed 32 % of the UK catch by volume and 38 % by value in 2005 (MFA 2006). The demersal sector particularly has suffered over the last 10 years. With a decline in key stocks, most notably cod, and the resulting limitations placed on those targeting them, landings and profitability have fallen.

Quotas for key demersal species fell dramatically. Between 1997 and 2007, cod quota for the North Sea and West of Scotland fisheries declined by 84 %, whiting by 65 % and plaice by 43 %. Demersal vessels operating in these areas have also been restricted to just 15 days at sea per month and other areas have been closed to fishing altogether or require specific permits (Aberdeenshire Council 2007). To reduce overcapacity in this sector, the Scottish Executive funded two rounds of decommissioning in 2001–2002 and 2003–2004, at a total cost of £ 56 million (Scottish Executive 2005). These schemes removed 165 over 10 m vessels from the Scottish demersal fleet (Scottish Executive 2006), most of them based in northeast Scotland.

Here, direct employment in catching fell by 55 % between 1995 and 2005, a loss of 1,770 jobs (Aberdeenshire Council 2007). Buckie fishing district was hardest hit with a 65 % loss of employment. However, despite the loss of employment in the catching sector, unemployment in northeast Scotland remained relatively low, 2 % in Fraserburgh and Macduff compared to the Scottish average of 2.8 % (Aberdeenshire Council 2005). Although official statistics of the destination of former fishermen are not available, anecdotal evidence suggests that many have been absorbed into the off-shore oil and gas industries.

Beyond the direct impact of restructuring on the fishing industry, there are high levels of multipliers meaning a much greater impact on the wider economy. A study using data from 1999, and so not covering the period of restructuring, found that Fraserburgh had a direct employment in fisheries, aquaculture and processing of 29 % (Scottish Executive 2002). Once the indirect employment in ancillary services such as net making, and the induced employment in local services supported by the spending of fishing households had been taken into account, Fraserburgh's fishery dependent employment rose to 57 %. Buckie's overall fishery dependent employment stood at 22 % (Scottish Executive 2002). The economic and employment impacts of restructuring have been large. However, dependence on fisheries concerns not only economic resources, but social and cultural resources which are also challenged by restructuring.

16.3 Social Issues in Fisheries Policy and Restructuring

Through the European Common Fisheries Policy (CFP), the restructuring described above has attempted to match the capacity of the European fleet with the health of commercial fish stocks (European Commission 2004). The socio-cultural dimensions of the industry have not traditionally been addressed by fisheries policy (Symes and Philipson 2009), but instead have been seen as an externality to be dealt with after policy decisions have been made (for example, see ESSFiN 1999). More recent policy developments have recognised the importance of taking these socio-cultural dimensions into account (Scottish Government 2010).

For example, at the European level the 2002 reform of the CFP saw the creation of Regional Advisory Councils (RAC) with the North Sea RAC creating a specific working group to provide advice on socio-economic impacts (NSRAC 2006). At the UK level, the 2004 report 'Net Benefits' recommended the creation of explicit social aims (Strategy Unit 2004). The response from the devolved administrations in 2005 set out several key objectives, one of which was "to tackle social exclusion and promote long term prosperity in communities traditionally dependent on the fishing industry" (DEFRA et al. 2005, p. 14). At the Scottish level, in the 'Sustainable Framework for Scottish Sea Fisheries', a sustainable fishing industry is seen as vital to maintaining viable communities around the coast. According to the Scottish report one of the key ways to achieve this would be through better decision making informed by interdisciplinary research that brings together biological and socio-economic specialisms. The report recognises a lack of adequate data on Scottish fishing communities, and commits to commissioning a fisheries dependency study (Scottish Executive 2005). At all three levels of policy—national, regional and local—there is a growing recognition of the need to take into account the social impacts of fisheries policy. There is, however, no clear definition of what these might be and how they could be assessed, although Urquhart et al. (2011) begin to set out an agenda for social science research in fisheries policy, outlining the need to increase the geographic scope of fisheries social science studies (which have been limited to small coastal communities on the Altantic fringe); the need for assessing the potential for co-management of marine resources; the need for an understanding of the contribution that fishing makes to local culture and sense of place for both local communities and the tourism industry; and finally, the need for an explicit attempt to make the social and cultural values of fishing more relevant to fisheries policy makers. Clearly, further research by and collaboration with the social sciences is called for.

The study of fisheries has traditionally been dominated by the biological and economic sciences and whilst the social science of fisheries is a growing field, it has not achieved the same influence over policy as bio-economic research (McClanahan et al. 2009; Symes and Hoefnagel 2010; Urquhart and Acott 2013). However, important contributions to the understanding of the socio-cultural dimensions of fishing and its relationship to restructuring have been made particularly in Newfoundland and Norway (see for example Power 2005; Binkley 2002, 2000; Women's Studies International Forum 2000). Social science research on fisheries in the UK has instead focussed on policy and management (see for example Brookfield et al. 2005; Daw and Gray 2005; Stead 2005; Symes 2005, 2001; Phillipson 2002). Nadel-Klein (2003, 2000, 1988) is a key exception; through sustained ethnographic research, she focussed on understanding Scottish fisheries from the perspective of individuals and households who work within it. More recently there have been a number of qualitative studies that explore the interaction between the socio-cultural dimensions of fishing and its relationship with the current period of restructuring from the perspectives of those living through it (see, for example, Reed et al. 2011; Acott and Urquhart 2012; Reed et al. 2013; Urquhart and Acott 2013; Ross 2013; Britton and Coulthard 2013; Zhao et al. 2013).

16.4 Identity and Place

The concept of identity has been widely used across the social sciences and produced a vast literature. Identity is utilised here in three specific ways: to demonstrate how individual and shared identities are constructed and why they are important; to explore how the idea of community is used in identity work; and to indicate how place is both created by and used in the performance of identities.

16.4.1 Individual and Collective Identities

Defined simply, identity is about "who we are and how we differ from those around us" (Jackson 2005, p. 391). Our identities are based on what makes us similar to or different from others, and so the concept can be used to encapsulate and define the social aspects of fishing that distinguishes it from other industries. Much of the individual identities expressed by the participants in this research draw on shared cultural identities that they hold, at least on the surface, in common with each other. These cultural identities perform important functions. They provide a sense of belonging, of having a place within the world. Having this understanding of oneself provides the cultural resources or maps to be able to place others and understand the world around you (Jenkins 1996). Maintaining a shared sense of identity ensures that the distinctiveness on which this identity is based, and the place in the world it provides, is not lost (Cohen 1985).

Identities are not inherent within a person or group but relational, based on processes of identifying these similarities and differences in relation to others (Jackson 2005; Hall 1996). Because they are not fixed within a person or group they are the outcome of active processes of negotiation, of identifying and placing others within the world as we understand it (Jenkins 1996). So whilst identities can be described as individual and/or collective, all are inherently social as they are based on interaction with others (Jenkins 1996). These negotiations of identity can be understood as performances. As identities are brought into being through performance and interpretation they can be understood as becoming, "a process never completed", as there is always the potential for further interpretation (Hall 1996, p. 2). And so as identities are contingent on interpretation and never completely achieved they must be worked at, or performed, in order to be maintained (Jenkins 1996).

People can simultaneously maintain both individual and collective identities through the use of symbols (Cohen 1985). Symbols do more than refer to something, be that an object or an idea, but allow the creation of meaning around them: "Symbols do not so much express meaning as give us the capacity to make meaning" (Cohen 1985, p. 15). In identity work this allows individuals within a group to refer to and value symbols of the same form, whilst understanding the symbols in a way that is meaningful to the individual. In this way the group is able to present a seemingly coherent identity without compromising the many varieties of its individual members.

16.4.2 Community

The idea of community can be understood as a symbol (Cohen 1985). It is a tool strategically used to express similarity and difference:

> People construct community symbolically, making it a resource and repository of meaning, and a referent of their identity.
> (Cohen 1985, p. 118)

The idea of community is used particularly when that which marks a collective identity as distinctive is under threat (Cohen 1985). Dalby and Mackenzie suggest that the idea of community does not necessarily exist in a particular form until it is threatened, "that in important ways they are formed and shaped by the opposition to [that which] is portrayed as threatening" (Dalby and Mackenzie 1997, p. 101).

The past plays an important role in the construction of collective identities, particularly in the use of community to defend against change. References to the past and the telling of stories surrounding people and places from times gone by are commonplace when reflecting on new circumstances and challenges (Cohen 1985). However, the use of the past is not due to a group's inability to face up to the present or future, of being unable to remove themselves from the past. Instead it is a strategic use of the past as a resource to make sense of and confront the present and future (Power 2005; Dalby and Mackenzie 1997; Hall 1996; Cohen 1985). It is used to help make the unfamiliar familiar, past experiences are drawn into the present to enable some sense to be made of it.

A shared past provides a sense of embeddedness and collective history, a sense of belonging and having a place within the world. Edwards suggests that this shared past is a connection to what makes a group distinctive, and so maintaining the past, through remembering and talking about it, is a tool for securing a future: "To lose the past would be to lose a present identity which could not, in turn, be projected ahead" (1998, p. 163). Shared stories or narratives of place are important in signifying belonging to that place (Bird 2002; Acott and Urquhart 2012).

16.4.3 The use of Place in Constructing Identities

Place is an important concept in the construction of individual and collective identities and is used here in two key ways. Firstly, place is created and given meaning through the performance of identities, and secondly place, and belonging to a place, is used in the construction of identities. Places are not benign objects, but are associated with particular meanings (Cresswell 2004). The meanings of place are constructed by the people claiming association with that place through the performance of their identities (Gregson and Rose 2000). As with identities, the symbolic nature of place is dependent on interpretation and so:

> Places are never finished but produced through the reiteration of practices—the repetition of seemingly mundane activities on a daily basis.
> (Cresswell 2004, p. 82)

With the potential for further interpretation of place, places only retain their meaning through its continual expression. The use of the past and the telling of shared stories to create collective identities are also used in the construction of place. The anecdotes and folklore recited by people creates a specific version of place imbued with carefully chosen meaning (Bird 2002). The places in this study become and remain 'fishing communities' through the performance of fishing identities, including through narrative.

Not only is place created through the performance of identity, it is also used in the construction of identities:

> In defining the discourses of inclusion and exclusion that constitute identity, people call upon an affinity with place or, at least, with representations of places, which in turn, are used to legitimate their claim to those places.
> (Ashworth and Graham 2005, p. 3)

An affinity with place is used to mark belonging. A long term relationship with a place often engenders a sense of feeling 'at home', of belonging, which forms part of identity (Convery and Dutson 2006). Having the local cultural knowledge to tell stories about this place signifies belonging to it (Bird 2002).

The idea of a 'fishing community' can be understood as an identity expressing symbol. It is used to express belonging to collective identities based on fishing and place, and to mark these identities as different from non-fishing people and places. Understanding and belonging to a 'fishing community' is part of collective and individual fishing identity. (Re)asserting the idea of fishing communities, through remembering and talking about them enables them to continue through periods of change: 'fishing communities' are brought into being through the performance of fishing identities.

16.5 Methodology

Within fisheries social science, sustained ethnographic research has been usefully and successfully applied to the contexts of Newfoundland, Norway and Scotland to understand the cultural aspects of the industry from the perspectives of those who live and work within it (Power 2005; Nadel-Klein 2003, 2000; Women's Studies International Forum 2000). The research underpinning this study draws on these approaches, employing semi-structured interviews to generate rich qualitative data. This style of interviewing allows participants to tell their own stories and express their unique perspectives on the issues raised (Bennett 2002; Smith 2001).

With no direct link to the fishing industry it was necessary to work through gatekeepers using a snowballing approach to generate potential research participants. This approach has many challenges, and can only be used successfully with a reflexive awareness of the role of these gatekeepers and the sample being produced (de Laine 2000; Cook and Crang 1995). However this proved to be the most successful way to build the connections and relationships needed to access these fishing communities. In total, 19 households took part over the period 2003 to 2006. All were either actively involved in the demersal sector (12), or had been in the case of the retired (3) and former fishermen (4). The chapter draws on a selection of passages from interviews across the range of participants.

Interviews were broadly biographical, with participants encouraged to give a brief life-history, before moving on to explore issues of change. Where possible interviews were conducted with both the husband and wife, and where practical a second interview was arranged at a later date. All interviews were conducted with the assurance of anonymity for participants, and so all names have been changed and some details blurred.

Carrying out research in fishing communities has particular challenges. On a practical level, it can be difficult to arrange interviews around the unique work patterns of fishermen, meaning a high level of flexibility is needed. The fishing industry is politically complex, and most noticeably divided along support for staying in or withdrawing from the CFP. The context in which the interviews were conducted was one where skippers who had applied for the decommissioning of their boats had recently received their confirmation offer and were deciding whether to accept. This generated much speculation within the community as to who would get what from the scheme. Working with fishing households at this time demanded great sensitivity and a reflexive approach.

16.6 Fishing Communities in Northeast Scotland

Ideas of fishing communities were frequently brought to life through the words and conversations of people interviewed. They were used primarily in a past tense, to construct idealised 'times gone by' in contrast to the places where they now live and work. Additionally the present day was assessed against the standards of the past to gauge whether their current situation could still be described as a 'fishing community'. While talking about their understandings of fishing communities, the participants identified a collection of symbols that for them make up a 'fishing community'. The symbols of boats, fishermen, a thriving economy and social events are used to illustrate what people mean by 'fishing community' and its use in the construction of fishing identities. Perhaps unsurprisingly, at the heart of all these ideas is the harbour and the role it plays in defining a place as a 'fishing community'. As Jack from Buckie neatly summed up after the interview:

> I couldnae imagine not living by the sea. I mean without a harbour, what's the centre of your community?[1]

16.6.1 Fishing Boats in the Harbour

When asked how places have changed, it was generally the loss of boats from the harbour that first came to mind with those who took part in the interviews. The presence of fishing boats is an outward, easily read sign that the place is an active 'fishing community'. When the number of boats using a harbour declines it impacts upon the economy, both around the harbour and in the towns. Alongside this economic impact, there is a symbolic impact. As the number of boats physically present in the harbour declines it illustrates the decline of the place as a 'fishing community' and of the industry more generally. This reinforces the growing negativity toward the industry felt by some fishermen who cite the loss of crew and lack of new entrants as other signs of decline.

Before fishermen began to work intensive trip patterns, there were specific times when the whole fleet came in to harbour, such as to land for Friday morning market, or to take their summer or Christmas breaks. At these times the true size of the local fleet could be seen, and they are remembered fondly by the people interviewed, as Peter who was in his 40s and from Cullen, describes:

> And Buckie there, Buckie on a Friday and Saturday night it was traffic jams, aye before, the harbour was full you could walk from one side, the boats was moored up from one end of the harbour, so you could walk from one end of the harbour to the other on them ken?

The notion of the boats filling the basins of the harbour came up time and time again. This narrative of fishing communities can be understood as part of the stories that signify belonging to these communities.

Now the harbour is much quieter. However, local boats do use it on certain occasions and when the fleet is in it demonstrates Buckie's continuing connection to the fishing industry and, for the people interviewed, its identity as a fishing community. For example, Buckie is the nearest large harbour for Adam who works mostly from Fraserburgh:

> *"So Buckie, I mean would you still consider Buckie to be a fishing community?"* "Aye there is see, well ken, Buckie is like our settling office, it's where we do all our business, and aye it is. There's a lot of boats from Buckie still, but they never come up because of the fishing grounds, ken. [...] But at Christmas the boats come up to Buckie because they are tied up for a fortnight just for ease, it's easier for maintaining your boat and running over

[1] The responses of the interviewees are reported in their vernacular form and several words may need translation into standard English viz: aye: yes; craic: chat (as in conversation); dinnae: do not; hae: have; haem: home; ken or ye ken: you know; mair: more; nae: not.

Fig. 16.1 Fishing fleet in for Christmas, Buckie. (Source: Author 1999)

Fig. 16.2 Pleasure boats at Lossiemouth marina, Lossiemouth. (Source: Author 2005)

> to it. So there is still a fleet, but even then there's nae the boats there used to be before. [...] So there's still a fleet, but in numbers I wouldnae like to say, it's probably 25% of what it was 25 year ago. When I started at the sea Buckie was full of boats every weekend and it's just 25 year and it's a big, big difference. But you do see it at Christmas, there is still a fishing fleet."

Although coming back to Buckie at Christmas is framed in terms of ease and economics, it also serves as a symbolic reassertion of Buckie as a fishing community (Fig. 16.1).

In Lossiemouth, as the fleet began to target fishing grounds further afield, the local harbour was used less and less. The largest basin in the centre of the town has been redeveloped into a marina for pleasure boats. This has been a source of conflict for local fishermen who see it as an inappropriate use of a fishing harbour, yet preferable to it going to ruin, as Charlie, a retired skipper describes (Fig. 16.2):

Fig. 16.3 Harbour Tearooms in former fish warehouses, Lossiemouth. (Source: Author 2005)

> We were against it at the start, you know, 'you're not turning that into a marina, it's a harbour'. But the harbour company had to do it, and it's better that than it going to ruin. And they built those flats in what used to be the stores.

James, a retired skipper who has lived in Lossiemouth all his life, eloquently describes the changes in the town and harbour:

> It's very sad to see a way of life die. The town of Lossie here, when I was young, every road led to the harbour. Everybody's work involved [the fishing] and now it's gone, just practically gone. Thankfully we haven't a derelict harbour because that's agonising, but we have in a sense because it's now a yacht marina, but it's a sad situation from the bustle of work and life that I knew. You could liken it to a mining town when a pit closes down, people have to move away [...] So it's a very different community today from what I grew up in.

The comparison James makes with mining communities is interesting. The harbour is understood in a similar way to the above ground workings found at the pit heads of mining towns. It is a visible, physical symbol that forms the focus for a town that distinguishes it as a place based on fishing.

Lossiemouth harbour is now a site of leisure and relaxation as opposed to the bustle of work and industry it once was. Along with the main basin being full of pleasure yachts, the stores along one pier of the harbour have been converted into flats. The larger stores opposite have been converted into 'The Harbour Tearooms', a café decked out in maritime themed objects, and gift shops selling model fishing boats, lighthouses and other maritime memorabilia to visitors (Fig. 16.3).

The last space in this row of buildings is occupied by the 'Lossiemouth Community and Fishing Museum'. Several of the retired fishermen from Lossiemouth are involved in the management of the museum and volunteer as staff. However, it is understood by some, such as Stuart who decommissioned his boat in 2002, as a sad situation that their livelihood has gone from being the centre of economic activity to an exhibit in a museum:

> I dinnae ken how many boats is left now, and that's before, I mean a lot's been offered decommissioning now. I don't know what will be left here at the end of the year [2003]. [...] If you go into the museum, you'll see the photos and [the boats are] all the way across

> the harbour. […] And this next lot [of decommissioning] will really put the nail in the coffin if they do away with the boats that have been offered decommissioning. […] It's just something to go and look at in the museum really, the fishing community, I would say. It's sad, but it's just the way it's gone.

16.6.2 Fishermen in the Towns

As well as a loss of boats from the harbours, participants raised concerns over the loss of fishermen from the towns and villages. Having fishermen living in a place is understood as an embodiment of the fishing identity of that place. But as the number of active fishermen declines, fishing community status is difficult to maintain. Jill describes this with reference to Cullen:

> When [my husband] first started going to sea there must have been dozens a skippers in Cullen. […] And now er there would be, let me see, there's …one, two, three, four… there would be about four skippers in Cullen… […] So it's nae really a fishing… er fishing community, whereas before, I mean every second man would have been a fisherman, erm so as far as that's concerned aye big changes. […] I suppose you would be lucky to get a dozen fishermen outta Cullen now, which is really quite amazing I think.

As the number of fishermen and people associated with the industry declines, fishing falls from view within the community, and is no longer seen as central to these places. Richard from Hopeman, who decommissioned his boat in 2002, describes how fishing has become invisible:

> Since I've come out of the boat I've not had much to do with the fishing, and if I didn't know the guys I wouldn't think there was a fishing community here.

The presence of boats and fishermen within these places contributes to the socialisation of younger generations into the industry. It is from these fishermen that youngsters learn how to perform fishing identities. As fishing slips from view it no longer surrounds young people at the centre of these places encouraging them to enter the industry, which contributes to the lack of new entrants. Charlie, a retired fisherman from Lossiemouth, describes the role played by the fishing industry in his life from an early age:

> When we left school we just made straight for the harbour, just sat watching people come in, we were there every night waiting for our father to come in, aboard the boats, we just loved boats, but they don't see that here now because there's no boats here, the boats from Lossie work the west coast or Peterhead.

Charlie remembers growing up surrounded by boats and fishermen, which encouraged and taught him about fishing and how to be a fisherman. He is referring to at least 60 years ago, but others see the same patterns happening more recently. Stuart, a skipper from Lossiemouth who was in his late 30s, describes his daughter's class at school:

> Obviously, when I was at primary school there was, say in a class of 30, at least a third of them, their fathers would have been fishermen. Now my daughter's in a class of 30 at the same school and there's not one!! There's not one of them that has got a father that's a fisherman. That's probably your biggest [change].

Fishermen are no longer the dominant occupational group in these places, and so the current generation are not growing up surrounded by or as immersed in the industry as they would have been in the past. As fishermen slip from the centre of these places, occupational identity could be devalued. However, as Adam from Portknockie describes, others fight against this identification of decline and it strengthens their commitment to the industry:

> It's getting to that stage, ken there's nae a lot of fishermen left, and folk have said ken, aye if they haven't seen you for a while, and they say 'are you still at the fishing' and you say 'aye', they're kinda shocked because they think it's finished! In a lot of folk's eyes it's finished already ken? So it's one of the things that makes me more determined to try harder, cos I enjoy it ken, and I can see it being sustainable.

16.6.3 Money in Towns

Fishing has traditionally been seen as the mainstay of not just the harbour and its surrounding businesses, but also the towns in which they were based. Although usually physically separate, with the harbour down at sea level and the towns 'up the brae' on the cliff tops, their fortunes have been inextricably linked. The boats drove the economy in and around the harbours and the pay the fishermen received from the boats drove the economies up in the towns. As Katherine from Fraserburgh said: "When the fishing's good, everyone gets a share, but when it's bad…". Participants suggested that because of its high dependence on the fishing Fraserburgh has suffered dramatically from the impacts of the recent restructuring of the industry, as Adam from Portknockie explains:

> And it has had an impact in their economy, because you drive through Fraserburgh now, ken afore it was bustling and now 20 % of the shops are boarded up ken having to close and it's nae a good thing to see. It's like a ghost town. […] Ken, they are feeling it a lot harder because there was a bigger percentage and it's happened quicker, ken? So it has had a major effect in that corner. […] Ken, you see the difference in the town, ken, unemployment's higher, shops are closing and the industry's shutting down, people were dependent, like they were nothing to do with fishing, like the shops, but they've had to shut down because there isn't the money going about the town. So it has affected them mair.

The changes in Fraserburgh were seen as happening rapidly. Store closures and the increase in the number of charity shops were symptomatic of the state of the local economy and so the fishing industry on which it is based. Fraserburgh was frequently referred to as a "ghost town". Where fishing was once, as one participant described it, "the backbone of places like this", it is no longer able to support them.

Fishermen and their wives were renowned locally for their ability to spend money. When I asked Jill from Cullen whether she had seen any changes in Fraserburgh she said:

> And [fishermen] were very affluent. You see they always say that farmers, when they make money they dinnae spend money, they're tight. But fishermen, and their wives!! [Laughing] Spend money and like ta hae nice things, erm… like to be dressed nice […] And to a certain extent some of it still goes on but, nae nearly so much as it used to be.

However, as the relative prosperity in the industry has declined, respondents referred to a fall in the level of consumer spending. When I asked what she thought the future for these places might be, Kate from Cullen described the scene in Fraserburgh:

> Well I think their economy is going to be badly affected, because there's one thing that's always been true of fishermen, if they've earned money they spent it, and that's a fact. Nobody spent better than them, especially in [Fraserburgh] there was a lot of competition in those two towns, so a lot of people have nae got sympathy for the fishermen because they portrayed an image possibly that was just over the top. And now the money's nae coming in, so it's nae being spent, so it's nae going through the local economy. The shops are closing down, [Fraserburgh is] full a charity shops, there just is nae… There used to be nice shops, but then there was the money going into the towns at that time.

Spending and conspicuous consumption has been part of the performance of fishing identity. Although fishing families had a reputation of liking to spend money it could be seen as well earned rewards for the hard work they had put in. Being able to buy things was a visible sign of success, as everything was bought with money that people had earned. Spending was part of the performance of being 'a good fisherman', it was a visible performance that could be read by others as symbolising hard work and successful fishing. It could also be seen as a barometer of collective success, of the industry and the fishing communities.

16.6.4 Social Events

As the number of fishermen has declined and those that are left are working more intensively and under more financial pressure, social events that were previously an important part of the fishing community have died out. As previously mentioned, in the past, Christmas was an important occasion in the fishing calendar. The whole fleet came back into harbour and tied up for a 2 week break. Christmas was associated with catching up with old friends and maintaining the social networks between fishermen that are hampered by being away at sea. This was facilitated through the 'Christmas Balls' that were organised by the various fish selling companies and agents all along the coast. Jack from Buckie who is in his early 30s, recalls the importance of these occasions for keeping in touch with other fishermen. For Jack the prosperity in the industry just 10 years ago was reflected in the Christmas celebrations:

> [You] had fishermen's balls at Christmas time and things that make the community, folk. The fishing, it's a funny thing, because you'll mebbe nae see a pal, he's on another boat and it's landing at a different time a you and you're missing him, and you'll mebbe nae see him for 2 or 3 month. Like I've got pals I've not seen since the back end of Christmas. But they'll all be in at Christmas, everybody comes haem at Christmas. So it just was a different time, it just, although it was only 7, 8 years ago… it just… it just… feels like a different time…

These events were not just a celebration of Christmas or the whole fleet being together, they also celebrated the achievements of individual fishermen, as Beth from Fraserburgh recalls:

> [T]he offices had always dinners and dances at the end of the year. And they was usually giving trophies for the best fishermen, ken, in the office and things like that. Well that's all gone, yes. Changes…

Fishing is also no longer at the heart of everyday social events like an evening down at the pub. George, a retired skipper from Buckie, describes the difference now that fishermen are no longer the dominant occupational group:

> Years ago if you went out in company, you had five or six couples going out, everyone was fishing. Now when you go out I sit there and say nothing, everyone else is involved with oil, eh? Whereas before it was nice, the fishing craic was always there, it's that type of job, it never, never leaves you, weekends and that; it was a great satisfying job. But now I'm sat there and it's oil rigs and working time directives and working this on and that off and I just cannae relate to that.

George expresses a feeling of exclusion, of no longer being able to relate to or take part in the social relationships that he, as a successful skipper, would once have been at the centre of.

16.7 Conclusions

Restructuring is an ongoing and seemingly perpetual feature of the fishing industry. However, little attention has been given to the socio-cultural impacts of this restructuring process. In recent years fisheries policy has indeed started to consider the role of social objectives. However, there remains much uncertainty as to how these should be conceptualised, measured and addressed. Here the concept of identity has been used to provide insight into the socio-cultural dimensions of fishing, and how they interact with the restructuring of the industry.

The idea of a 'fishing community' has been conceptualised in two ways. Firstly it is brought into being through the performance of individual and collective fishing identities within a place. Secondly the claim to belong to a fishing community is an important symbol in the construction of fishing identities. The notion of fishing communities was frequently used by participants who brought them to life through the stories they told. Their construction of fishing communities was based on shared ideas of what makes a 'fishing community'.

Fishing communities are understood as having an active harbour used by fishing boats, which signifies a place as a fishing community. However, the harbours along this coast are increasingly being used for other purposes—for heritage and leisure, or other industries. Fishermen living in a place are the embodiment of the idea of a 'fishing community' that can be maintained after the boats have gone. Participants recalled a time when "every second man would have been a fisherman" and how this provided important social functions in symbolising a place as a fishing community and socialising the younger generations into the industry. However, the number of fishermen is declining. The industry once drove a vibrant economy in these places which signified the success of individual fishermen and the wider industry. The economy now reflects the decline in the industry. The social events associated with the industry were sites of performance of fishing identities and an expression of 'fishing community'. These events celebrated the success of individuals, the local fleet and the prosperity in the industry as a whole. Now fishing is no longer at the centre of social occasions and some fishermen feel a sense of isolation within broader social relations. The 'fishing community' has become invisible within the wider 'community'.

The performances of fishing identities based on the symbols outlined above have been challenged by the restructuring of the fishing industry. Without these performances, the places along this stretch of coast are no longer seen as 'fishing communities' in the way that they once were. This reinforces the loss of fishing identity as people can no longer lay claim of belonging to a 'fishing community' which had previously been an important symbol of fishing identity. Whether the present and future generations of people living in the coastal settlements of northeast Scotland are able to readjust to a new equilibrium in the region's fishing industry sufficient to reclaim their fishing identities will be an intriguing question for future investigation.

Acknowledgements An earlier version was presented at the European Society of Rural Sociology's XXIIIrd conference held in Vaasa, Finland in August 2009. It is based on doctoral research funded by the Economic and Social Research Council and undertaken at Newcastle University (Williams 2008). The content of this chapter would not have been possible without the generosity of the people who participated in the interviews and to them I express my sincere thanks.

References

Aberdeenshire Council. (2005). Profile: Fraserburgh and Macduff, Aberdeenshire Council: Aberdeen. http://www.aberdeenshire.gov.uk. Accessed 20 July 06.

Aberdeenshire Council. (2007). Fishing industry statistics, Aberdeenshire Council: Aberdeen. http://www.aberdeenshire.gov.uk. Accessed 3 Feb 2007.

Acott, T., & Urquhart, J. (2012). Marine fisheries and sense of place in coastal communities of the English Channel. *Final report prepared as part of the INTERREG 4a CHARM III project*. Chatham: University of Greenwich.

Ashworth, G. J., & Graham, B. (2005). Senses of place, senses of time and heritage. In Ashworth, G. J. & Graham B. (Eds.), *Senses of place: Senses of time*. Aldershot: Ashgate.

Bennett, K. (2002). Interviews and focus groups. In Shurmer-Smith, P. (Ed.), *Doing cultural geography*. London: SAGE Publications Ltd.

Binkley, M. (2000). 'Getting By' in tough times: Coping with the fisheries crisis. *Women's Studies International Forum*, *23*(3), 323–332.

Binkley, M. (2002) Set Adrift: Fishing Families. University of Toronto Press: Toronto.

Bird, S. E. (2002). It makes sense to us: Cultural identity in local legends of place. *Journal of Contemporary Ethnography*, *31*(5), 519–547.

Britton, E., & Coulthard, S. (2013). Assessing the social wellbeing of Northern Ireland's fishing society using a three-dimensional approach. *Marine Policy*, *37*, 28–36.

Brookfield, K., Gray, T., & Hatchard, J. (2005). The concept of fisheries-dependent communities: A comparative analysis of four UK case studies: Shetland, Peterhead, North Shields and Lowestoft. *Fisheries Research*, *72*, 55–69.

Cohen, A. P. (1985). *The symbolic construction of community*. London: Routledge.

Convery, I., & Dutson, T. (2006). *Sense of place*, Project Report University of Central Lancashire: Newton Rigg.

Cook, I., & Crang, M. (1995). *Doing ethnographies, concepts and techniques in modern geography no. 58*. Norwich: Institute of British Geographers.

Cresswell, T. (2004). *Place: A short introduction*. Oxford: Blackwell.

Dalby, S., & Mackenzie, F. (1997). Reconceptualising local community: Environment, identity and threat. *Area, 29*(2), 99–108.

Daw, T., & Gray, T. (2005). Fisheries science and sustainability in international policy: A study of failure in the European Union's Common Fisheries Policy. *Marine Policy, 29*, 189–197.

De Laine, M. (2000). *Fieldwork, participation and practice: Ethics and dilemmas in qualitative research*. London: SAGE.

DEFRA, SEERAD, WAG and DARDNI (2005). *Securing the benefits: The joint UK response to the Prime Minister's strategy unit net benefits report on the future of the fishing industry in the UK*. London: DEFRA.

Edwards, J. (1998). The need for a 'bit of history': Place and past in English identity. In N. Lovell (Ed.), *Locality and belonging*. London: Routledge.

European Commission (EC). (2004). *About the common fisheries policy*, EC. http://ec.europa.eu/fisheries/cfp_en.htm. Accessed 17 May 2006.

European Social Science Fisheries Network (ESSFiN). (1999). *Final report*, ESSFiN: Universities of Aarhus and Hull.

Gregson, N., & Rose, G. (2000). Taking Butler elsewhere: Performativities, spatialities and subjectivities. *Environment and Planning D, 18*, 433–453.

Hall, S. (1996). Introduction: Who needs 'Identity'? In S. Hall & P. du Gay (Eds.), *Questions of cultural identity*. London: Sage Publications.

Jackson, P. (2005). Identities. In P. Cloke, P. Crang, & M. Goodwin (Eds.), *Introducing human geographies*. London: Hodder Arnold.

Jenkins, R. (1996). *Social identity*. London: Routledge.

Marine Fisheries Agency (MFA). (2006). *United Kingdom sea fisheries statistics 2005*. London: Marine Fisheries Agency.

McClanahan, T. R., Castilla, J. C., White, A. T., & Defeo, O. (2009). Healing small-scale fisheries by facilitating complex socio-ecological systems. *Reviews in Fish Biology and Fisheries 19*, 33–47.

Nadel-Klein, J. (1988). A Fisher laddie needs a fisher lassie: Endogamy and work in a Scottish fishing village. In J. Nadel-Klein & D. L. Davis (Eds.), *To work and to weep: Women in fishing economies*. St Johns: Institute of Social and Economic Research, Memorial University of Newfoundland.

Nadel-Klein, J. (2000). Granny baited the lines: Perpetual crisis and the changing role of women in Scottish fishing communities. *Women's Studies International Forum, 23*(3), 363–372.

Nadel-Klein, J. (2003). *Fishing for heritage: Modernity and loss along the Scottish coast*. Oxford: Berg.

North Sea RAC (NSRAC). (2006). *Socio-economic protocol for NSRAC*, unpublished working document NSRAC.

Pettersen, L. T. (1996). Crisis management and household strategies in Lofoten: A question of sustainable development. *Sociologia Ruralis, 36*(2), 236–248.

Phillipson, J. (2002). *Widening the net: prospects for fisheries co-management*. Newcastle upon Tyne: Centre for Rural Economy.

Power, N. G. (2005). *What do they call a fisherman? Men, gender, and restructuring in the Newfoundland fishery*. St Johns: Institute of Social and Economic Research, Memorial University of Newfoundland.

Reed, M., & Courtney, P., et al. (2011). The social impacts of England's inshore fishing industry. *Final report* (p. 75). London: DEFRA.

Reed, M., & Courtney, P. et al. (2013). Beyond fish as commodities: Understanding the socio-cultural role of inshore fisheries in England. *Marine Policy, 37*, 62–68.

Ross, N. (2013). Exploring concepts of fisheries 'dependency' and 'community' in Scotland. *Marine Policy, 37*, 55–61.

Royal Society of Edinburgh (RSE). (2004). *Inquiry into the future of the Scottish fishing industry*. Edinburgh: The Royal Society of Edinburgh.

Scottish Executive. (2002). *Scottish economic report: June 2002: The regional employment contribution of the fisheries sector to the Scottish economy*. Edinburgh: Scottish Executive.

Scottish Executive. (2005). *A sustainable framework for Scottish sea fisheries*. Edinburgh: Scottish Executive.

Scottish Executive. (2006). *Scottish sea fisheries statistics 2005*. Edinburgh: Scottish Executive.

Scottish Government. (2010). *The future of fisheries management in Scotland: Report of an Independent Panel*: Edinburgh.

Smith, S. (2001). Doing qualitative research: From interpretation to action. In M. Limb & C. Dwyer (Eds.), *Qualitative methodologies for geographers: Issues and debates*. London: Arnold.

Stead, S. M. (2005). Changes in Scottish coastal fishing communities-Understanding socio-economic dynamics to aid management, planning and policy. *Ocean and Coastal Management, 48*, 670–692.

Strategy Unit. (2004). *Net benefits: A sustainable and profitable future for UK fishing*. London: Strategy Unit.

Symes, D. (2001). The future of Europe's fisheries: Towards a 2020 Vision. *Geography, 86*(4), 318–328.

Symes, D. (2005). Altering course: Future directions for Europe's fisheries policy. *Fisheries Research, 71*, 259–265.

Symes, D., & Hoefnagel, E. (2010). Fisheries policy, research and the social sciences in Europe: challenges for the 21st century. *Marine Policy, 34*, 268–275.

Symes, D. and J. Phillipson (2009). Whatever became of social objectives in fisheries policy? Fisheries Research 95 (1), 1-5.

Urquhart, J., & Acott, T. (2013). Constructing 'The Stade': Fishers' and non-fishers' identity and place attachment in Hastings, South-East England. *Marine Policy, 37*, 45–54.

Urquhart, J., Acott, T., Reed, M., & Courtney, P. (2011). Setting an agenda for social science research in fisheries policy in Northern Europe. *Fisheries Research, 108*, 2-3, 240–247.

van Ginkel, R. (2001). Inshore fishermen: Cultural dimensions of a maritime occupation. In D. Symes & J. Phillipson (Eds.), *Inshore fisheries management*. Kluwer: Dordrecht.

Williams, R. (2008). Changing constructions of identity: Fisher households and industry restructuring, Ph.D Thesis, Newcastle University.

Women's Studies International Forum Special Issue. (2000). Gender and resource crisis in the North Atlantic Fisheries 23(3).

Zhao, M., Tyzack, M., Anderson, R., & Onoakpovike, E. (2013). Women as visible and invisible workers in fisheries: A case study of Northern England. *Marine Policy, 37*, 69–76.

Chapter 17
Perspectives about the Sea in the Azores: Respecting Narratives that Sustain Inshore Fishing Communities

Alison L. Neilson, Rosalina Gabriel, Ana Moura Arroz and Enésima Mendonça

Keywords Azores islands • Narratives • Photo-elicitation • Traditional fishing communities • Sustaining culture • Knowledge construction • Environmental justice

17.1 Introduction

A relationship with the ocean has long shaped the lives of Portuguese people. They have a long history of "discovering the world" by sea, through explorers such as Fernão de Magalhães (1480–1521) and Vasco da Gama (1469–1524). They rely on the ocean for food and consume more fish per capita than any other country in the European Union (56 kg/person/year vs EU average 22 kg)[1]. It is said that there are over 1000 ways to cook bacalhau, dried cod, which is a staple in the kitchen and *fiel amigo* (faithful friend). The sea is notably represented in art, literature, and other expressions of culture and celebration. It is, therefore, easy to expect that in Portugal you will find a diverse range of perspectives about the sea and that the ocean will figure prominently in narratives of national, individual and community identity. Despite this, not all of these perspectives are granted equal weight in education and within political decision-making.

The Azores is an autonomous region of Portugal, consisting of nine small volcanic islands in the Atlantic Ocean, midway between Lisbon and Newfoundland

[1] European Atlas of the Sea http://ec.europa.eu/maritimeaffairs/atlas/index_en.html.

A. L. Neilson (✉) · R. Gabriel · A. M. Arroz · E. Mendonça
Azorean Biodiversity Group, University of the Azores, 9700-042 Angra do Heroísmo, Azores, Portugal
e-mail: aneilson@uac.pt

R. Gabriel
e-mail: rgabriel@uac.pt

A. M. Arroz
e-mail: aarroz@uac.pt

E. Mendonça
e-mail: efmendonca@uac.pt

J. Urquhart et al. (eds.), *Social Issues in Sustainable Fisheries Management*,
MARE Publication Series 9, DOI 10.1007/978-94-007-7911-2_17,

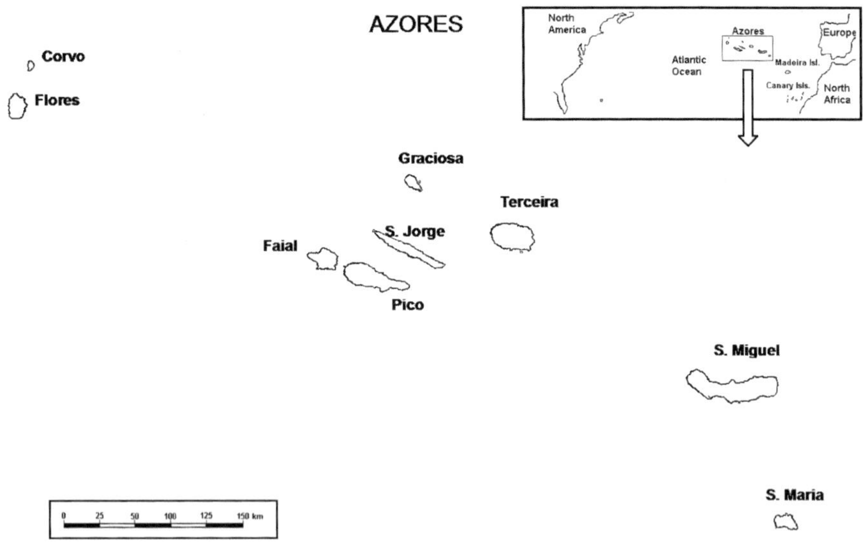

Fig. 17.1 Map of Azores Islands, Portugal. (Copyright 2005 by Isabel R. Amorim, adapted with permission)

(Fig. 17.1). Immigrants from various parts of mainland Portugal, Spain, France, and Flanders and Moorish prisoners and Sephardic Jews have, among others, colonized the islands since the beginning of the 15th century. Today, whale-watching has taken over from the previously economically vital activity of whaling, and the waters splash with divers, surfers, spear-fishers, kayakers, bathers and, of course, fishers. From the smallest island of Corvo (17 km^2) to the largest of São Miguel (747 km^2), there are few places in the Azores from where you cannot see the ocean. Around 240,000 people call the Azores home, and the population varies with intermittent currents of travellers, departing emigrants and returning retirees (SREA 2006).

There are 667 km of coastline in the Azores with numerous seamounts in the surrounding ocean that are remarkably biodiverse. Large open-sea pelagic fish, such as tuna and swordfish, blue jack mackerel and conger as well as spiny and common lobster and clams are important species caught in the Azores and make up 6 % of the total catch of Portugal (2009 data, Iborra Martin 2011). Although the Azorean Exclusive Economic Zone (EEZ) is almost one million km^2, the depth of much of the area limits the potential fishing grounds to seamounts, banks and shallow areas around the islands. The regional fishing fleet is primarily considered to be small-scale and is dominated by small, old wooden vessels of low power. These traditional boats make up 90 % of the fleet and are responsible for the vast majority of Azorean landings. Azorean fisheries are primarily artisanal in that fishers use traditional, labour-intensive, passive fishing gear (Carvalho et al. 2011). The small boats go out to sea and return within a day, while the fewer larger boats sail and fish between islands for multiple days (Pinho and Menezes 2009).

The regional government of the Azores was given local control over their fisheries in 1974 although this autonomy was limited by the separation of powers between the islands, mainland Portugal and the European Union (Lewis and Williams 1994; Murray 2012). Since the implementation of the Common Fisheries Policy, Azorean fishers have been on the defence against a push to open local waters to exploitation by trawlers (which was locally outlawed in 1995) and have faced a lack of support from mainland Portugal in policing the waters for illegal fishing, chiefly Spanish and French longliners (Faustino 2010). In 2002, protests from Azorean fishers regarding the government's handling of fisheries had a strong influence on the regional elections and in 2009, Azorean fishers took the Portuguese Ministry of Defence to court over its failure to patrol the 200 nm Exclusive Economic Zone around the Azores to prevent illegal fishing. The fishers won the case although as of late 2012 no compensation has been received.

In this chapter, we explore the ways in which various narratives about the ocean interact and how some ideas and some people appear to have a stronger voice when discussing marine sustainability, while others are often invisible or their voice is not considered worthy. When discussing research methodology, the context of the authors' lives on the islands is important to consider in terms of how we do research and present this narrative. All the authors have been living in the Azores for various lengths of time. One was born and raised of Azorean parents, two came from mainland Portugal over twenty years ago and another came from Canada to reside in the Azores five years ago. We are part of families, and professional and personal networks, within a "small connected community" (cf. Damianakis and Woodford 2012) and so we are insiders in the communities we are studying. But we are not fishers, nor from fishing families; so we are also outsiders. However, the concept of insider and outsider may be too simplified and compartmentalised to be accepted uncritically as we try to engage with our own understanding about the narratives we see and hear (see Besley and Peters 2009 for a discussion of Lyotard's 1984 language games). Indeed, research that focuses on the social construction of knowledge requires the researchers to recognise their own social positions and processes of meaning making.

This chapter is structured by themes on identity, respecting the ocean, environmental justice, education and sustainability, with each section including the words of research participants as well as the words of other authors who have written about these themes and have hence influenced our inquiry and the presented narrative. Firstly, though, we outline the methodological approach taken in this study.

17.2 Listening with Respect

The narrative presented here explores the ways that experiences, stories and perspectives of inshore fishing communities of the Azores islands inform the way knowledge of the sea is constructed. It asks the reader to consider the connections between (and implications of) ways of knowing and the ways of living with the sea. We present this work as narrative because of its strength for disrupting practices that privilege invisible normative ways of knowing (Neilson 2008; Stone-Mediatore 2003), and because many environmental problems in particular are "produced, reproduced and intensified… by the ways in which *we* live our 'storied lives'" (Gough 1993). Besley and Peters (2009) outline the adoption of narrative as a way of showing our identity and to "emphasize the link between narrative self-understandings, imagination, the role of the narrative arts in shaping our political life". Framing, metaphors and narratives are important for meaning making (Lakoff 2010) and are particularly important when challenging dominant views that may have been taken as common sense (Knoespel 1991; Stone-Mediatore 2003), as well as inviting critical reflection on the very story being told. Compartmentalizing knowledge-making from knowledge and knowers, using the techniques of empiricism (Richardson 1990), is a way in which some perspectives get excluded from policy and education. Presenting the work in a narrative way also reflects the holistic way that Azoreans understand the ocean.

Taking an ethnographic and cross-cultural approach to this work involved "being reconnected to complex, multiple perceptions, and lived realities of the environment" (Neilson 2008, p. 137) and required much practical work—travelling, learning language and listening attentively to other peoples' stories, always critical of the flow of power (Stone-Mediatore 2003). Our work includes research led by us or assisted for numerous scholars on multiple islands of the archipelago as well as education, both in the formal school system and community development in fishing communities. While our understanding is based on broader ethnography and autoethnographies (Ellis and Bochner 2003) of our lives and work in the Azores, all quotations from research participants in this chapter are from one research project, EDUMAR, which ran between 2008 and 2010. For this specific project, we began by reading and directly contacting fishers, organizations such as the association of wives of fishers and fleet owners, sailing clubs, whale boat teams, museums and marine tourism operators, in order to find key informants as well as gather photo images identifying current and past work and pastimes located on or related to the sea. For this chapter, we predominantly share the words of people connected to fishing communities, and although at times we specify fishers in our arguments, we do not label speakers as such since the label is contested in these very communities about which we write. Our research attempts to highlight perspectives that are missing from policy and education without reinforcing dualisms or reifying identity. Quotations are labeled with first or last names, sometimes preceded by the honorific forms, Senhor or Senhora, depending on what is culturally appropriate. Participants asked to be identified by their actual names. We believe that this has increased our local accountability and emphasizes that real flesh-and-bone people are affected by such policy-directed words such as "capacity reduction" (EU Common Fisheries Policy). Sons and daughters, songs and meals, homes and boats and work and environment all contribute to culture and community, and these referents of real-life should share the page with words like policy and sustainability.

The same normative practices that might create expectations on the part of readers that this text should be compartmentalised into the opinions of expert and non-expert, results and discussion, also affect the people who agreed to participate in this work. We informed participants of our desire to hear their perspectives and used photo elicitation and focus groups, methods particularly effective in overcoming potential barriers of language and other social and cultural differences between the participants and researchers (Doyle 2001; Madriz 2003). We visited current and former fishing villages on four islands and facilitated focus groups with elderly retired people in *Casas do Povo* (community centres). We offered a selection of photographs to start conversations within groups of three to five people, specifically inviting participants to also consider images that we did not, or could not, photograph. Facilitation involved following the conversation as directed by those speaking, with occasional questioning to ensure we had the necessary details of their narratives. As Beilin (2005) suggested would happen, we saw that photos allowed participants to engage with complex and changing meanings. The photos triggered memories, nuances and ambiguities, which challenged as well as built rapport and helped to reduce researcher misinterpretation (Hurworth 2003). In addition to the primary research team, two other Azoreans helped facilitate some focus groups. The primary data includes over 25 h of interviews with people associated with fishing and whaling, and tourists; and 10 h of interviews with whale watching tour operators and museum personnel, marine biologists, sailors and kayakers. Narratives were read according to the way participants defined the sea and their own experiences (Clandinin and Connelly 2000). We engaged in this analysis via reflexive writing and discussions about how our own identities and experiences were influencing which themes we identified from the transcripts, which sections we focused on and which ones were ignored. Since these interviews were undertaken and analysed, we have further engaged with fishers and fishing associations on all nine islands; these interactions contribute to our continuing understanding of the local narratives particularly in the context of wider political discourse and in particular the EU's Common Fisheries Policy (Neilson et al. 2012).

17.3 The Ocean and Identity

As may be expected most of the people who live by and work on the sea and tourists who travel to the Azores from around the world for whale watching, spoke to us about the significance of the ocean in their lives and identity. Even Portuguese who do not work on the sea told us that being near the sea was important to their happiness: "my body wanted the sea, the beach, the smell of the sea … I need to feel the sand…" (Teresa). Although the sea figures centrally in the narratives of most of the Portuguese who spoke with us, we heard about an especially keen and deeply embedded sense of sea within the Azorean psyche. Azoreans and even people who moved to settle on the islands spoke about being *in* the sea rather than being near the sea: "because I think that living in the Azores is sea. It's almost like living in the sea"[2] (Cláudia). For Azoreans, the ocean is full of history and stories; it is not an empty wilderness where people are not welcome. Families have the ocean in their blood and their blood in the ocean. Women wept when telling us about children

[2] "Porque acho que viver nos Açores é mar. É quase igual a viver no mar" (Cláudia).

and husbands lost to the sea. But they also told us that fishing families are part of a natural system in which the ocean has a need to be fed. In this way, the influence of the ocean does not stop abruptly on the shore, but flows through much of their lives and is deeply embedded in and part of the fishing communities.

Azoreans describe themselves as people of the sea: "I think also the fact that we live on an island, also makes us turn to the sea"[3] (Edite). The people who identify themselves as fishers do so with strong pride, such as Sr. Dutra who claims, "the sea, for me, is everything. I am nothing without the sea. The sea is everything to us."[4] This pride and identity seem to be connected to their island location, far from any major landmass, and their fluency in the natural cycles of the local sea and weather. As Ritinha says: "The sea is survival for all people, all humankind. The sea is a treasure."[5] Azoreans do not dominate the sea, but, for them, the sea is familiar and normal[6] (Edite). Sometimes they thrive, sometimes they struggle, but they have the physical and emotional abilities to live with the sea and the sea shapes the Azorean.[7]

> I could not live for much time away from the sea. I could not manage well when, for example, I was in the army. I remember once I was a little long without going to the sea and I had to go near the sea because I was not feeling good, I was surely ill. I cannot live long away from the sea.[8] (Sr. Madruga)

Working in a garage overlooking the port of São Mateus (Terceira), untangling lines with one hand while adding a hook, then sticking it into the wooden side of a square box, a "gamela", informs the stories some Azorean women tell of the ocean. Stringing up fishing lines along the road in Velas (São Jorge) reminds the men of the hungry dolphin that tried to steal their catch as they pulled their lines quickly into the boat. The way the Azoreans understand the ocean is intimately connected to the small wooden boats that they take to the open sea, to the rocks they use to weigh the end of their fishing lines, and those things that allow them to be free men and women. They do not live as primary wage earners, with differentiated work and free time: "the goal of the family enterprise is ultimately to be able to remain self-employed. It is a means that is its own end" (Højrup 2003, p. 23). They rely on family and other community members whose involvement in "fisheries" is sometimes invisible within analytical structures that rely implicitly on the universality of the logics of wage working. They go out to the sea when the weather, the movements of

[3] "Eu acho que também o facto de vivermos numa ilha, também nos faz ligar ao mar." (Edite).

[4] "O mar para mi é tudo. Eu sem o mar não sou nada. O mar é tudo para a gente." (Sr. Dutra).

[5] "Pois o mar é a sobrevivência de toda a pessoa, pessoas humanas. O mar é uma riqueza." (Ritinha).

[6] "É o nosso dia a dia. Acaba por ser importante." (Edite).

[7] "A man who has a life in the sea, has life… it creates, it creates the person. It is an addiction. I have been going to the sea for 70 yrs" Portuguese: "Um homem que tem a vida do mar tem aquele… cria, aquilo cria na pessoa. É um vício. Eu já vou para a banda dos 70" (Sr. Silva).

[8] "Eu não podia viver assim algum tempo afastado do mar. Não conseguia e o tempo que tive, por exemplo, quando tive no exército. Recordo-me que uma vez fiquei mais algum tempo sem ir para o mar e às tantas tive que ir mesmo para um sítio próximo se não já não me estava a sentir bem, a adoecer seguramente. Eu não consigo viver muito tempo longe do mar." (Sr. Madruga).

fish or hunger either allows or demands that they go. "Fishing", however, is made up of more than the action of going out to sea or even preparing the gear on land; it is also about the recipes and the sharing of the meals of *chicharros* (see Sousa and Medeiros 2012). The term *fishing* may not, however, tell us enough about the lives of all people, in these coastal communities, particularly women who are engaged in the wider activities of accounting for boats, cleaning boats or even loading the "gamelas". These women, despite their vital involvement in the industry, will often not describe themselves as "doing fishing".

By acknowledging that the sea is part of the fishers, we also acknowledge that the fishers are part of the sea. In suggesting this, we need to be careful about how we discuss this relationship, for although environmental philosophies such as deep ecology suggest that it is important to recognize and reaffirm that humans are part of nature, the dominant social paradigm assumes that human society is separate from nature (Zimmerman et al. 1998). Historically, indigenous peoples, labourers and others who worked or lived outdoors have often been portrayed as part of nature and, thus, as less-than-human and inferior to the "civilized" or "educated" which has been used to justify generations of colonialism (McClintock 1995; Said 1978). We must take care, therefore, not to idealize or impose a convenient but false homogeneity onto fishers or onto the sea. People in coastal communities have a diverse mix of interests, experiences and histories and, most importantly, have the ability and the right to define themselves and their traditions as culture changes and adapts via intercultural encounters. In a recent meeting of Azorean fishers and local and visiting foreign scientists about policy issues, fishers spoke of feeling excluded from discussions by top-down processes in both the islands' government and the European Union (Bulhão Pato et al. 2011). Through our study we suggest that broader inclusion of people and their perspectives within the processes of fisheries management and conservation can lead to better and more equitable efforts toward sustainability in coastal fishing communities resonating with the ideas of a "partnership ethic" (Merchant 1997) and promoting "peopled seascapes" (Shackeroff et al. 2009).

17.4 Respect for the Ocean

As in anthropologies of coastal fishing communities throughout other parts of Europe, we heard that fishing is life and imparts an individual and cultural identity far greater than that of a mere job. Details differ but the stories we heard resonate with those recorded elsewhere (e. g. Carbonell 2012; Knudsen 2009; van Ginkel 2007). There is great complexity within the stories told by the fishers, as well as by a broad range of people from various countries and walks of life that we interviewed. The discussions touched on numerous concepts and discourses, although some words, such as 'respect', were used by nearly everyone. Indeed, fishers and sailors who work on the open ocean regularly referred to the need to respect the sea as an immediate necessity for their own well being. Additionally, many coastal hazards, including tsunamis, floods, landslides, earthquakes and volcanoes have been a recurring reality for the people living in the Azores (Calado et al. 2011), especially those communities located along the particularly vulnerable areas of the coast. We heard many stories of family members being swept out to sea while collecting limpets on shore and of storms destroying buildings and killing livestock. Although most were reluctant to discuss the specifics of past dangers, some fishers acknowledged that they had much experience with bad weather, accidents and boat failures.

> But a life of the sea is very dangerous. It is beautiful for those who take sea journeys, it is very beautiful, but for those who are on the sea alone, the sea is very dangerous.[9] (Sr. Correia)

> But it's dangerous. It has to be respected. It's like I say. When you lose respect for the sea, do not go there anymore. Do not have a fear of the sea, respect it. There are many surprises for us.[10] (Lionel)

Some of the reluctance to speak about dangers at sea may be part of informal but strong taboos which are reported amongst cultures and people who work in dangerous conditions (van Ginkel 2007). To talk of dangerous near-misses at sea causes great anxiety for the people who wait on shore not knowing if their loved ones are safe when the weather turns bad. The relationships that develop between people and the sea seem to be immediate and direct, which is also highlighted by the way fishers in the Azores spoke of the fish that they caught and their positions in relation to one another within the larger ecosystem.

> Although, I often look at a fish, at the fish that I caught, and I look at him, and he had to die for me to live. And that's it, life in the end is also showing this truth. So I can survive and all those who eat fish, it is necessary that the fish die. That's it, and we are all part of this chain. So, the biggest eat the little ones.[11] (Sr. Madruga)

> So the sea is a way of surviving for everyone, for all human beings. The sea is a treasure, as the earth is. The earth and the sea are two kinds of influence on people's lives, but for me the sea is more important. It is due to the fact that one gets more food from the sea than from the earth. Mainly it is fish that we have to eat, fish as well as meat, but fish is healthier than meat and also thinking of nature. The way as the fish is raised in the sea, it is another way of looking at things different from what happens on earth.[12] (Ritinha)

---

[9] "mas a vida do mar é muito perigosa. É bonita para quem vai passear, é muito bonita, mas para quem está tratando só do mar é muito perigosa." (Sr. Correia).

[10] "Mas, é perigoso. Tem que se respeitar. É como o senhor me dizia. Quando perder o respeito pelo mar, não vou lá mais. Não é ter medo do mar, é respeitá-lo. Dá muitas surpresas à gente." (Lionel).

[11] "Embora, muitas vezes eu olhe para um peixe, para aquele peixe que pesco e olho para ele, e ele teve que morrer para que eu viva. E é isto, a vida no fundo também passa por esta verdade. Para que eu consiga sobreviver mais todos aqueles que comem peixe, é preciso que o peixe morra também. Pronto, e todos nós fazemos parte desta cadeia. Enfim, os maiores comem os mais pequenos." (Sr. Madruga).

[12] "Pois o mar é a sobrevivência de toda a pessoa, pessoas humanas. O mar é uma riqueza, como a terra é. É as duas influências que tem para a vida das pessoas é o mar e a terra, mas para mim o mar tem mais. Derivado, o sustento é mais tirado do mar do que da terra. Principalmente do peixe, a gente tem que comer peixe, como também como a carne mas, o peixe é mais saudável do que a carne e também para efeitos de natureza. O mar como é criado o peixe, como é criado tem outra maneira de ver diferente da terra." (Ritinha).

The fishers we met spoke with pride about being providers and helping feed the community. They were also concerned about protecting the natural resource that provided their livelihood. Indeed, one of the first marine reserves to be established was created and maintained voluntarily by fishers around Corvo Island (Abecasis 2012). Yet the role of fishers in protecting the resource can often be overlooked and, indeed, the narratives of non-fishers in our study did not make any acknowledgement that fishers could have respect for the ocean and the fish they catch. There is a global environmental discourse that dictates that humans can only be seen as alien to the ocean and their presence is always harmful (King 2005).

A common phenomenon at work here is the erasure from the official tourism narrative of the fishers who directly gather food and resources from the ocean (Fife 2006; King 2005). We did a simple analysis of images used in tourism brochures of the Azores and found that humans were rarely shown in the photos[13]. Those photos that did include people showed tourists and locals participating in festivals with no photos showing fishers or whalers. The photos we used to elicit discussion in our focus group research showed people engaged in various fishing activities, yet few tourists spoke about people with reference to how they understood the sea even when they were offered photos that contained people. During one such interview, a couple sitting on the wharf in Horta (Faial) said they had not been eating fish during their stay in the Azores because they were concerned about large fishing trawlers, which they assumed were part of the Azorean fishing fleet. Horta has a busy marina with rows of water tourism operators, a sailing club, dry dock for small boats and the passenger ferry terminal. Fishing boats, which bring in daily catch, are moored close to the ferry dock. This area has open access to local and tourists alike with no entrance gates or signs of restricted access. There are no physical barriers to keep the tourists away from the fishing activities. The fishers and fishing activity nevertheless appear invisible to the tourists. The strength of this storied invisibility is great considering that fish are being unloaded regularly from small boats under the noses of the tourists and government-issued brochures and posters about sustainable fisheries in the Azores are displayed prominently, including in some of the whale-watching booths in the marina.

Despite this apparent invisibility of fishing, most people we spoke to mentioned emotional and aesthetic responses to being on the ocean. Some people were afraid and spoke of not liking to be on the ocean, but many spoke about a strong positive connection. This aesthetic and emotional response was connected to ideas of respect primarily for tourists. While fear seems to be an element of respect within the narratives of fishing, it jars with tourism narratives. For some, fear is based on a lack of knowledge and respect: "It is like a therapy, because the dolphins give us something, and the ocean teaches also….so the people and I learn to change the fear to respect" (Angelika). Respect is a prominent concept powering tourism narratives, but it may only flow on the surface of the water in which we swim, sail

[13] 40 out of 163, 24.5% photos had humans in"destinazores.com," 96 pp.; 29 out of 145, 20% photos has humans in "Azores the Living Nature Guide, 208 pages; 1 out of 37, 2.7% photos has humans in "Guide for Macronesia Treasures," 90 pages.

or whale watch. It encompasses the aesthetic of an empty and romanticized ocean and does not make contact with anyone or any activities that construct the ocean as a "peopled seascape" (Shackeroff et al. 2009) or challenge distinct boundaries between what is and what is not ocean.

Tourists and other non-fishers, like fishers, have complex relationships with other places and a capacity to understand complex and ambiguous situations and multiple perspectives, even if the dominant stories limit their ability to do so while engaged in normative tourism experiences. It seems unlikely that a mad dash to find a whale within a narrow window of time allows tourists to see or know the culture, history and multiple stories that exist within those same breaking waves. Nevertheless, tourism can actively seek to bridge differences and privilege the local people by facilitating interactions between visitors and locals. The official code of safety and ethics for walking trails in the Azores,[14] published by the regional government, invites walkers to consider the existing livelihoods of locals by asking them to close gates to keep domestic animals in their pastures and suggests they greet local people as they walk the trails that cross their lands. This small consideration can create a lived story in which people and their activities belong to the land and challenge the powerful discourse of "wilderness" and empty space.

A similar opportunity exists as part of the local whaleboat regatta culture. Some of the people we interviewed from whaleboat racing teams were marine scientists from the local university. Their experiences of racing whaleboats brought them into direct contact with the heritage of whaling, but may also have given them an opportunity to live their meanings of the sea through a paradigm different from that of their marine biology experience. The marine scientists who are also whaleboat racers learn experientially different perspectives toward the sea, which could be heard in their narratives about the sea and in the way that they spoke about fishers with more empathy and respect than other non-fishers.

17.5 Constructing Knowledge and a Question of Environmental Justice

There is not one simple truth about the sea. Rather, people perceive the sea differently based on their different experiences of the world. We come to know the world initially by experiencing it as real and we construct knowledge gained through later experiences in ways that complement our original knowledge (Berger and Luckmann 1966). By choosing or creating concepts of reality that match our existing concepts—cumulative continuity—and choosing social interactions with people who share the same concepts—interactive continuity—we reinforce our world or our concept of "the ocean" as the truth (Fraser 2001). Although there is not any one truth that is *truer* than the others, some stories block other stories and create a world in which only some stories can be told as the official script and other stories are

[14] http://www.trails-azores.com/etica.php.

ignored in media, school and public policy discourse. Language, culture and type of knowledge play important roles in this process as well. Philip Steinberg, in *The Social Construction of the Ocean* (2001), presents a detailed account from 1450 to the present of the way the ocean has been represented legally, artistically, in policy, in advertisements, and in commercial and military histories. He explores how human activity is affected by the processes, as well as the storytelling, of global narratives such as industrial capitalism.

The sea itself can play a part in creating the story based on how it acts on people. In our interviews some people said that the sea 'craves human life'[15], and they talked about the ocean as a live beast that can be angry: "we went to Porto dos Biscoitos and the sea was very angry and took persons, and two died there, in the sea" (Srª Borges) [16]. Religion and spiritual beliefs have an impact as well. We heard how certain saints have protected communities from the hunger of the sea. This story of the sea is full of people and culture. It contains life and death, and good and evil. We also heard a very different story mainly by people going on the sea for leisure and tourism—activities in which people generally experience pleasant weather and water. These people talked as if the ocean was an empty space, calm, neutral and free of people. It is important to ask about the relative power of different stories of the ocean, and acknowledge that there are implications of accepting one story over another. For instance, if educators and tour operators construct the ocean as empty of human endeavour, they would have little reason to include the stories or perspectives of fishers in their work. When a student or tourist sees garbage in the sea, are they concerned for the livelihoods of the fishers as part of the ocean ecosystem? We have heard blame directed primarily at the fishers as the likely culprits of contaminating an otherwise virgin space and in need of punishment or education. The global infrastructure of air travel, international hotel reservations, or even the technology of semi-rigid boats and life preservers, all part of many whale watcher's experience of the sea in the Azores, somehow is exempt from any implication in this adulteration.

Environmental injustice (Bullard 1994; Shiva 1994) refers to the ways some people are disproportionately harmed by environmental hazards (such as fishers dealing with garbage in the ocean) while others receive unjust proportions of environmental benefits. Environmental colonialism and environmental refugees are related concepts (see Carbonell 2012). In the Azores, the topography of the islands limits the potential spaces for people to build homes. In Rabo de Peixe on São Miguel Island, many of the poor fishers live in houses that are located on unstable cliffs immediately adjacent to the sea and in danger from storms. Many homes have fallen into the sea and currently people are being relocated from these cliffs by the government (Gonçalves 2012). These houses were originally built on the coast because the safer land inland was arable and too expensive for fishers. We

[15] "The sea eats a person every day, it eats a foot of land or it eats a person." "O mar todos os dias come uma pessoa, come um palmo de terra ou come uma pessoa." (Srª. Maria).

[16] "… fomos para o porto dos Biscoitos e o mar estava muito revoltado e levou pessoas, que duas morreram lá no mar" (Srª Borges).

heard many stories in our interviews about the sea crashing into homes, killing the family's pigs and chickens, and endangering people. While a focused study comparing Azoreans of different socioeconomic status is needed to more fully understand environmental justice issues in these islands, it is quite clear from our interviews that poverty has placed many of our interviewees in living and working situations in which they are in danger of harm from the sea in ways that other Azoreans were not.

The varying relationships with the environment affect individuals and cultures, hence influencing how different people perceive the environment. When formal education and informal learning draw only from a limited scope of perspectives, environmental injustice can be perpetuated through the silencing of voices that might otherwise call for an end to injustice. Environmental justice, perception and education are, therefore, interrelated. Deconstruction of knowledge sources and discourses are important for ensuring environmental justice within both formal and informal education. Who we listen to and how we listen are important to what narratives inform learning. Indigenous scholars (Battiste 2005; Smith 1999) challenge us to acknowledge multiple privileges and to avoid reproducing these through education and research. Indeed, Thomas King reminds us that "stories [are] medicine, that a story told one way could cure, that the same story told another way could injure" (2003, p. 92).

Like others interested in the perspectives of coastal fishing communities (see Knudsen 2009 for Turkey; King 2005 for Australia), we explore the stories at play and tease apart the power dynamics and implications of being in relation with these narratives. We did not specifically target gender as an issue to explore in our focus group interviews. However, other research done in the islands clearly shows that the invisibility of women in fisheries casts doubts on gender equity (Sempere and Sousa 2008). This adds complexity and uncertainty to our earlier statement about some women choosing to avoid the label of "fisher". Sousa (2011) outlined that although women play an important role in the fisheries sector, they are invisible in two ways: first, while engaged as fish workers preparing gamelas and working in canning, the public does not see them even though they may be paid and listed in official statistics as working in the fisheries; and, second, when undertaking the often unpaid tasks that are part of a family business, such as doing the accounting and paperwork. We began this chapter with the perspectives of people from the fishing communities in the Azores. However, when we interviewed people who used the sea for leisure (whale watching, kayaking, swimming, etc), they sometimes complained about the fishers not wanting to share the same ocean space with them. In a statement about feeling that they were intruding on the fishers' space, a kayaker implied that the fishers were irresponsible with finances and that fish were free for the taking and that fishers had no need and did not have to take responsibility for populations of fish.

> They are the kings of the sea, is not it? The sea is just for them. Just for them. And we often paddle, we're intruding in their space… the fisherman did not generate much of its richness…As we grow a plant to grow and we see it, they were there fishing for a fish that did

not have to be feeding him, did not have to take care of him, put him in an aquarium, it had nothing. They get there and get it for free. They will receive things for free. (Sr. Barros).[17]

17.6 Narratives that Educate

Although there are a growing number of educators whose practice is shaped by critical pedagogy and environmental justice (Russell et al. 2000), formal education that focuses on the science of the ocean (see Tran 2009; Tran et al. 2010) is dominated by reductionist thinking and unchallenged assumptions. For instance, since the 1970s overcoming "misconceptions" in science has been the classic approach to science teaching, although it has been increasingly criticised for not representing how science is actually practised (Vosniadou 2012). This approach to teaching means that a normative way of understanding the ocean is promoted. For instance, facts are taught as if there was only one way to know the ocean. Anyone not understanding or experiencing the ocean in the same way may also not understand or support normative conservation efforts. Clearly, complex biological and social systems that operate in the ocean and within ocean management schemas are difficult for science as well as social science teachers to handle in formal school systems, since learning is subdivided according to subjects, not by holistic approaches to issues. Furthermore, tourists and tour operators who told us that they were interested in conserving the ocean also said that fishers needed to be educated and implied that fishers were a main cause of problems in the ocean. This can lead to a bias in environmental advocacy (Clover and Hill 2003) and other types of informal education such as tourism, which may prove unjust to fishers. Of course, science teachers will draw from science of the ocean in their teaching. However, if scientists who regularly work on coastal fisheries issues acknowledge the importance of the knowledge and perspectives of the fishers (Grant and Berkes 2007; Davis et al. 2004), teachers, tourists and others could benefit from doing the same.

In their study about children's understandings of the environment, Cordeiro et al. (2012) argue that children from the Azores rarely mentioned the sea or marine life. Similarly, Pacheco et al. (2012) found that kindergarten teachers in the Azores felt that they needed more information concerning the sea. These studies further suggest that although children seem to have little knowledge of marine life near their shores, they have complex, sometimes contradictory and often deliberately hidden understandings of and connections with their marine home. Children in other fishing communities sometimes hide great pain caused by the denial of how deeply

[17] "Eles são os reis do mar, não é? O mar existe só para eles. Só para eles. E nós muitas vezes a remar, estamos a intrometer-mos no espaço deles…o pescador não gere muito a sua riqueza…. Enquanto nós cultivamos uma planta e vemos ela a crescer, eles ali pescam um peixe que, não tiveram que estar a alimentá-lo, não tiveram que cuidar dele, metê-lo dentro de um aquário, não tiveram nada disso. Eles chegam lá e recebem aquilo de graça. Eles vão recebendo as coisas de graça." (Sr. Barros).

immersed they are in the sea amidst the collapse of the fisheries ecosystem and the economy that directly affects their families (Howard 2007).

Perhaps an understanding of how Azorean fishers know the sea would contribute to future efforts to sustain fish and coastal fishing cultures. Fishers' understanding of the nature of ocean processes, as "dynamic, unpredictable, complicated, disordered, and chaotic" (van Ginkel 2007, p. 27) is similar to the views of many ecologists (Olsson and Folke 2001). For example, in our interviews, fishers directly spoke about ocean food-chains and they commented on the changes they have seen in the number of fish over their lifetimes. Sr. Borba, one of the research participants, pointed out that the idea of building new ports on some of the islands may be helpful for landing boats, but would destroy fish habitat. Undoubtedly there are differences among fishers from different countries, but we clearly see that the fishers in the Azores share similar types of experiences and knowledge to coastal fishers elsewhere. For instance, they live with uncertain and changing conditions and fish availability, and pride themselves on their ability to provide food to their communities. Neves-Graça (2002), in her work exploring the transition from whale hunting to whale watching in the Azores, found that the perspectives of former whalers to be similar to those of deep ecologists, that whalers had provided much expertise to scientific researchers, but that the expertise of whalers was largely ignored and not considered relevant for tourism, conservation or education. In our interviews, we also heard of in-depth knowledge of whale behaviour by former whalers from the island of São Jorge. They spoke about the lunacy of "killer nets" that local fishers had recognized as causing serious problems for the sustainability of fisheries:

> They die like this, this size [shows small size with his hands], they are arrested in the trolls, they [fishes] are brought dead to the surface. They don't take them to the port, they send them to the sea. It destroys the fauna. They come to the coast and spread nets that kill everything that catch. It's everything.... Those nets are killing nets. But why do they do that? In a few years that will not be good. Maybe for me there is no problem; maybe for my son there is no problem, but for the sons of my son I would like to be here and see what we will have.[18] (Sr. Borba)

These same men questioned why universities would rely on books rather than fishers to understand fishing. Researchers who are concerned with small-scale fishing communities are beginning to document how the current neoliberal management of the European Common Fisheries Policy, while espousing the importance of healthy communities, is often not based explicitly on the wellbeing of these communities and can actually contribute to pushing them out of fishing (Høst 2010).

While some readers may indeed feel in tune with the ocean as described by our Azorean informants, these narratives from the coastal fishing communities will be understood through a particular lens which is mediated by an individual's

[18] "Elas morrem assim deste tamanho, vêm presas nos trolls, chegam cá cima mortas. Eles não as levam para terra, deitam-nas no mar. Destrói a fauna. Eles vêm para a costa deitar redes que mata tudo o que vai lá. É tudo. Eles estão a deitá-las rente às pedras a matar vejas, sargos, tudo. Aquelas redes são redes assassinas. Mas, porque é que fazem isso? Daqui a anos, não vai servir, talvez para mim já não. Talvez para o meu filho mas, para os filhos do meu filho, eu gostava de poder estar cá a ver o que era." (Sr.Borba).

own experiences, or lack of experiences, of the ocean as well as various dominant metaphors prominent in mass media, schooling and other social and political influences. The terms "sustainability" and "sustainable fisheries" have been pervasive in rhetoric, documents and speech in forums addressing issues of fisheries policy and governance. In this respect, we can consider two important underlying concepts, the "Tragedy of the Commons" (Hardin 1968) and "over-fishing", as we seek to invite a reconsideration of local people, their views of the ocean and the way these may ensure a future for inshore fishing communities.

Garrett Hardin (1968) argued that only private ownership of a resource would overcome the urge to be selfish and over-exploit to the ultimate destruction of the commons: a tragedy. His assertions were challenged because he ignored communal systems that prevented over-exploitation, and he failed to consider how colonial rule destroyed other long sustained communal land and marine-tenure systems (Fenny et al. 1990). Critics of the neoliberal assumptions underlying the ideas of Hardin (1968) and the EU Common Fishing Policy abound (Bromley 2009; Lam and Pauly 2010), yet we have heard this metaphor used regularly in various academic and public forums and included without challenge even in writings that inherently question its validity (cf. Pierce et al. 2012). While we do not claim that coastal fishers are impeccable in their daily actions, this metaphor incorrectly and unfairly centres the blame on the fishers, which can dissuade policy makers, educators, tour operators and others from seeking the perspectives of fishers.

Similarly powerful is the idea of "over-fishing". It is difficult to argue convincingly that the very real possibility of over-exploitation of fish is not a problem and we do not want to negate this concern. Local fishers, as well as biologists, are clear that fish populations have decreased considerably. In a recent meeting of researchers that focused on the sea, a statement about overfishing in the Azores being the cause of depletion of fish stocks went unchallenged. This is surprising because artisanal fishing that occurs in the Azores is promoted by these same research/government partnerships as a sustainable practice. EU fishing regulations are promoted officially as the way to make European fishing efforts sustainable, yet European populations continue to eat more fish than are caught in European waters, consuming fish from African and Asian waters while continuing to blame the fishers for over-exploitation of the oceans. King (2005) wrote about similar incidents of incongruence when a fisheries biologist in Australia chastised a fisherman for catching the same species of fish that the biologist himself regularly enjoyed eating. Many different species of fish exist in the ocean, but around the world, the same fish are sold commercially in restaurants and shops. During a conference about fisheries and fishing communities held recently in the Azores, the fishers pointed out that during our lunch in the university cafeteria, we were served fish caught in foreign waters potentially thousands of kilometers away. The alternative concept of "over-consumption" redirects the focus to our own complicity and to questions about the damaging effects of other activities such as oil exploration and drilling, marine traffic, military exercise and other exploitive activities on the sea. The story of "overfishing" provides strong temptation to grasp a simpler argument than wading through the complicated concepts and processes of centralization, capitalization and "marketization" of neoliberal management systems, all protected by powerful interests (see Høst 2010).

17.7 Narratives that Respect Sustainability

In this chapter, we have tried to place Azoreans at the centre of the themes discussed to explore the way in which their narratives may be taken up by policy and education. We have also considered our own role in these processes of knowledge creation and transmission. As researchers and educators we have chosen to disrupt some of the normative processes that we believe inadvertently work against the holistic narratives and the sustainable living practices of inshore fishers. In creating policies for local and global management of the sea and fisheries, various narratives are listened to, created and lived, including stories of facts and experts. Some of the most powerful underlying stories tell us that there are experts who should make decisions, "Folks trained in public policy, science, economics, and law….may believe that if you just tell people the facts, they will reason to the right conclusion" (Lakoff 2010, pp. 72–73). The Common Fisheries Policy is being debated and revised now in 2012; fish populations continue to be threatened and Europeans continue to eat marine fish. When policy is made to regulate fishing, some people will have to stop fishing either because it is no longer legal for them to do so, or because it is no longer feasible for other reasons. We believe it is time to look more carefully and critically at the narratives at play, especially those promulgated by the communities who are most affected by the changes in policy and environmental conditions.

Without critical reflection on multiple and conflicting narratives, we can easily make assumptions and understand the perspectives of other people in ways that do not fit accurately. As we have explained, the stories of the ocean as an empty wilderness, tragedy of the commons and overfishing, among others, are powerful fetishes and rarely challenged. Understanding other people is a useful goal if we hope to have broader participation within efforts to create sustainable futures for fish and fishing communities. Some narratives have more power than others and thus have direct implications on how societies create rules for using the ocean. The idea that respect for the ocean must include sailing for pleasure and spending time on the beach but not pulling up fish with hooks, serves to undermine the respect for and the authority of fishers as experts on the ocean.

In reviewing the ethics behind the changing management approach to fisheries, Carolyn Merchant (1997) called for a partnership ethic in which the needs of both fish and people are privileged. Such an ethic "holds that the greatest good for human and nonhuman communities is in their mutual living interdependence" (Merchant 1997, p. 29). She included consideration for cultural and biological diversity, relationships and obligations. The most direct way to include cultural diversity in management approaches is to include a diversity of people who have different perspectives of the sea and fisheries, particularly local fishers (Neves-Graça 2004). However, if the people who are in charge of management do not recognize a lack of diversity, then there will be no reason for them to address it. In looking at the various narratives and perspectives on the ocean, we believe that it is useful to challenge dominant and normal stories in order to be willing to listen more carefully to lesser heard, but important, alternative stories from the very coastal communities where sustainability is sought. In reviewing the concept of education for sustainability and the efforts of the United Nations Decade of Education for Sustainable Development, Gadotti (2008) advocates educating for other possible worlds. He follows Paulo Freire's call for educating to dissolve power, for awareness of power issues, to "unalienate" and "defetishize", to make visible what has been hidden through oppression, to give voice to people who have been ignored (or silenced) and to include what has been missing.

Considering how to ensure sustainability for such a complex system as the marine environment may be one of the biggest challenges facing human kind. In addition to the myriad of unknown ecological factors, there are strong economic forces that seek to increase profit at the expense of equity between people or conservation of fish. These same forces build narratives that aim to make this intention invisible or innocuous. By exploring the narratives that sustain inshore fishing communities, we help to disrupt the idea that these economic stories are natural and inevitable. We disrupt the idea that inshore fishing communities are helpless in the face of unstoppable stories of individual transferable quotas (ITQs) and other narratives of economic management. These narratives are powerful, partially in their ability to make themselves invisible, rather than be exposed as stories that can be told in alternative ways. Therefore, experiential learning may be helpful for ensuring that the perspectives of fishers are fully included in fisheries management. Rather than telling scientists, politicians or tourists that fishers are embedded in the sea, perhaps, similarly to the Azorean scientists who race in the whale boats, they could benefit from opportunities to experience the sea in a more embedded way. Whether narrating differently or living different narratives, we seek to disrupt narratives that have unfairly targeted fishers as the problem to be solved in management issues and instead see fishers as vital partners in creating sustainable management systems which could benefit both fish and people.

Acknowledgments Funding for this research came from The Regional Director of Science, Technology and Communications, Government of the Azores, M2.1.2/F/007/2008 Project Support of Post-doctoral Fellows of DRCT/FRCT.

References

Abecasis, R. C. (2012). *Understanding local views on marine conservation: a case study in the Azores*. Palestra. Conhecer o mar dos Açores II, fórum científico de apoio à decisão (Knowing the Ocean of the Azores, forum II for science for policy). Biblioteca Pública e Arquivo Regional João José de Graça, Horta, Pt. 9 a 10 de Julho de 2012.

Battiste, M. (2005). Post-colonial remedies for protecting Indigenous knowledge and heritage. In P. Tripp & L. Muzzin. *Teaching as activism. Equity meets environmentalism*, (pp. 224–232). Montreal and Kingston: McGill-Queen's University Press.

Beilin, R. (2005). Photo-elicitation and the agricultural landscape. *Visual Studies, 20*(1), 56–68.

Berger, P. L., & Luckmann, T. (1966). *The social construction of reality: A treatise in the sociology of knowledge*. New York: Doubleday.

Besley, A. C. T., & Peters, M. A. (2009). *Critical educational studies and the narrative turn: A poetics of resistance?* Presentation Philosophy of Education Society of Australia, PESA, December 3–6, 2009, Honolulu, Hawaii. http://www2.hawaii.edu/~pesaconf/sun/html. Accessed 24 Oct 2012.

Bromley, D. W. (2009). Abdicating responsibility: The deceits of fisheries policy. *Fisheries, 34*(6), 280–302.

Bulhão Pato, C., Neilson, A., & Sousa, L. (2011). *Exploring the wealth of coastal fisheries: Listening to community voices*. Unpublished final report, 21–24 October 2011 Angra do Heroísmo and Ponta Delgada, Portugal. http://conferencewealthofcoastalfisheriespt.blogspot.pt/.

Bullard, R. D. (1994). Overcoming environmental racism. *Environment, 36(4), 10–20,* 39–44.

Calado, H., Borges, P., Phillips, M., Ng, K., & Alves, F. (2011). The Azores archipelago, Portugal: Improved understanding of small island coastal hazards and mitigation measures. *Natural Hazards, 58*(1), 427–444.

Carbonell, E. (2012). The Catalan fishermen's traditional knowledge of climate change and the weather: A distinctive way of relating to nature. *International Journal of Intangible Heritage, 7,* 61–75.

Carvalho, N., Edward-Jones, G., & Isidro, E. (2011). Defining scale in fisheries: Small versus large-scale fishing operations in the Azores. *Fisheries Research, 109,* 360–369.

Clandinin, D. J., & Connelly, F. M. (2000). *Narrative inquiry: Experience and story in qualitative research.* San Francisco: Jossey-Bass Publishers.

Clover, D. E., & Hill, L. H. (2003). Learning patterns of landscape and life. In L. H. Hill & D. E. Clover (Eds.), *Environmental adult education: Ecological learning, theory, and practice for socioenvironmental change* (pp. 89–96). San Francisco, CA: Jossey-Bass.

Cordeiro, L. M. M., Arroz, A. M., & Gabriel, R. (2012). O ambiente sob o signo da infância: Diversidade de conceitos e expectativas face ao futuro. In R. Gabriel (Ed.), *Abordagens do ambiente em contexto educativo* (pp. 177–289). Cascais, Pt: Principia.

Damianakis, T., & Woodford, M. R. (2012). Qualitative research with small connected communities: Generating new knowledge while upholding research ethics. *Qualitative Health Research, 22*(5), 708–718.

Davis, A., Hanson, J. M., Watts, H., & MacPherson, H. (2004). Local ecological knowledge and marine fisheries research: The case of white hake (Urophycis tenuis) predation on juvenile American lobster (Homarus americanus). *Canadian Journal of Fisheries and Aquatic Sciences, 61*(7), 1191–1201.

Doyle, A. (2001). *The colliery aesthetic: Cultural responses at the end of industry.* Unpublished doctoral dissertation, Durham University, UK.

Ellis, C., & Bochner, A. P. (2003). Autoethnography, personal narrative, reflexivity. Researcher as subject. In N. K. Denzin & Y. S. Lincoln (Eds.), *Collecting and interpreting qualitative materials* (pp. 199–258). Thousand Oaks, CA: Sage.

Faustino, P. (2010). Ministério da Defesa acusado de "má-fé" em processo judicial de vigilância do mar. *Diário de Notícias*, 26 de Janeiro de 2010.

Feeny, D., Berkes, F., McCay, B. J., & Acheson, J. M. (1990). The tragedy of the commons: Twenty-two years later. *Human Ecology, 18*(1), 1–19.

Fife, W. (2006). National parks and disappearing people: Romantic landscapes and the urbanization of rural spaces in Newfoundland. In X. S. R. Campos & X. M. Santos Solla (Eds.), *Galicia and Terranova and Labrador, Comparative Studies on Economic, Political and Socio-cultural Processes*. Proceedings of the International Congress, Universidade de Santiago de Compostela, October 28–31, 2002.

Fraser, L. A. (2001). *The development of commitment to the common good: The case of adults leaning about global issues*. Unpublished doctoral dissertation, University of Toronto, Canada.
Gadotti, M. (2008). Education for sustainability: A critical contribution to the Decade of Education for Sustainable Development. *Green Theory and Praxis: The Journal of Ecopedagogy, 4*(1), 15–64.
Gonçalves, C. D. S. (2012). *Realojados de Rabo de Peixe: Sentimento de Comunidade e Sentimento de Segurança*. (Relocation of Rabo de Peixe: Feeling of community and sense of security). Unpublished master's thesis, University of the Azores, Portugal.
Gough, N. (1993). Narrative inquiry and critical pragmatism: Liberating research in environmental education. In R. Mrazek (Ed.). Alternative paradigms in environmental education research. Troy, OH: North American Association for Environmental Education. www.edu.uleth.ca/ciccte/naceer.pgs/pubpro.pgs/alternate/14.Gough.rev.htm. Accessed 18 Sep 2000.
Grant, S., & Berkes, F. (2007). Fisher knowledge as expert system: A case from the longline fishery of Grenada, the Eastern Caribbean. *Fisheries Research, 84,* 162–170.
Hardin, G. (1968). The tragedy of the commons. *Science, 162,* 1243–1248.
Højrup, T. (2003). *State, culture and life-modes. The foundations of life-mode analysis*. Aldershot: Ashgate.
Høst, J. (2010). *A neoliberal catch: Access rights and the clash of coastal lifemodes*. Presentation at the World Small-Scale Fisheries Congress, Bangkok, Thailand, October 18–22, 2010.
Howard, P. (2007). The pedagogy of place: Reinterpreting ecological education through the language arts. *Diaspora, Indigenous, and Minority Education, 1*(2), 109–126.
Hurworth, R. (2003). Photo-interviewing for research. *Social Research Update, 40.*
Iborra Martin, J. (2011). *Fisheries in Portugal*. Brussels: Policy Department on Structural and Cohesion Policies, European Parliament.
King, T. (2003). *The truth about stories, a native narrative*. Toronto: House of Anansi Press Inc.
King, T. J. (2005). Crisis of meanings: Divergent experiences and perceptions of the marine environment in Victoria, Australia. *The Australian Journal of Anthropology, 16*(3), 350–365.
Knoespel, K. J. (1991). The emplotment of chaos: Instability and narrative order. In N. K. Hayles (Ed.), *Chaos and order. Complex dynamics in literature and science* (pp. 100–122). Chicago: The University of Chicago Press.
Knudsen, S. (2009). *Fishers and scientists in modern Turkey. The management of natural resources, knowledge and identity on the Eastern Black Sea Coast*. New York and Oxford: Berghahn Books.
Lakoff, G. (2010). Praxis forum. Why it matters how we frame the environment. *Environmental Communication, 4*(1), 70–81.
Lam, M. E., & Pauly, D. (2010). Who is right to fish? Evolving a social contract for ethical fisheries. *Ecology and Society 15*(3), 16. http://www.ecologyandsociety.org/vol15/iss3/art16/. Accessed 10 June 2011.
Lewis, J., & Williams, M. (1994). Regional autonomy and the European communities: The view from Portugal's Atlantic Islands. *Regional Politics and Policy, 4*(2), 67–85.
Madriz, E. (2003). Focus groups in feminist research. In N. K. Denzin & Y. S. Lincoln (Eds.), *Collecting and interpreting qualitative materials* (pp. 363–388). Thousand Oaks: Sage.
McClintock, A. (1995). *Imperial leather: Race, gender and sexuality in the colonial contest*. New York, NY: Routledge.
Merchant, C. (1997). Fish first! The changing ethics of ecosystem management. *Human Ecology Review, 4*(1), 25–30.
Murray, F. (2012). *The European Union and Member State territories: A new legal framework under the EU Treaties*. New York: Springer.
Neilson, A. L. (2008). *Disrupting privilege, identity and meaning: A reflexive dance of environmental education*. Rotterdam, The Netherlands and Taipei City, Taiwan: Sense Publishers.
Neilson, A. L., Cardwell, E., & Bulhão Pato, C. (2012). Coastal fisheries in the Azores, Portugal. A question of sovereignty, sustainability and space. In K. Schriewer, & T. Højrup (Eds.), *European fisheries at a tipping-point* (pp. 464–505). Murcia: Cátedra Jean Monnet Universidade de Murcia.

Neves-Graça, K. (2002). Revisiting the tragedy of the commons: Ecological dilemmas of whale watching in the Azores. *Human Organization, 63*(3), 289–300.

Neves-Graça, K. (2002). *A whale of a thing: Transformations from whale hunting to whale watching in Lajes do Pico*. Unpublished doctoral dissertation, York University, Toronto, Canada.

Olsson, P., & Folke, C. (2001). Local ecological knowledge and institutional dynamics for ecosystem management: A study of Lake Racken watershed, Sweden. *Ecosystems, 4,* 85–104.

Pacheco, C., Sousa, F., & Gabriel, R. (2012). As crianças ainda são imaturas – Perspetivas dos educadores de infância sobre a abordagem de termáticas ambientais nos jardins de infância dos Açores. In R. Gabriel (Ed.), *Abordagens do ambiente em contexto educativo* (pp. 177–289). Cascais, Pt: Principia.

Pierce, G., Pita, C., Santos, B., & Seixas, S. (2012). Sustainability of fisheries. In F. J. Goncalves, R. Pereira, W. Leal Filho & U. Miranda Azeiteiro (Eds.), *Contributions to the UN Decade of Education for Sustainable Development* (pp. 325–367). Frankfort: Peter Lang.

Pinho, M. R., & Menezes, G. (2009). Pescaria de demersais dos Açores. *Boletim do Núcleo Cultural da Horta*, (18), 85–102.

Richardson, L. (1990). Narrative and sociology. *Journal of Contemporary Ethnography, 19*(1), 116–135.

Russell, C. L., Bell, A. C., & Fawcett, L. (2000). Navigating the waters of Canadian environmental education. In T. Goldstein & D. Selby (Eds.), *Weaving connections: Educating for peace, social and environmental justice* (pp. 196–217). Toronto: Sumach Press.

Said, E. (1978). *Orientalism*. New York, NY: Vintage Books.

Sempere, M. J., & Sousa, R. (2008). *Estamos cá. Existimos. As mulheres na pesca nos Açores. Ponta Delgada*. Portugal: UMAR-Açores.

Shackeroff, J. M., Hazen, E. L., & Crowder, L. B. (2009). The oceans as peopled seascapes. In K. McLeod & H. Leslie (Eds.), *Ecosystem-Based Management for the Oceans* (pp. 33–54). Washington, D.C.: Island Press.

Shiva, V. (1994). *Close to home: Women reconnect ecology, health and development*. London: Earthscan publications Ltd.

Smith, L. T. (1999). *Decolonizing methodologies: Research and indigenous peoples*. New York: Zed Books.

Sousa, L. (2011). *Women in fisheries in the Azores. How they become visible*. Presentation at 'It's Not Just About the Fish' Social and cultural perspectives of sustainable marine fisheries. Greenwich: UK 4–5 April 2011.

Sousa, L., & Medeiros, J. (2012). *O chicharro à nossa mesa. Ponta Delgada*. Portugal: UMAR-Açores.

SREA (2006). *Anuário Estatístico 2006*. Região Autónoma dos Açores. Açores, pp 276.

Steinberg, P. E. (2001). *The social construction of the ocean*. Cambridge: Cambridge University Press.

Stone-Mediatore, S. (2003). *Reading across border: Storytelling and knowledges of resistance*. New York, NY: Palgrave.

Tran, L. U. (2009). Children and adults' understanding of ocean and climate sciences. Report to the national research council to review the noaa education program in the ocean literacy principles and climate literacy principles documents. http://www7.nationalacademies.org/bose/NOAA%20Teaching%20and%20Learning.pdf.

Tran, L. U., Payne, D. L., & Whitley, L. (2010). *Research on learning and teaching ocean and aquatic sciences*. Special Report #3. National Marine Educators Association. http://coexploration.org/oceanliteracy/NMEA_Report_3/NMEA_2010-6-Learning.pdf.

van Ginkel, R. (2007). *Coastal cultures. An anthropology of fishing and whaling traditions*. Apeldoorn, NL and Antwerpen, BE: Het Spinhuis Publishers.

Vosniadou, S. (2012). Reframing the classical approach to conceptual change: Preconceptions, misconceptions and synthetic models. In B. J. Fraser, K. G. Tobin, & C. J. McRobbie (Eds.), *Second international handbook of science education (vol. 1, pp. 119–130)*. New York: Springer.

Zimmerman, M. E., Callicott, J. B., Sessions, G., Warren, K. J., & Clark, J. (Eds.), *Environmental philosophies. From animal rights to radical ecology*. Upper Saddle River, NJ: Prentice Hall.

Part VI
Conclusion

Chapter 18
Towards an Understanding of Social Issues in Sustainable Fisheries Management

Tim G. Acott, Julie Urquhart and Minghua Zhao

Keywords Sustainable fisheries management • EU Common Fisheries Policy • Institutional sustainability • Knowledge • Methodologies • Ecosystem approach to fisheries

18.1 Introduction

At the outset this book presented the idea that sustainable management of fisheries is a complex and difficult area to negotiate. Through the example of the evolution of the Common Fisheries Policy in Europe in Chap. 1, it was argued that politics, power and governance play key roles in influencing management debates and discourses and it demonstrated the obstacles to bringing about reform in well established systems. The assertion was made that despite increasing international interest in sustainable development and natural resource management, when it comes to fisheries, the primacy of achieving biological and economic sustainability is likely to continue to dominate, with social issues being overlooked, despite clear evidence to suggest that sustainable fisheries will only be achieved through the integration of biological, social and economic issues (FCR 2000; Forst 2009). This edited collection contributes to making visible a broad range of stakeholders and social issues in order to balance the dominant biological/economic discourses in fisheries

T. G. Acott (✉) · J. Urquhart
School of Science, University of Greenwich, Central Avenue, Chatham Maritime, Chatham, Kent ME4 4TB, UK
e-mail: T.G.Acott@greenwich.ac.uk

J. Urquhart
e-mail: j.urquhart@greenwich.ac.uk

M. Zhao
Greenwich Maritime Institute, University of Greenwich, Old Royal Naval College, Park Row, Greenwich, London SE10 9LS, UK
e-mail: M.Zhao@greenwich.ac.uk

J. Urquhart et al. (eds.), *Social Issues in Sustainable Fisheries Management*, MARE Publication Series 9, DOI 10.1007/978-94-007-7911-2_18,

management. The contributions explore different aspects of the social and cultural dimensions of sustainability and are organised thematically to reflect five 'issues' in fisheries: governance and co-management, local ecological knowledge and scientific evidence, sustaining fisher livelihoods, gender roles and, finally, social identity and cultural heritage. The diversity of content in these pages demonstrates that unpacking the social and cultural perspectives of marine fishing draws in a range of issues, approaches and epistemologies. To move towards a sustainable fisheries management paradigm entails understanding more about the social and cultural dimensions of sustainability. In this final chapter, four dimensions are put forward as a way of framing and articulating social and cultural issues. This is not an exhaustive list, but represents a signal of the type of broad interdisciplinary thinking that is needed to move towards sustainable fisheries in the future.

18.2 Four Dimensions of Socio-Cultural Issues

Returning to the sustainability triangle presented in the introductory chapter, four additional dimensions of sustainability can be added: institutional, discourse (knowledge), translation and methodological (Fig. 18.1). These dimensions have grown out of the content and ideas presented in this book. As such they are a selective illustration of issues, drawn together as a result of an international conference 'It's Not Just About The Fish' held at the University of Greenwich in 2011. The areas of concern that are described highlight some key points that need addressing in order to facilitate the transition to a sustainable fisheries management paradigm. In order to do this the social and cultural dimensions of sustainability must be addressed, however these terms can be vague and lack specific meaning. Our intention here is to begin to unpack this socio-cultural dimension and stimulate the debate on its relevance and relationship to fisheries management. The four new dimensions have been located towards the social/cultural axis of the sustainability triangle as they comprise important elements in understanding social/cultural sustainability. Consideration of each point on the triangle is important in developing a sustainability strategy for fisheries with sustainable development achieved through balancing the environmental, economic and socio/cultural elements. The following sections draw out some key aspects of these four sustainability dimensions. However, it remains for future work to consider in more detail how these dimensions intersect with the environmental and economic aspects.

18.3 Institutional

As early as 1994 Charles presented the sustainability triangle and placed institutional sustainability at its centre. He saw the importance of maintaining suitable financial, administrative and organisational capabilities over the long term as central

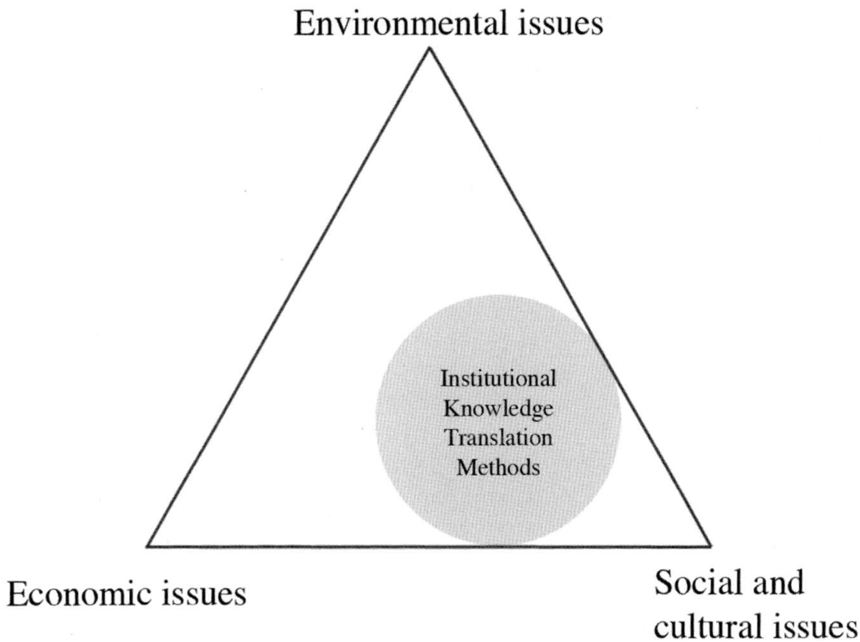

Fig. 18.1 Dimensions of sustainability for fisheries management

to the idea of sustainability. However, in moving fisheries towards a more sustainable perspective there are other institutional issues that need consideration (the term institution is being used here in a broad sense to refer to sets of rules/norms rather than just organisations). First, what is the scope of fisheries management? Should fisheries management be sector specific or should broader consideration be given to multiple stakeholders and sustainable communities more generally by adopting an area-based approach? Symes (Chap. 2) considers this issue in relation to 'wicked problems', where he describes a category of problems that are difficult to define and separate from other wider issues. He recommends a broad governance approach with stakeholders fully engaged in the governance process. Drawing on ideas of resilience and interactive governance he sees fisheries not as a self-contained policy area but one that is intimately associated with environmental conservation and social/community issues.

Similarly, although from a very different perspective, Gallizioli in Chap. 4 agrees that little attention has been paid to those coastal communities where fishing is still important. Despite the problems of defining fisheries dependant communities, Gallizioli describes funds that have been made available in the EU to regions dependant on fisheries. In particular the importance of offering opportunities for diversification is considered through PESCA and subsequently Fisheries Local Action Groups (FLAGs). Gallizioli makes the concluding point that fishing only has a future as part of a flourishing coastal community and adds, somewhat sadly, "that today the image of being a fishing community has a value exceeding that of the landed catches" (p. 84). This theme of diversification and the scope of fisheries management is continued by Budzich-Tabor (Chap. 10), who looks in more detail at opportunities offered by the FLAGs, and Morgan et al., who look specifically at issues of diversification (Chap. 9). These authors look for solutions/opportunities by considering the intersection of fisheries and other stakeholders and the possibilities for creating new ways to conduct business. Fitzpatrick (Chap. 3) considers the importance of governance benchmarking and draws out the importance of a detailed contextual understanding to develop the right level of stakeholder participation. A central concern in these papers is the extent that a broader area-based approach should be utilised (as opposed to sector specific), i.e. thinking about fisheries as a part of a broader coastal community and the extent that fishing be considered alongside other employment and economic development opportunities.

In addition, it is also important to consider how decisions are made, and to understand who is included and excluded from the decision-making process. This is dealt with directly by Bigney Wilner (Chap. 5) who suggests that there is the risk of creating an institutional void if attention is not given to the power relations at play in the policy making process. She asserts that all stakeholders need to be given a voice and not just those representing dominant discourses or those with the most power. Martindale (Chap. 15) also adds to this thesis by revealing some of the discussions from different stakeholders around the importance of the *Ribble* as a regeneration project. He makes a convincing argument that asks what norms are influential in shaping relationships between heritage and contemporary fishing practice. He describes preconceived ideas of the relationship between fishing and heritage in terms of a sharp divide that includes clear boundaries, social distinction and different economic priorities. A fluid picture is painted where the boundaries between contemporary fisheries and heritage are connected and contribute to a sense of place. These issues of decision-making are closely related to the second dimension under consideration here, that of discourse/knowledge production and whose knowledge is privileged.

18.4 Discourse and Knowledge

A number of the chapters in this volume consider the idea of knowledge production and associated ideas of dominant discourses. Bigney Wilner (Chap. 5) and Neilson et al. (Chap. 17) perhaps deal with these issues in the most direct way. Bigney Wilner looks at three distinct policy discourses: food safety, the tragedy of the commons and subsistence, and local economies. Neilson et al. consider the ocean from the perspective of fishing communities in the Azores and how these narratives are taken up in policy and education and makes the case that, through an understanding of narrative, we can learn something about who is included and who is excluded within dominant management paradigms. Bigney Wilner describes how the discourses associated with integrated management in the Annapolis basin in Nova Scotia allow us to understand how issues and problems are framed by different stakeholders. The idea of integrated management is, therefore, placed within a broader discourse context where the distribution of power between different stakeholders is of central importance.

Mackinson and Wilson (Chap. 7) describe the significance of different types of knowledge production by looking at the sharing of knowledge between scientists and fisherman. They argue for governance structures in which the sharing of knowledge between different stakeholders is a key element and the value of fishers' local ecological knowledge is recognised. Blanchard et al. (Chap. 6) approach the importance of knowledge from a slightly different perspective. The relationship between attitudes, values, knowledge and ethics is explored and an argument is constructed about the importance of educational tools to strengthen stewardship and conservation perspectives. Knowledge also underpins Martindale's contribution as he charts the difficult terrain of drawing on old skills to make visible relationships between heritage and contemporary fishing communities. In a similar vein, in the chapter by Acott and Urquhart (Chap. 14) 'sense of place' is used as a vehicle to explore the intimate associations between fisheries and terrestrial communities. The importance of understanding relationships created between fishers' knowledge, practice and identity is central to their thesis Williams (Chap. 16) explores how fishing identities are bound up with shared symbols and understandings of the fishing industry.

Britton (Chap. 8) uses a wellbeing approach and draws conclusions about the importance of participation, power and co-management. Her analysis points towards how small-scale fishers are frequently marginalised from the decision making process. She advocates a multi-stakeholder approach that facilitates listening and learning from others although an outright acceptance of community-based management is cautioned against.

All three chapters that deal with women in fisheries touch upon ideas of discourse and knowledge in different ways. Frangoudes and Pascual-Fernández (Chap. 12) describe the factors leading to the formation of women's organisations in fisheries while Zhao et al. (Chap. 13) chart the case of numerous individuals working in different parts of the industry in England. Although the terminology of discourse is not used in these papers, they are fascinating accounts of an under-represented group struggling against a dominant paradigm and finding ways to intervene so that the visibility of women participating in the industry is increased. Using a quite different approach, Vervaele (Chap. 11) gives a personal glimpse into the lives of Flemish women involved in fisheries. Experiencing the narratives that emerge from her account helps to reveal insights into the experiences of the women living in the world of fisheries. The three papers on women in fisheries collectively start to form a discourse detailing the vital role women play but at the same time how their contribution is often undervalued and unrecognised.

In many of the book chapters we can see a tension between different types of knowledge production and how that knowledge gets heard in different discourses. 'Invisibility' is a word used throughout a number of the contributions in this book. This idea takes us into the third dimension of social/cultural sustainability, translation.

18.5 Translation

Marine fishing is an activity that connects a largely invisible undersea world with terrestrial communities. The process of taking fish from the sea sets in motion a wide range of relationships and impacts. The impacts considered important are, at least partly, reflected in the dominant discourses or power relations that are operating. Thus, within fisheries management, biological and economic impacts are privileged over social and cultural issues. However, many of the chapters that make up this volume describe different ways that the activity of fishing becomes translated into a range of socio-cultural impacts in terrestrial communities and how they might be made more visible in a policy-making context. This is clearly evident in Martindale's (Chap. 15) contribution where he skillfully reflects on the ability of the lugger *Ribble* to challenge ideas of industry and heritage. He draws on ideas of "nostalgia to stimulate imagination and innovation"; in essence he is entreating us to explore how a heritage-led regeneration project can offer new perspectives on the relationships between fisheries and contemporary society.

Acott and Urquhart (Chap. 14) ask us to think about how marine fishing is translated into numerous different place-based themes. These draw in subjective domains of identity, memory and inspiration. But they go further than this and show us how marine fishing is also translated into the material environment though buildings, boats, street decorations etc. They also remind us that these are not just academic musings; understanding the extent of these translations has important implications for tourism, recreation and social cohesion.

The women in fisheries papers also describe three different accounts of how marine fishing is translated in the lives of women involved in the industry. The often heartfelt pictures that emerge from these tales are clear reminders of what can happen when groups are rendered invisible in an industry and the actions that can be taken to resolve that.

Budzich-Tabor's (Chap. 10) paper provides a descriptive account of how Axis 4 European funding is encouraging the translation of fishing activity in different ways. Examples of edible seaweed production, the use of mobile phone text messages to increase the visibility of the fishers' catch and opportunities for diversifying into tourism. Each of these is an account of how, with appropriate funding, the process of fishing can be translated into different material circumstances.

Thinking about translation as one dimension of a sustainable fisheries paradigm moves the debate beyond a dominant economic/biological paradigm. It forces consideration of much broader questions about the reach and influence of marine fishing. From this perspective fishing is seen as a process that connects the undersea world to terrestrial environments and communities. It can be argued that one element of a sustainable fisheries management paradigm is understanding the relationships that emerge from the process of fishing regardless of sector/area boundaries. To achieve this, appropriate research methods are needed that can capture a broad range of social issues. This leads us onto the fourth dimension, methodological.

18.6 Methodological

The previous three dimensions have illustrated the range of issues that emerge from a consideration of the social and cultural dimensions of marine fishing. However, in addition to the actual issues themselves, thought must also be given to research approaches used to investigate the issues. It has been argued at various points in this book that fisheries management is dominated by a positivist economic/biological paradigm. It might be argued that this is perhaps appropriate where the object of concern is management of fish populations (although decades of this approach has not resulted in sustaining fish populations). However, if fisheries management is increasingly about understanding people as well as the biology of fish populations, then a broader methodological toolkit is needed.

Within the social sciences there are many different research traditions that can be drawn on to understand people and the environment. Within this book we have presented a range of different approaches that, in their own way, contribute to making visible some of the relationships created by the process of marine fishing. The research methods employed range from formal quantitative approaches (e.g. multi-criteria decision analysis used by Morgan et al. Chap. 9) through to qualitative in-depth interviewing and photo elicitation (Neilson et al. Chap. 17) Part of developing a sustainable fisheries management paradigm is understanding that incorporating people into an area-based approach is a complex and messy affair. Methodological pluralism is needed where policy makers and managers draw on a variety of information and data sources. Accepting qualitative data alongside quantitative data can be a challenge for decisions makers and a key task for the future is to understand more about the transition of research knowledge to policy in practice.

18.7 Future Challenges

The contributions in this book reveal some of the complex social and cultural issues associated with marine fishing in coastal communities. Institutional problems are highlighted with regard to fisheries management and policy, leading to social problems. Emphasis is placed in many cases on the integration of local knowledge and social values being incorporated into decision-making frameworks. Developing a sustainable fisheries paradigm includes thinking about marine fishing as a process that sets in motion numerous complicated impacts that draw together marine ecosystems and society in a series of relationships. The challenge for the future is how to incorporate social and cultural dimensions into marine fisheries management in the move towards a sustainability paradigm for fisheries.

In recent years perhaps the clearest indication of this emerging is the increasing interest in an ecosystem approach to fisheries. Arguments for the inclusion of social, economic and institutional factors in ecosystem management for fisheries are clearly articulated by the FAO (De Young et al. 2008). These FAO guidelines seek a pragmatic balance between understanding the importance of a broad inter-sectoral approach to fisheries and consideration of those issues that are the responsibility of the fisheries manager. However, it is believed that “the more integrated or cross-sectoral the approach taken is, the more likely the attainment of sustainable development goals” (De Young et al. 2008, p. 4).

There is, however, no uniform agreement on how to achieve an ecosystem approach for fisheries (Morishita 2008) and there are impediments to its implementation. In a review of progress towards an ecosystem approach in Europe, Jennings and Rice (2011) suggest that one problem is the “low specificity and incompatibility of environmental, social and economic objectives” (p. 125). Paterson et al. (2010) suggest that an ecosystem approach to fisheries is a concept that draws together biological conservation and fisheries management. In this context fisheries management is considered as “management that focuses on fishing activities and targets resources in order to satisfy societal and human needs” (p. 782). They call for a new transdisciplinary approach to fisheries management that integrates social and natural science perspectives. Moving towards an ecosystem approach for fisheries perhaps involves thinking about fisheries as webs of interconnected biological, economic, social and cultural elements (Charles 2001). Christensen et al. (2007) suggest a lack of interdisciplinarity results in the failure of sustainability programmes. However, a key challenge for the future will be finding ways to create governance mechanisms that facilitate broad interdisciplinary/cross-sectoral approaches (see Gavaris 2009 for a discussion on management planning and an ecosystem approach). Social science can make an important contribution here, particularly in terms of understanding the broad discourses that frame management decisions. Highlighting the distribution of power within systems, understanding how some elements of a system are valued over and above other elements, and perhaps most importantly, reflecting on interconnections between fishery and non-fishery systems. Perhaps many of these suggestions lie outwith the realm of ‘fisheries management’ but a key goal would seem to be to develop approaches that reveal the connectivity between different systems (within and beyond fisheries) and then consider appropriate governance mechanisms and indicators (Jennings 2005).

Social Issues in Sustainable Fisheries Management has focused on trying to highlight how different social science perspectives can help make visible a range of issues associated with understanding the social and cultural dimensions of marine fisheries management. These have been contextualised with reference to institutional, discourse/knowledge, translation and methodology sub-categories which highlight future challenges for the integration of social issues into sustainable fisheries management including: how to incorporate a broad range of stakeholders into the decision-making process and overcome the problem of dominant discourses and power relations that prevail; determining whose knowledge is valid and making visible the voices of under-represented stakeholders; how to translate the often intangible socio-cultural values into policy speak; and developing integrative approaches for combining quantitative and qualitative data in the decision-making process.

Doing so results in perhaps a more messy world than quotas and maximum sustainable yields. Yet it is this messy world that is the home of fishers and their families; people that are living in a world that industry and regulation has created. This is not a sanitised existence of rational economic people. This is an existence governed and sculpted by the intersecting worlds of institutions, marine ecosystems and heartfelt passions. This is a world where scientific knowledge and local economic knowledge are juxtaposed, a world where an understanding of power and hegemony can reveal the relative influence of different stakeholders. It is hoped that this volume can be a useful addition to the on-going debate about how fisheries management should take account of the social issues in sustainable fisheries management. We believe that this book is, therefore, an important contribution to moving towards a sustainable fisheries management paradigm, both from a theoretical and issue-based perspective but also in terms of how we practically go about trying to understand social issues in fisheries management.

References

Charles, A. T. (1994). Towards sustainability: The fishery experience. *Ecological Economics, 11,* 201–211.

Charles, A. T. (2001). *Sustainable fishery systems.* Oxford: Blackwell Science.

Christensen, V., Aiken, K. A., & Villanueva, M. C. (2007). Threats to the Ocean: On the role of ecosystem approaches to fisheries. *Social Science Information, 46*(1), 67–86.

De Young, C., Charles, A., & Hjort, A. (2008). *Human dimensions of the ecosystem approach to fisheries: An overview of context, concepts, tools and methods.* FAO Fisheries Technical Paper. No 489. Rome: FAO.

FCR. (2000). *An assessment of the socio-economic costs and benefits of integrated coastal zone management.* European Commission: 61. Firn Crichton Roberts Ltd, G.S.o.E.S.U.o. Strathclyde.

Forst, M. F. (2009). The convergence of Integrated Coastal Zone Management and the ecosystems approach. *Ocean and Coastal Management, 52*(6), 294–306.

Gavaris, S. (2009). Fisheries management planning and support for strategic and tactical decisions in an ecosystem approach context. *Fisheries Research, 100,* 6–14.

Jennings, S. (2005). Indicators to support an ecosystem approach to fisheries. *Fish and Fisheries, 6,* 212–232.

Jennings, S., & Rice, J. (2011). Towards an ecosystem approach to fisheries in Europe: A perspective on existing and future directions. *Fish and Fisheries, 12,* 125–137.

Morishita, J. (2008). What is the ecosystem approach for fisheries management? *Marine Policy, 32,* 19–26.

Paterson, B., Isaacs, M., Hara, M., & Moloney, C. L. (2010). Transdisciplinary co-operation for an ecosystem approach to fisheries: A case study from the South African sardine fishery. *Marine Policy, 34,* 782–794.

Index

J. Urquhart et al. (eds.), *Social Issues in Sustainable Fisheries Management*,
MARE Publication Series 9, DOI 10.1007/978-94-007-7911-2,

Printed by Printforce, the Netherlands